# The vegetation of the Subantarctic islands Marion and Prince Edward

# Geobotany 3

Series Editor
M. J. A. WERGER

DR W. JUNK PUBLISHERS THE HAGUE – BOSTON – LONDON 1982

N. J. M. GREMMEN

# The vegetation of the Subantarctic islands Marion and Prince Edward

DR W. JUNK PUBLISHERS THE HAGUE - BOSTON - LONDON 1982

*Distributors*

*for the United States and Canada*

Kluwer Boston, Inc.
190 Old Derby Street
Hingham, MA 02043
U.S.A.

*for all other countries*

Kluwer Academic Publishers Group
Distribution Center
P.O. Box 322
3300 AH Dordrecht
The Netherlands

**Library of Congress Cataloging in Publication Data**

Gremmen, N. J. M.
The vegetation of the Subantarctic islands, Marion and Prince Edward.

(Geobotany; 3)
Bibliography: p. 129
Includes index.
1. Botany – Marion and Prince Edward Islands – Ecology.
2. Plant communities – Marion and Prince Edward Islands – Ecology.
I. Title. II. Series.
QK429.P74G74 581.969'9 81-13646
AACR2

ISBN 90 6193 683 7 (this volume)
ISBN 90 6193 895 3 (series)

Cover design: Max Velthuijs



PRINTED IN THE NETHERLANDS

*Pringlea antiscorbutica.*

# Preface

The first scientific exploration of Marion and Prince Edward Islands took place in 1873, one century after their discovery, when the Challenger-expedition landed a party on Marion Island. The next visit of scientists to the islands was in 1939, when members of the Bougainville-expedition landed on Marion Island. In the years after the establishment of the meteorological station on Marion Island in 1948 some biological and geological observations were made (see e.g. Rand 1954, 1955a, b, 1956; Trusswell 1965; La Grange 1962), but systematic study of the islands only started with a biological and geological expedition in 1965–66 (Zinderen Bakker et al. 1971). Since then a number of biological expeditions to Marion and Prince Edward Islands have been made.

The present study forms part of the Marion Island Bioenergetics and Mineral Cycling Programme, initiated and directed by Prof. Dr. E. M. van Zinderen Bakker. The author spent two summers on Marion Island as a member of the 4th and 5th biological expeditions to the island (November 1973 – June 1974 and November 1974 – June 1975). Prince Edward Island was explored during a visit of one week in May 1974.


**Acknowledgements**

Many people have contributed in one way or another to the realisation of this study, and only a few are mentioned below. The fact that the names of the numerous others who have contributed to this work are not mentioned does not make my gratitude to them less sincere.

Prof. Dr. E. M. van Zinderen Bakker, former director of the Institute for Environmental Sciences, University of the Orange Free State, Bloemfontein, South Africa, was the initiator of this study. I want to express my sincere gratitude to him for giving me the opportunity to undertake this study, for his stimulating guidance and encouragement in the early stages of the work, and for his continuing interest and comments thereafter.

Valdon R. Smith, of the same institute, gave me his unrelenting support, both during our stay on Marion Island in 1973–74, and in later stages of our work. His lasting, active interest in the progress of this work has been very stimulating.

I am grateful to all members of the 4th and 5th expeditions to Marion Island and to the members of the 30th and 31th relief teams to the Marion Island weather station for their companionship and their interest in my work.

The South African Department of Transport sponsors the biological research programme on the islands, and provides transport to and from the islands, as well as accommodation, food and clothing to the members of the expeditions.


N. J. M. Gremmen

# Contents

CHAPTER ONE

# Introduction

Marion and Prince Edward Islands are remote specs of land in the vast ocean surrounding the Antarctic continent (Fig. 1). They are extremely oceanic, in a geological, geographical, climatological as well as a biological sense. The islands are of volcanic origin, about 500 000 years old (Verwoerd 1971). Their climate is characterized by cool temperatures (about 5 °C) which show little variation, by very high precipitation and by strong winds (Schulze 1971). The islands' flora and fauna are characterized by a low number of species. To colonize the islands these species have been selected on their ability to cross wide expanses of ocean and to establish themselves successfully on the islands. Consequently the biota is very impoverished in comparison with continental areas, since many groups of organisms which are very successful in the surrounding continental areas are lacking on the islands (cf. e.g. Holdgate 1967c; Holdgate & Wace 1961; Wace 1960, 1966; Bramwell 1979).

The present work is the first comprehensive account of the vegetation of a Subantarctic area using the floristic-sociological approach to vegetation study of the Braun-Blanquet School (cf. Westhoff & Maarel 1973, 1978; Westhoff 1979). Several authors (e.g. Wace & Holdgate 1958; Huntley 1971; cf. Westhoff & Maarel 1973, 1978) have remarked upon the difficulty or even impossibility of using species-fidelity in characterizing plant communities of cold-temperate or Subantarctic islands, due to the species-poverty of the flora and the wide ecological amplitude of many species. Studies in azonal species-poor communities, such as salt marshes (Beeftink 1965, 1966, 1968) and communities of higher water plants (Hartog & Segal 1964; Segal 1965), have shown that, although difficulties are encountered, e.g. in the determination of the relationships between communities that show very little floristic similarities, the application of the Braun-Blanquet approach results in a very useful classification. At the outset of this study it was believed that an investigation into the vegetation of Marion and Prince Edward Islands using the methods of the Braun-Blanquet approach would yield a detailed inventory of plant communities as well as information on their ecology, which could serve as a basis for further scientific research on the islands.

CHAPTER TWO

# Description of the environment

## 2.1 Geographical and oceanographic position

The islands Marion (46° 54′ S, 37° 45′ E) and Prince Edward (46° 38′ S, 37° 57′ E) are small specs of land in the vast Southern Ocean (Fig. 1). The remoteness of these islands – which measure 290 and 44 km$^2$ respectively and lie 22 km apart – is illustrated by the distance to the nearest land, the Crozet Islands, which are situated 945 km to the east, while the nearest continent, Africa, lies 1800 km NNW of the islands. The distance to the Antarctic continent is 2300 km.

The surrounding ocean has a strong influence upon the ecosystems of the islands. Not only is the climate of the southern oceanic regions strongly influenced by the distribution of watermasses with different temperatures, but oceanographic boundaries also have biogeographic significance (cf. Deacon 1964; Winterbottom 1971). A diagrammatic representation of the main meridional currents and the distribution of water masses in the Southern Ocean is given in Fig. 2. Cold Antarctic surface-water flows northward from the boundary region of east and west wind drifts at about 65° S. Cold water with a comparatively high salinity forms the Antarctic Bottom Current, moving northward from the continental shelf. This northward-flowing water is replenished by a southward current of warmer water at a depth of 2000–3000 m (Deacon 1959, 1960, 1964).

The temperature of the Antarctic surface-water gradually increases in a northerly direction. At about 50° S a sharp rise in surface temperature occurs within a short distance. Here Antarctic surface-water, with temperatures ranging from 1–2 °C in winter to 4–5 °C in summer, meets Subantarctic surface-water, which is c. 2 °C warmer. These watermasses do not mix. The Antarctic surface-water sinks below the Subantarctic water. The Antarctic Convergence seems to occur where the warm, deep southward current rises most steeply over the colder Antarctic Bottom Current. The position of the Antarctic Convergence is remarkably stable: it is always found within about 100 km from its mean position (Deacon 1959, 1960, 1964). The temperature gradient at the surface shows another sharp increase at about 40° S. Here the Subantarctic surface-water, with a temperature of c. 10 °C in winter and 14 °C in summer, meets the water of the subtropical seas, which is some 4 °C warmer (Fig. 2). This boundary region is called the Subtropical Convergence. It has a much less stable position than the Antarctic Convergence (Deacon 1964).

At present the Antarctic Convergence is situated about 250 km south of Marion and Prince Edward Islands. During recent geological times the position of this oceanographic boundary has shifted considerably. Some 18 000 years ago the convergence apparently was lying to the north of the islands, leaving them surrounded by cold Antarctic surface-water (Zinderen Bakker 1969, 1971b, 1973b, 1978; Hays et al. 1976). This went together with a lowering of the air temperature, resulting in an extensive glaciation of Marion Island (Zinderen Bakker 1969, 1970, 1971b, 1973b, 1976, 1978; Hall 1978a, b; see also par. 2.2).

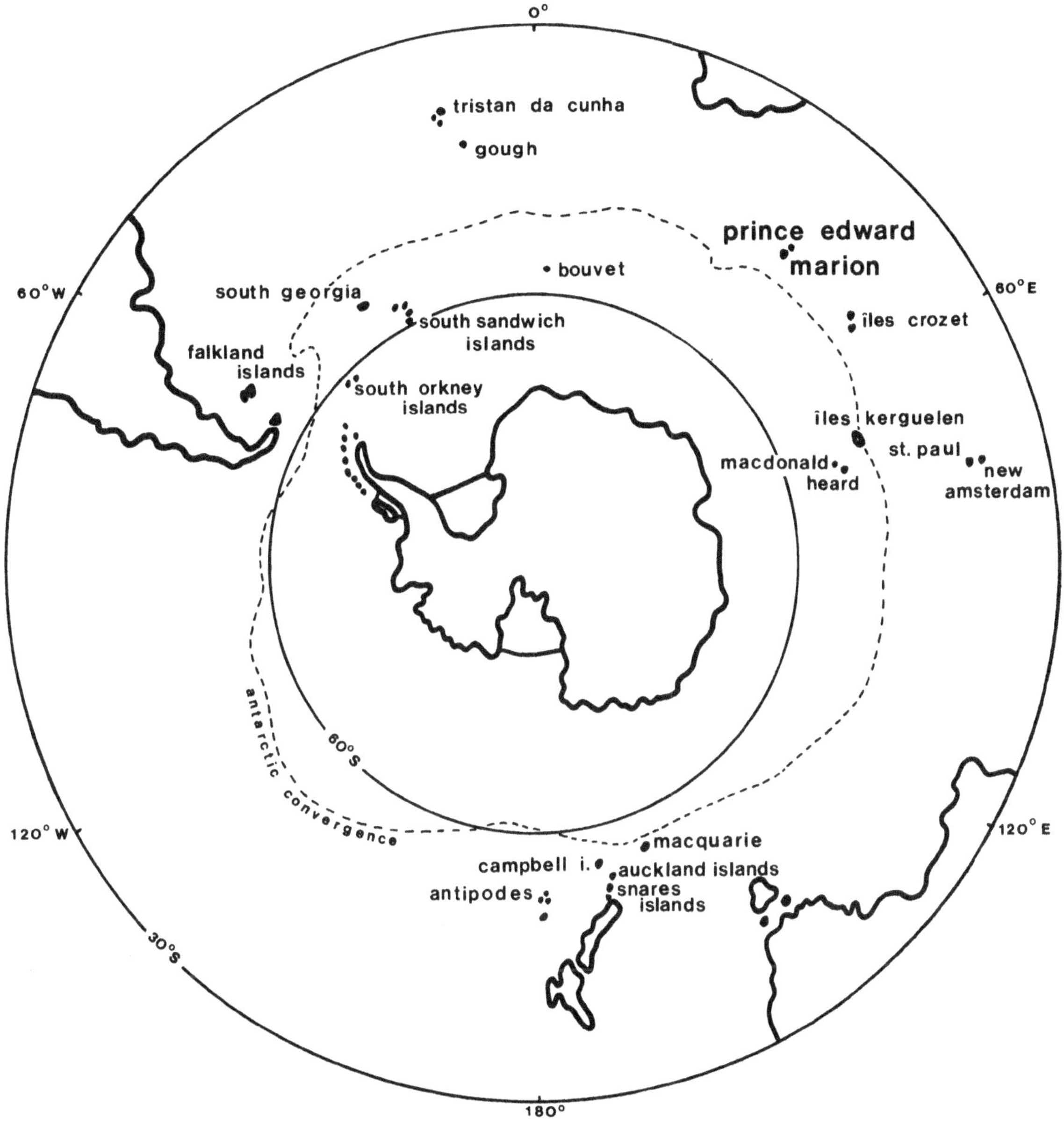


*Fig. 1.* Geographical map of the Southern Ocean.

**2.2 Geology**

Marion and Prince Edward Islands are situated near the centre of the West Indian Ocean Ridge, which is a continuation of the Mid-Atlantic Ridge and connects the latter with the Indian Ocean Ridge system (Girod & Nougier 1972). Most of the islands associated with these ridges have recently shown volcanic activity, e.g. Bouvet (Baker & Tomblin 1964) and Tristan da Cunha (Gass 1963; Dickson 1965a, b). For several thousands of years Marion and Prince Edward Islands apparently have not shown volcanic activity, but in 1980 an eruption took place at Kaalkoppie on Marion (V. R. Smith, personal communication).

During the short visits of the Challenger (1873)

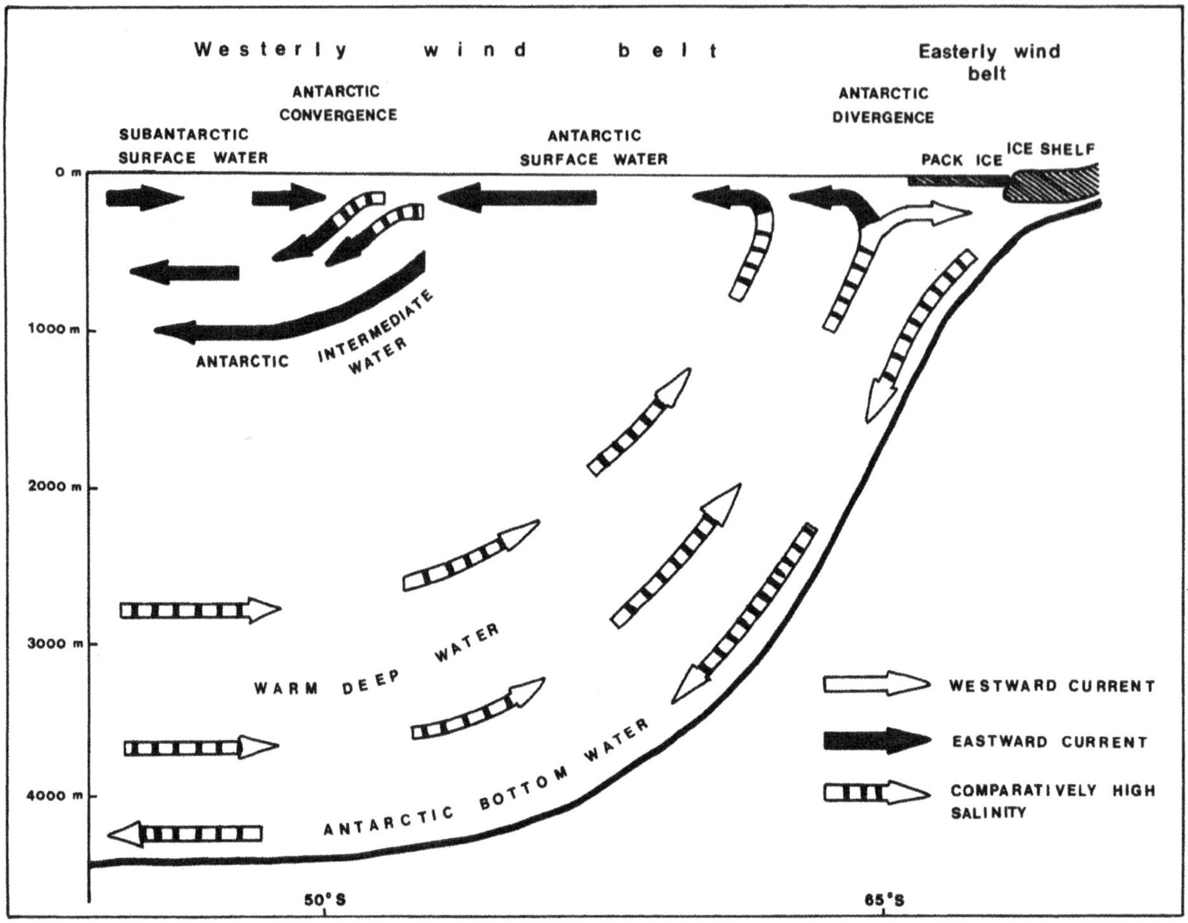


*Fig. 2.* The meridional circulation of water in the Southern Ocean (after Deacon 1964 and Zinderen Bakker Sr. 1967).

and the Bougainville (1939) expeditions some information on the geology of the island was collected (Renard 1889; Moseley 1892; LaCroix 1923, 1940). A comprehensive geological survey was undertaken during the South African Biological and Geological Expedition to Marion and Prince Edward Islands in 1965/1966. The results have been published by Verwoerd (1967, 1971), Kable, Erlank & Cherry (1971), McDougall (1971) and Snape & Retief (1971). The following is largely based on accounts by Verwoerd (1967, 1971).

Marion and Prince Edward Islands are the summits of coalescing shield volcanoes of late Quaternary origin. Their age is estimated at 0.5 million years. Two distinct series of lavas can be discerned (Fig. 4), associated with two periods of volcanic activity.

Fine grained, compact basalts with a grey colour resulted from the first volcanic stage. Potassium-argon dates show ages for lava flows of this type of between c. 276 000 and 100 000 years. Stratified tuff and pyroclastic deposits associated with this first volcanic stage are present but not prominent. Eruptive and effusive activity during this period took place at a number of sites on the islands.

The greater part of the islands is covered with deposits which originated from a more recent period of volcanic activity (Figs. 5 and 6). The lavas of this second volcanic stage are generally black in colour and strongly vesicular. In chemical composition they do not differ markedly from the old grey lavas. Potassium-argon dates show that black lava was erupted c. 15 000 years ago. Different methods of consolidation of the lava resulted in variation in structure of the lava flows. Flows consisting of solid lava covered with clincker fragments (aa flows) are

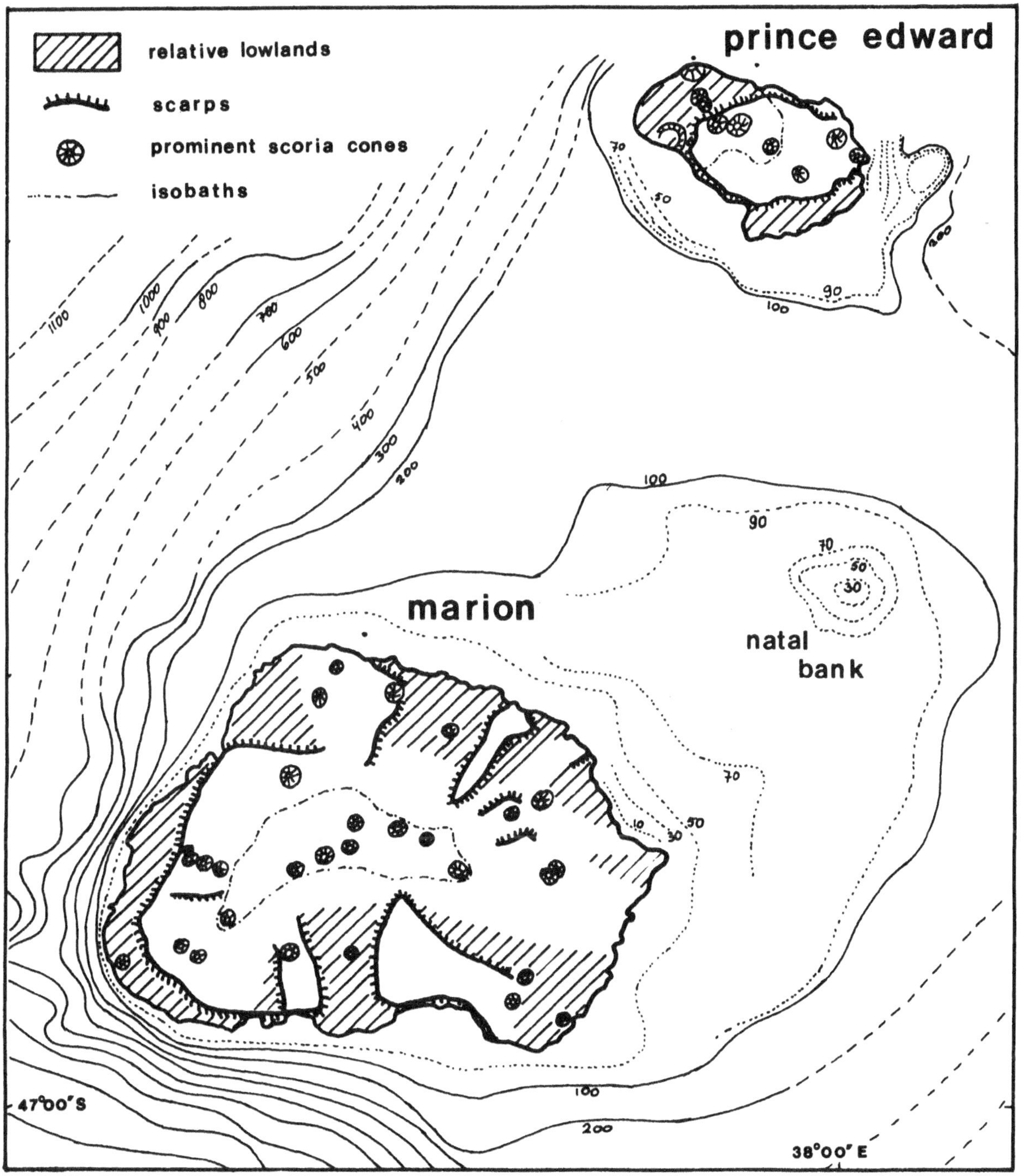


*Fig. 3.* Physiography of Marion and Prince Edward Islands and bathymetry of the surrounding ocean (after Verwoerd 1971).

*Fig. 4.* Black lava flow (left) and a grey lava area near Kildalkey Bay, Marion Island.

the most common type. The loose lava fragments are piled into heaps often up to 20 m high. Block lava covers a large part of the islands. These flows are built of loose, ragged lava blocks up to 1 m in diameter, generally arranged into steeply sloping, parallel ridges. Pahoehoe lava, characterized by its even surface, is not very common. Lava tunnels, caves and other structures caused by large gas bubbles are frequently encountered in flows of this type.

The numerous cones of red and black pyroclastic material occurring on the islands (Figs. 6, 7, 8) have been the main centres of activity during the second volcanic stage. Tuffaceous deposits associated with this stage are very rare.

Marion Island shows a conspicuous radial pattern of grey lava ridges and plateaus, with the valleys and plains covered by black lava. This is the result of radial faulting which has occurred between the two stages of volcanic activity. No such faulting is apparent on Prince Edward (Fig. 3).

*Glaciations of Marion Island*

At present, permanent snow and ice covers are restricted to the central highland of Marion Island, above c. 1000 m a.s.l. The occurrence of striated platforms, smoothly rounded grey lava outcrops and the general topography of the grey lava areas (Fig. 5) strongly suggest an extensive glaciation of the island in the past. No black lava seems to have been subjected to ice-covering. This dates the glaciation of Marion Island between c. 100 000 and 15 000 years ago. A study of glacial landforms and deposits showed that the island was subjected to three periods of glaciation, all of the Würm and Riss age, each with stades and interstades (Hall 1978a, b). Palynological research revealed the last stages of the last ice age (Schalke & Zinderen Bakker 1967, 1971; Zinderen Bakker 1967b, 1969, 1970, 1971a, b, 1973b). Hall (1978) concludes that a minimal temperature decrease of 3.5 °C during the glaciations was reached. This is in agreement with

*Fig. 5.* Grey lava area near Kerguelen Rise, Marion. Note the cushions of *Azorella selago*. The vegetation is the Andreaeo acutifoliae – Racomitrietum crispuli hymenophylletosum peltati.
*Fig. 6.* View from Skua Ridge over part of the north-eastern coastal plain of Marion Island towards Junior's Kop, one of the c. 130 scoria cones on the islands.

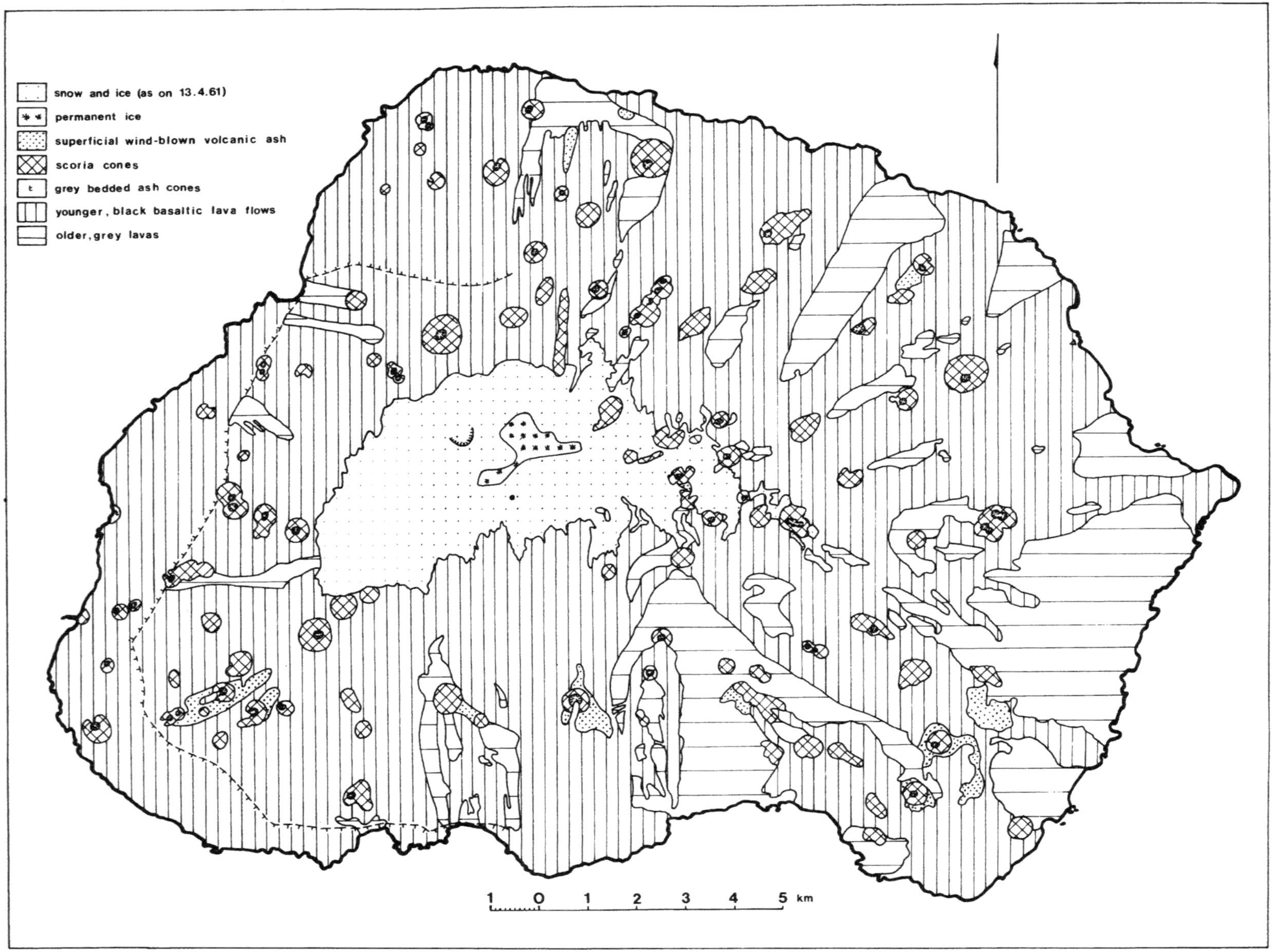


... of Marion Island (after Verwoerd 1971)

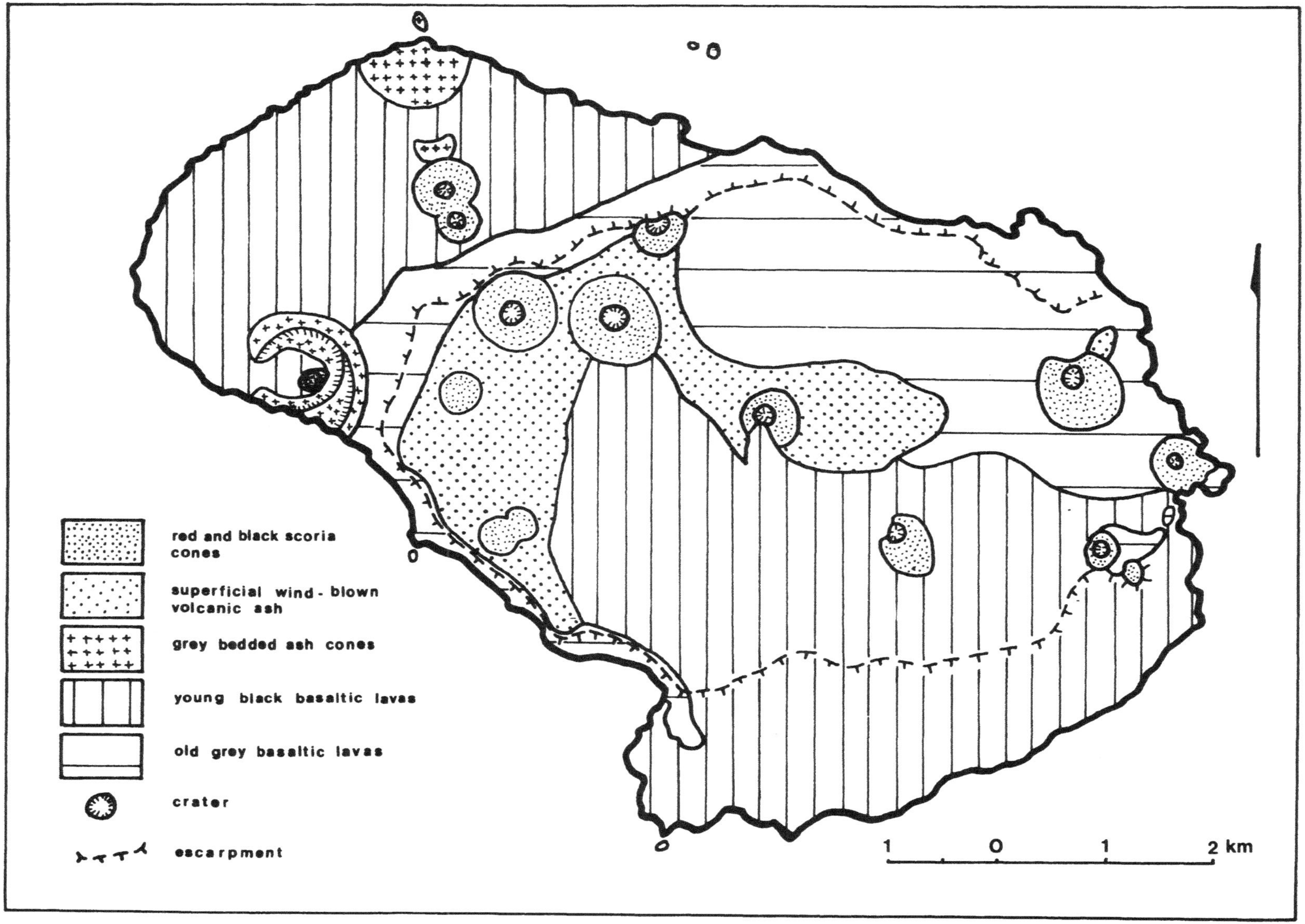


*Fig. 8.* Geological map of Prince Edward Island (after Verwoerd 1971).

values suggested by Zinderen Bakker (1973b) and Hays et al. (1976). A temperature decrease of this magnitude implies a shift of the Antarctic Convergence to the north of the islands (Zinderen Bakker 1973b; Hall 1978a, b).

The glaciers do not appear to have covered Marion Island completely. In the ice-free areas cold-resistant species probably survived during the ice ages (cf. Zinderen Bakker 1970). A similar situation apparently existed on Kerguelen (cf. Bellair 1965; Young & Schofield 1973).

### 2.3 Topography and hydrology

Marion Island consists of a central highland, sloping down in all directions towards the sea. This central highland has the form of a plateau, at c. 1000 m above sea level and with a number of higher volcanic cones. The highest of these reaches 1230 m. The surface of the highland is formed by unconsolidated scoriae and volcanic ash, with few rocky outcrops. Most of the central highland is permanently covered by ice and snow.

From the central highland the land slopes down more or less steeply towards the sea. This 'island slope' forms the greater portion of the island. It is divided into valleys and plains separated by plateaus and ridges in a radial pattern which is especially conspicuous on the northern and eastern sides of the island. The ridges and plateaus almost always consist of grey lava. Black lava generally covers the less elevated areas. Ridgetops and valleys are often separated by steep cliffs of over 100 m high. This pattern seems to have resulted from radial faulting (Verwoerd 1971).

On the northern and eastern sides of Marion Island the island slope descends more or less gradually towards the sea. On the western, and part of the southern sides, the island slope abruptly ends in an escarpment of 200–300 m high, which in turn descends steeply to a low coastal plain. This escarpment runs parallel to the coast. It seems to be the result of marine erosion and apparently formed the coastline of the island before the second period of volcanic activity. The surface substrate of the escarpment now largely consists of black lava. The coastal plain is 1–2 km wide and also consists entirely of black lava. It is elevated up to about 50 m above sea level.

Approximately 130 scoriae cones mark the centres of eruption of the second volcanic stage. Many of these are well preserved, often showing circular craters and sometimes containing a craterlake.

The coastline is generally formed by cliffs. Where the old, grey lava meets the ocean, marine erosion produced high cliffs, at some places reaching several hundreds of metres above sea level. Beaches are common along grey lava coasts. Except for a narrow sandy beach at Ship's Cove (Fig. 9), all beaches consist of boulders. Grey lava areas produced a smooth coastline, often with large, curved bays. Black lava coasts are very irregular, with numerous capes, inlets, caves and off-shore rocks. The coastal cliffs of black lava areas generally do not reach a height of more than 20 m and beaches occur at only a few places along these coasts.

The highest part of Prince Edward Island is formed by the central block, with a summit of 672 m. To the east the central block slowly slopes down towards a coastal plain. On the western side it terminates abruptly in nearly vertical cliffs of c. 400 m high. At the foot of this escarpment lies a coastal plain. To the north and south the central block ends in precipitous cliffs, up to 500 m high, descending into the sea.

Marion Island possesses a poorly-developed radial drainage system. (Verwoerd 1971). Underground drainage occurs to a considerable extent and is greatly facilitated by the porous structures of black lava and pyroclastic deposits. Grey lava flows are more impermeable. Consequently, the aboveground drainage system in these grey lava areas is more developed. Only a few of the c. 35 streams on Marion Island are perennial, notably the Van den Boogaard River and the Diving Petrel Stream. After four days without rain, the Van den Boogaard River was observed to flow strongly 1.5 km inland, but not much water reached the sea through its course at that time. Considerable drainage of water from the rivercourse into the ground appeared to take place. On Prince Edward Island only a few small streams are found, none of which carries water permanently.

Lakes and ponds are not important in the islands drainage system. Large lakes (1000–30 000 $m^2$) are generally confined to the impermeable grey lava areas. Their depth ranges from 0.5–2.5 m

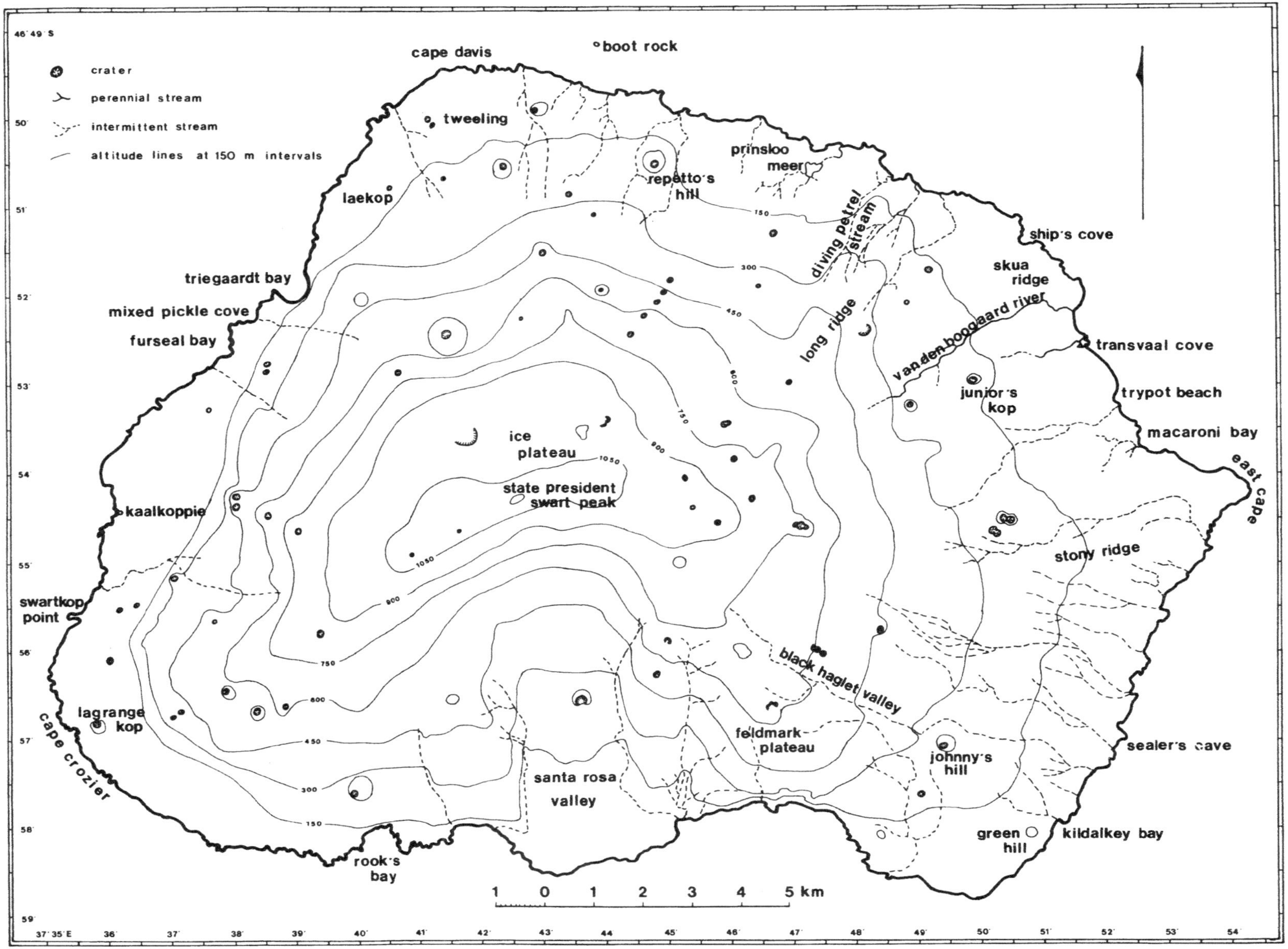


*Fig. 9.* Topographical map of Marion Island (after Langenegger & Verwoerd 1971).

(Grobbelaar 1974). In mire areas bog ponds are common. They are mostly small in size.

Water movement through the soil is locally concentrated in drainage lines. In some places at the bottom of grey lava slopes water emerges to form small springs. Generally the catchment area providing the water for these springs is large enough to assure a continuous water supply. Below the springs water tracks are often found, which in most cases do not possess a free water channel. These watertracks generally run out into a stream, but in some cases the water disappears again below the surface.

## 2.4 Climate

The islands possess a cool, extremely oceanic climate, characterized by an average annual temperature of c. 5 °C with little diurnal and seasonal variation, by a very high precipitation and humidity, and by strong winds, blowing predominantly from westerly directions. Weather conditions on the islands are strongly influenced by the numerous depressions that travel from north-west to south-east through the area. The passing fronts cause high cloudiness and precipitation (Vowinckel 1954).

The Meteorological Station at Marion Island, established in 1948, is maintained by the South African Weather Bureau. It is situated on the north-eastern side of the island at Transvaal Cove, 23 m above sea level, c. 100 m away from the coast (Fig. 9). Its geographical position is Lat. 46°52'34" S, Long. 37°51'40" E (Langenegger & Verwoerd 1971). Annual reports on meteorological

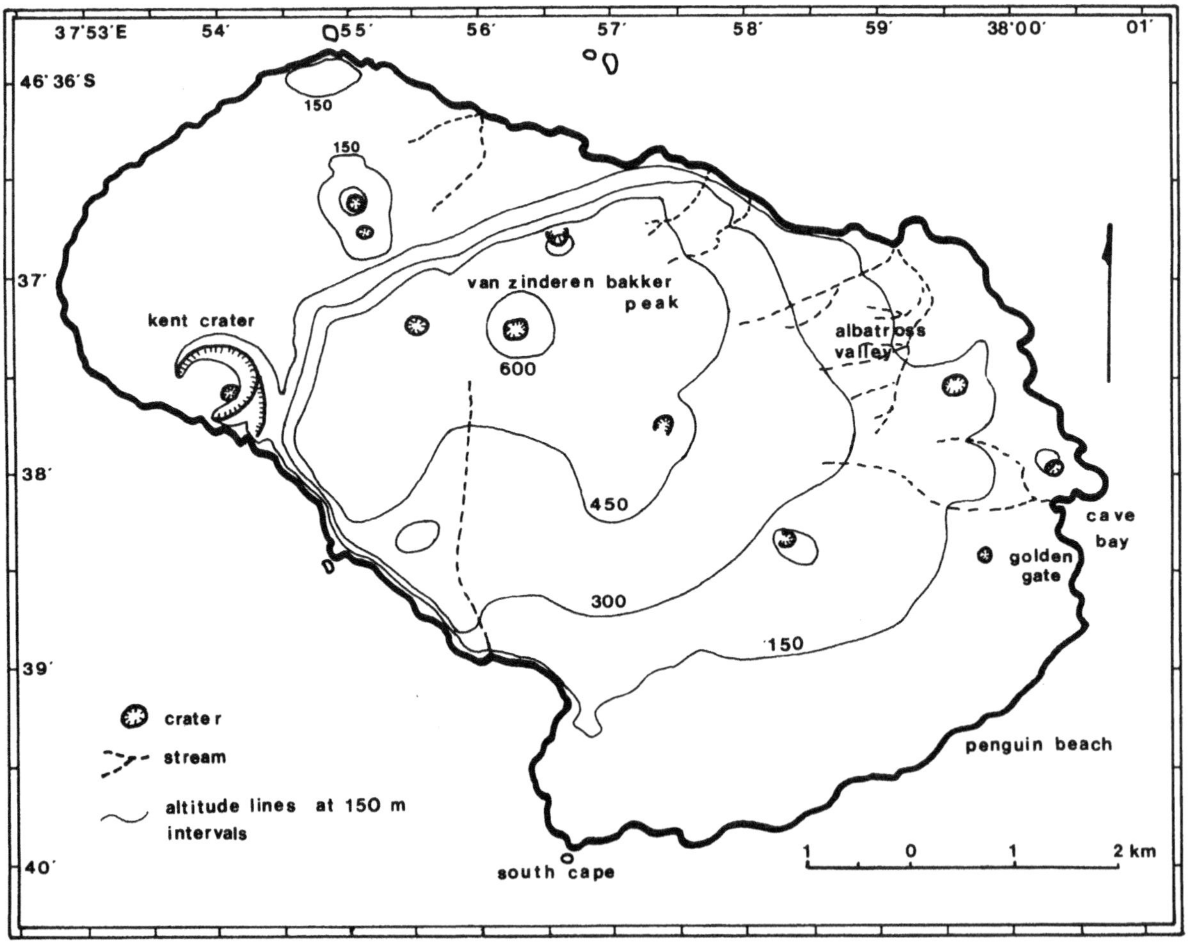


*Fig. 10.* Topographical map of Prince Edward Island (after Langenegger & Verwoerd 1971).

data are published by the Weather Bureau. Summarized data have been provided by Fabricius (1957) and Schulze (1971). For the computation of the climatic statistics given in this chapter, summarized data from Schulze (1971) were used as well as detailed data for the years 1963–1973, excluding 1971 (Weather Bureau 1967a, b, 1968, 1970, 1971a, b, 1972, 1973, 1975a, b).

*Sunshine*

In Table 1, 23 years of data on sunshine duration are summarized. Days with full sunshine are very rare. During winter, Marion Island receives a smaller part of the astronomically possible sunshine than in summer, but no corresponding differences in cloudiness are observed (Table 2). Part of this variation is due to the methods of computation of the possible sunshine duration, which assumes an unobstructed horizon (Weather Bureau 1973). At the weather station the central mountain area of the island obstructs the horizon in south-westerly, westerly and north-westerly directions, reaching up to c. 5° above the horizon.

*Cloudiness*

The sky over Marion Island is generally overcast or nearly so. During a 15-year period 70% of the observations showed a cloud cover of 6–8 oktas (Schulze 1971). Cloud observations are summarized in Table 2. The evenings are somewhat less cloudy than the mornings and afternoons, especially during winter.

Part of the amount of clouds observed is of orographic origin. Because of orographic effects the average cloud cover on the western side of Marion Island is probably higher than at the weather station (Schulze 1971). No meteorological observations are available from this part of the

*Table 1.* Sunshine duration at the Marion Island Weather Station (24 years: 1948–1973, excluding 1967 and 1971).

| | Sunshine duration | | Frequency of days with | | | | |
|---|---|---|---|---|---|---|---|
| | Hrs · day$^{-1}$ | % of possible | 0 | 1–10 | 11–49 | 50–89 | 90–100% |
| | | | of possible sunshine | | | | |
| J | 5.0 | 32 | 4 | 4 | 15 | 8 | 0 |
| F | 4.9 | 35 | 4 | 3 | 12 | 8 | 0 |
| M | 3.8 | 31 | 6 | 5 | 12 | 8 | 0 |
| A | 2.8 | 26 | 8 | 5 | 11 | 7 | 0 |
| M | 2.4 | 25 | 9 | 5 | 11 | 7 | 0 |
| J | 1.8 | 21 | 10 | 5 | 10 | 4 | 0 |
| J | 2.0 | 22 | 10 | 5 | 11 | 5 | 0 |
| A | 2.5 | 25 | 8 | 6 | 12 | 6 | 0 |
| S | 3.1 | 27 | 7 | 4 | 13 | 6 | 0 |
| O | 4.5 | 33 | 4 | 4 | 13 | 9 | 0 |
| N | 5.1 | 35 | 3 | 4 | 15 | 8 | 0 |
| D | 5.2 | 33 | 4 | 4 | 15 | 8 | 0 |
| Y | 3.6 | 30 | 76 | 54 | 149 | 84 | 2 |

*Table 2.* Mean monthly cloud cover at the Marion Island weather station in oktas at 0800, 1400 and 2000 h S.A.S.T. (1959–1973, excluding 1971).

| | J | F | M | A | M | J | J | A | S | O | N | D | Y |
|---|---|---|---|---|---|---|---|---|---|---|---|---|---|
| 0800 h | 6.4 | 6.5 | 6.3 | 6.4 | 6.2 | 6.3 | 6.4 | 6.2 | 6.3 | 6.2 | 6.4 | 6.4 | 6.3 |
| 1400 h | 6.3 | 6.4 | 6.2 | 6.4 | 6.2 | 6.3 | 6.5 | 6.4 | 6.5 | 6.3 | 6.3 | 6.3 | 6.3 |
| 2000 h | 6.2 | 6.2 | 5.8 | 5.8 | 5.4 | 5.6 | 5.8 | 5.8 | 5.9 | 5.8 | 6.1 | 6.1 | 5.9 |

island, but the presumption is supported by the experiences of field parties in this area.

*Temperature*

Due to their small landmass, the relatively low incidence of sunshine and the nearly permanent winds, the temperature regime of the islands is closely related to that of the surrounding ocean (cf. Kessler 1978). The islands are situated c. 200 km north of the Antarctic Convergence, where the Antarctic surface water sinks below the Subantarctic surface water which is c. 2 °C warmer (Deacon 1964).

The average sea surface temperature at Transvaal Cove ranges from 4.0 °C in August and September to 6.1 °C in February and March. The extremes observed during a 15-year period were 2.1 °C and 8.0 °C. The mean monthly air temperature follows closely the temperature of the sea water, being c. 1.5 °C higher in summer and c. 1 °C lower in winter (Schulze 1971).

The seasonal and diurnal variation in air temperature on Marion Island is very small, as is shown by the thermoisopleth-diagram (Fig. 11). This diagram shows lower summer temperatures than those given by Zinderen Bakker (1978). This is due to the fact that three of the four years of which the latter author used the temperature data in the compilation of his diagram had warmer summers than the average. Very similar patterns are shown by thermoisopleth-diagrams for Macquarie (Troll 1955) and Kerguelen (Trol 1944). The warmest period on Marion occurs nearly two months after the sunshine maximum; this clearly indicates the strong influence of the surrounding ocean on the temperature regime of the islands. The absolute extremes measured at the weather station are -6.8 °C and 22.3 °C. Generally the temperature range on one day does not exceed 6 °C. An average of 55 days with frost occur annually at sea level. Although frost may occur in any month of the year at the weather station, generally a frost-free period of 2–4 months occurs during summer. The number

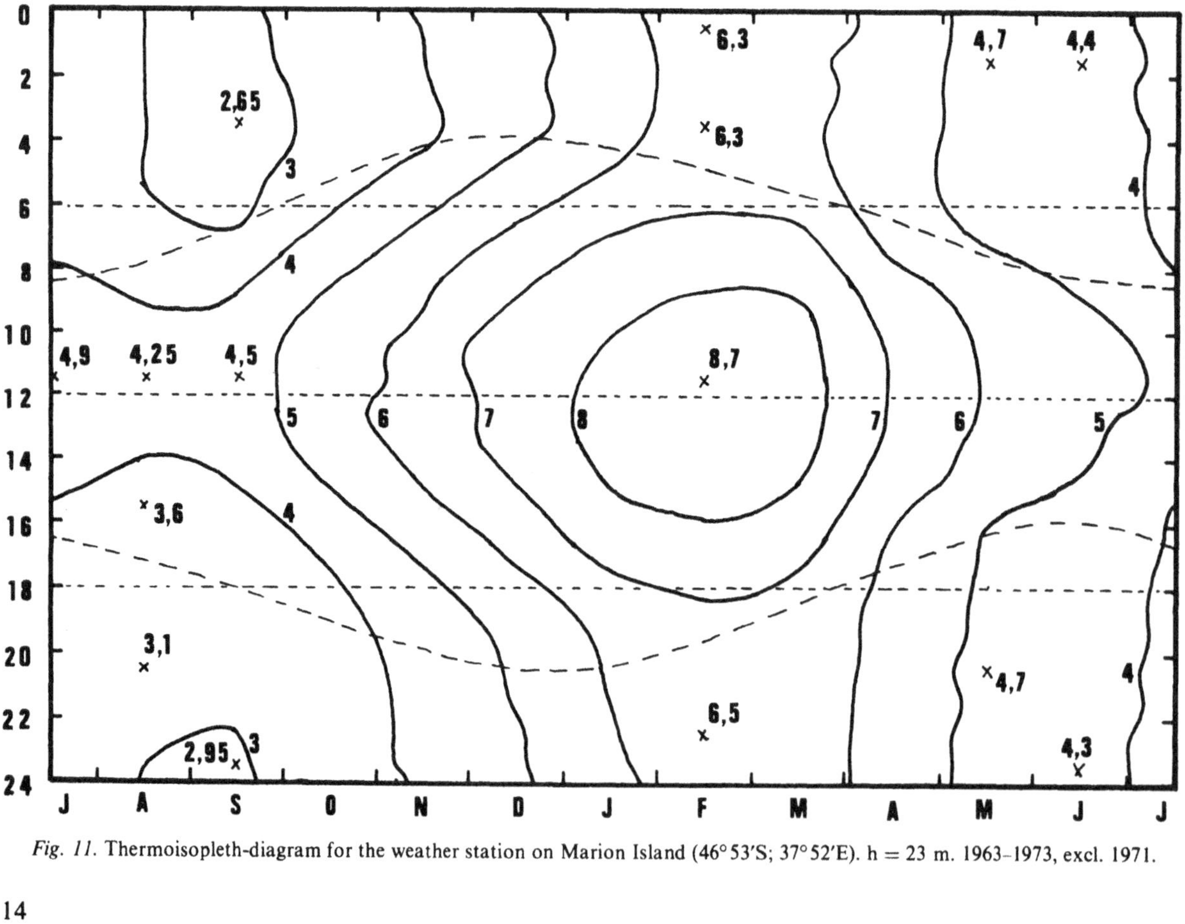


*Fig. 11.* Thermoisopleth-diagram for the weather station on Marion Island (46° 53′S; 37° 52′E). h = 23 m. 1963–1973, excl. 1971.

of days with frost increases with altitude: Zinderen Bakker (1978) estimates that at 250 m above sea level frost occurs on about 50% of the nights. On only 90 days/year the temperature rises above 10 °C (Table 3).

From 0–2000 m above sea level the average temperature lapse rate in the free atmosphere is c. 4.5 °C/1000 m in summer. In winter a temperature decrease of 4.0 °C/1000 m is found. The 0°-isotherm fluctuates between c. 600 m above sea level in winter to c. 1900 m in summer (Schulze 1971).

*Relative humidity*

The average relative humidity on Marion Island is somewhat higher than 80%. Diurnal and seasonal variation is small. Only on rare occasions does the humidity fall below 50% and then generally only for a few hours. Values of less than 40% are exceptional (Schulze 1971; Weather Bureau 1967a,b, 1968, 1970, 1971a,b, 1972, 1973, 1975a,b). It should be noted that these are screen humidity values and, especially during windy conditions, the actual humidity at plant canopy level may be lower.

*Precipitation*

Some data on precipitation for a period of 22 years are summarized in Table 4. There is a distinct seasonal variation in precipitation. April and May are the wettest months with a precipitation of more

*Table 3.* The number of days on which the maximum temperature at the Marion Island weather station is less than 10.0 °C and the minimum temperature less than 0.0 °C (1961–1973, excluding 1971).

| | J | F | M | A | M | J | J | A | S | O | N | D | Y |
|---|---|---|---|---|---|---|---|---|---|---|---|---|---|
| Days with max. temp. ≤10.0 °C | 16.1 | 12.8 | 15.6 | 20.9 | 26.2 | 26.9 | 29.5 | 29.6 | 28.2 | 24.8 | 23.0 | 21.5 | 274.5 |
| Days with min. temp. ≤0.0 °C | 0.1 | 0.2 | 0.1 | 1.4 | 4.8 | 5.3 | 8.3 | 11.9 | 11.1 | 7.1 | 3.9 | 0.6 | 55.1 |

*Table 4.* Summary of some precipitation observations at Marion Island (1951–1973, excluding 1971).

| | Precipitation in mm | | | | Frequency of days with precipitation | | | |
|---|---|---|---|---|---|---|---|---|
| | Average | Adj. av.* | Min. | Max. | >0.1 mm | >1.0 mm | Snow | Fog |
| J | 231 | 226 | 137 | 350 | 26 | 20 | 2.9 | 3.8 |
| F | 212 | 227 | 103 | 333 | 23 | 19 | 1.8 | 4.3 |
| M | 226 | 222 | 149 | 397 | 24 | 19 | 2.6 | 5.6 |
| A | 238 | 242 | 142 | 327 | 27 | 21 | 4.5 | 4.1 |
| M | 247 | 243 | 100 | 428 | 28 | 23 | 8.4 | 4.2 |
| J | 219 | 222 | 104 | 461 | 28 | 23 | 9.6 | 3.9 |
| J | 224 | 220 | 148 | 319 | 28 | 23 | 11.6 | 4.8 |
| A | 191 | 187 | 87 | 250 | 28 | 22 | 13.0 | 3.4 |
| S | 190 | 193 | 113 | 359 | 27 | 20 | 13.4 | 2.4 |
| O | 175 | 172 | 108 | 288 | 25 | 19 | 9.6 | 2.6 |
| N | 197 | 200 | 88 | 316 | 25 | 20 | 7.7 | 2.4 |
| D | 225 | 221 | 166 | 300 | 26 | 22 | 6.4 | 3.2 |
| Y | 2575 | 2575 | 2109 | 2992 | 313 | 251 | 90.7 | 44.7 |

* Average precipitation adjusted for months of equal length.

than 240 mm each, whereas October on average receives 172 mm. Although the amount of precipitation for each individual month may differ considerably from the average, the yearly extremes lie within 20% of the average. About 50 days/year have less than 0.2 mm of precipitation. Dry spells of more than a few days are rare.

Snow may occur during any month of the year, but is most frequent during winter, when one out of every three days receives snow. The lowland areas are covered with snow only for short periods in winter, but the highest part of Marion is permanently covered with snow. In summer the snow-line is found at about 1000 m above sea level, in winter at about 600 m (Zinderen Bakker 1971a).

Hail is observed on 15 days/year and fog occurs on an average of 45 days.

*Wind*

Marion and Prince Edward are situated within the belt of West winds between 40 and 60° S. These westerlies are interrupted by numerous depressions passing from north-west to south-east. During about 60% of the time the wind blows from directions between west and NNW (Fig. 12).

The average wind speed at the weather station is 6.7 m. $sec.^{-1}$. Winds blowing from the westerly quarter are on the average stronger than those from other directions (Schulze 1971). The meteorological station is sheltered considerably from winds from directions between south-west and north-west (Taljaard 1957). Thus the average wind speed at exposed sites may be much higher than at the station. According to Huntley (1970), at high elevations winds are considerably stronger than at sea level.

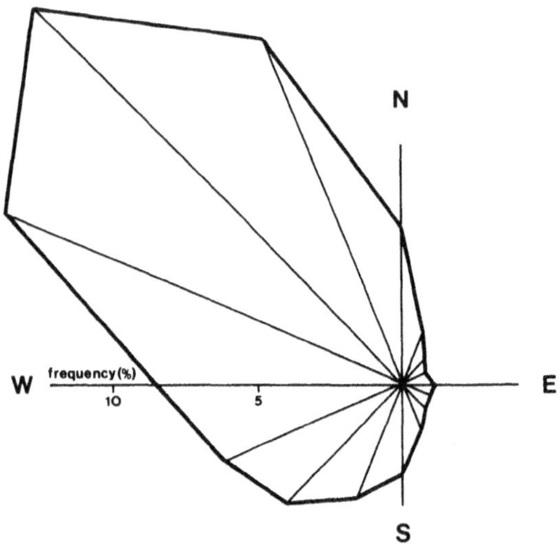


*Fig. 12.* Wind direction frequency at the Marion Island meteorological station (data from Weather Bureau 1967a, 1971a, 1973, 1975a, b).

There is a small seasonal variation in wind speed, with on average somewhat higher velocities in winter than in summer. The frequency of calm conditions shows a maximum during summer, while gales are most frequent in winter (Table 5).

Schulze (1971) has pointed out a clear diurnal rhythm in wind speed at Marion Island during summer. The average wind velocity is highest at noon (c. 7.7 m. $sec.^{-1}$) and is only 6.5 m. $sec.^{-1}$ during the night. This pattern, caused by heating of the island during the day and cooling at night, is much less clear in winter.

*Climate classification*

The cool, extremely oceanic climate of Marion Island does not fit well into the climate classification made by Walter & Lieth (1967). Walter et al. (1975) mention a Subantarctic climate zone when describing the cold temperate or boreal climate type, but from their description it is apparent that this type differs considerably from the Marion climate. Walter (1970) classifies the climate of the Subantarctic regions as Arctic. Fig. 13 shows the climate diagram (cf. Walter 1955; Walter & Lieth 1967; Walter 1970; Walter et al. 1975) of the Marion Island weather station. Comparison of this diagram with those given in Walter & Lieth's world atlas reveals only few stations with similar climates, e.g. Staaten Island and Evangelistas on the southern tip of South America, the Subantarctic islands Kerguelen and Macquarie, and some mountain stations in New Zealand, such as Chateau Tongariro. In Walter & Lieth's system the climate of Marion Island can be classified as VIII–IX (oceanic), i.e. an extremely oceanic climate with colder summers than the cold temperate climate (VIII) and lacking the cold winters of the latter. At the same time, however, the climate is much warmer than the Arctic climate type (IX), especially in winter.

*Table 5.* Frequency of wind speed classes at the Marion Island weather station in 0/00 (1963–1973, excluding 1966 and 1971).

| | Wind speed in m. sec$^{-1}$ | | | | | | | | | | |
|---|---|---|---|---|---|---|---|---|---|---|---|
| | 0.0 -1.0 | 1.1 -1.5 | 1.6 -3.3 | 3.4 -5.4 | 5.5 -7.9 | 8.0 -10.7 | 10.8 -13.8 | 13.9 -17.1 | 17.2 -20.7 | 20.8 -24.4 | 24.5 -32.7 |
| J | 11 | 3 | 10 | 14 | 17 | 13 | 9 | 5 | 2 | 0 | 0 |
| F | 15 | 2 | 8 | 12 | 16 | 10 | 8 | 4 | 2 | 0 | 0 |
| M | 16 | 3 | 9 | 14 | 15 | 12 | 8 | 3 | 2 | 1 | 0 |
| A | 15 | 2 | 10 | 13 | 16 | 11 | 9 | 4 | 1 | 0 | 0 |
| M | 12 | 3 | 10 | 14 | 17 | 14 | 9 | 4 | 2 | 0 | 0 |
| J | 12 | 3 | 8 | 11 | 13 | 15 | 13 | 5 | 2 | 1 | 0 |
| J | 9 | 3 | 8 | 13 | 16 | 14 | 12 | 5 | 2 | 1 | 0 |
| A | 8 | 3 | 10 | 14 | 16 | 13 | 11 | 5 | 3 | 1 | 0 |
| S | 8 | 3 | 9 | 15 | 19 | 12 | 10 | 5 | 2 | 0 | 0 |
| O | 12 | 3 | 9 | 13 | 16 | 14 | 11 | 5 | 2 | 0 | 0 |
| N | 9 | 3 | 9 | 17 | 17 | 12 | 10 | 5 | 2 | 0 | 0 |
| D | 12 | 4 | 11 | 16 | 16 | 13 | 9 | 3 | 2 | 0 | 0 |
| Y | 141 | 34 | 108 | 162 | 193 | 152 | 120 | 54 | 23 | 5 | 2 |

A second climate classification made especially with regard to biological phenomena is given in the map of seasonal climates of the earth by Troll & Paffen (1966, see also Troll 1966). Marion Island lies within zone 14, the zone of highly oceanic, sub-polar climates with moderately cold winters (coldest month -8 ° to +2 °C), poor in snow, and with cool summers (warmest month +5 ° to +12 ° C; annual fluctuation less than 13 °, often less than 10 ° C). This climate type occurs in a belt between c. 50 and 60° S. In the northern hemisphere it is restricted to a few areas, notably the Aleutan Islands and southern Iceland (Troll & Paffen 1966).

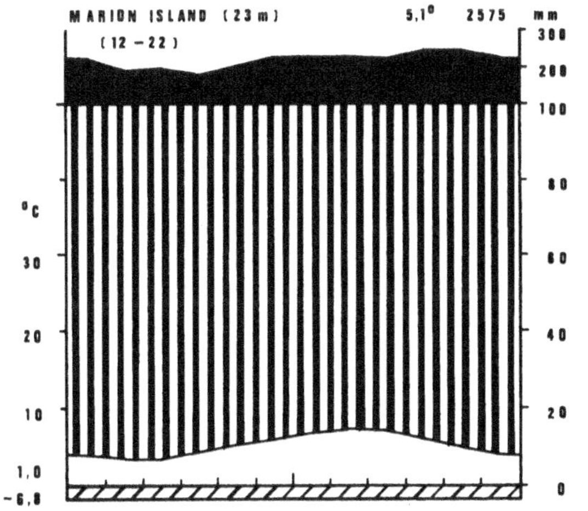


*Fig. 13.* Climate diagram of the weather station at Marion Island.

## 2.5 Soils

Classification of the soils of the islands according to international systems is difficult, due to the immaturity of the soils, the negligible effect of parent material on soil development and the strong influence of small variations in topography (Smith 1976c).

The cool, extremely oceanic climate strongly favours the formation of peat. On large parts of the lowland areas, especially where drainage is impeded, peat deposits occur ranging in thickness from a few cm to more than 4 m. The peat may be resting directly on parent rock, but usually a layer of loamy clay separates the two (Fig. 14, Table 6).

At well-drained sites which are not markedly exposed (e.g. most lowland slopes) the soil generally consists of 5–30 cm of plant litter and humus overlying a layer of loamy clay (Fig. 15, Table 7). The depth of these slope soils generally lies between 50 and 130 cm.

At strongly exposed sites, especially at high

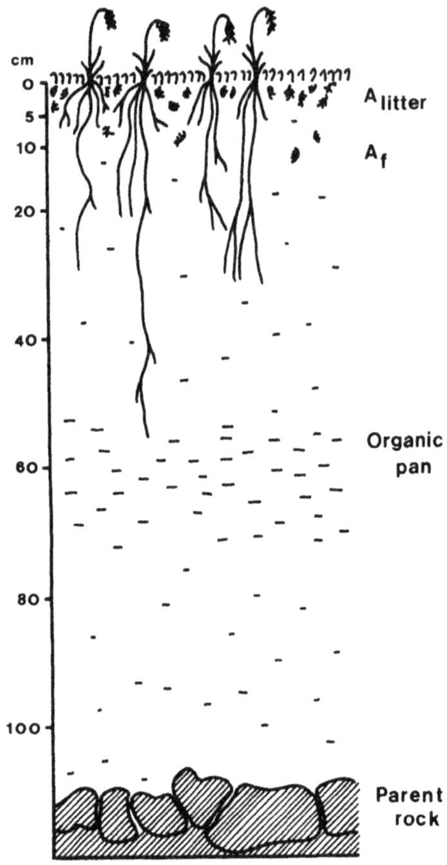


*Fig. 14.* Soil profile of a peat deposit on Marion Island (after Smith 1976c).

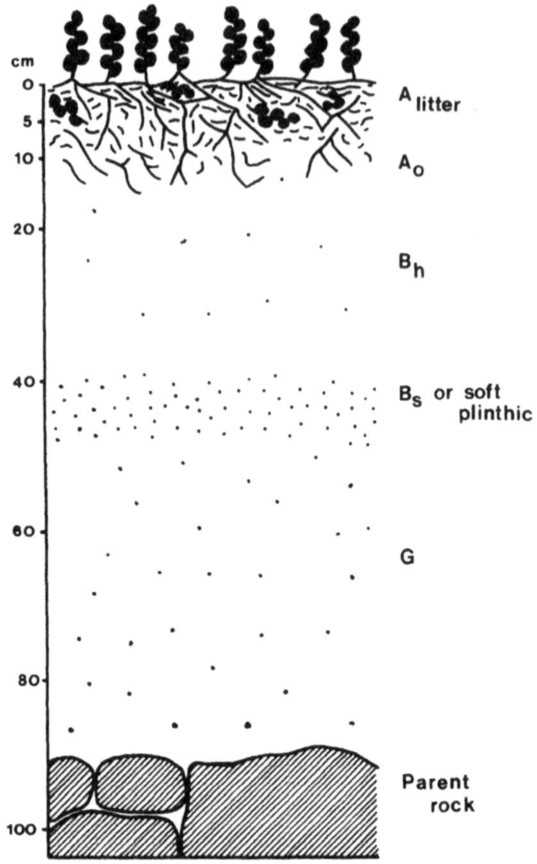


*Fig. 15.* Profile of the soil of a lowland slope (after Smith 1976c).

*Table 6.* Chemical data of the profile of peat soil illustrated in Fig. 14 (after Smith 1976c).

| Depth (cm) | pH | Loss at 105 °C (%) | Org. C (%) | Tot. N (%) | Tot. P (%) | C.E.C. | exch. $Ca^{2+}$ | exch. $Mg^{2+}$ | exch. $Na^{+}$ | exch. $K^{+}$ |
|---|---|---|---|---|---|---|---|---|---|---|
| | | | | | | (milli-equivalents · 100 $g^{-1}$) | | | | |
| 0–5 | 4.2 | 1931 | 45.9 | 2.30 | 0.66 | 42.7 | 2.4 | 6.3 | 2.2 | 1.6 |
| 5–20 | 4.4 | 1403 | 46.4 | n.d. | n.d. | n.d. | n.d. | n.d. | n.d. | n.d. |
| 20–40 | 5.0 | 365 | 30.3 | 0.76 | 0.52 | 28.4 | 1.7 | 0.8 | 0.7 | 0.1 |
| 50–70 | 4.9 | 218 | 24.6 | 0.27 | 0.60 | 15.9 | 2.2 | 0.9 | 0.9 | 0.2 |
| 90–100 | 4.6 | 477 | 10.4 | 0.67 | 0.54 | 13.0 | 3.3 | 1.3 | 0.8 | 0.2 |

*Table 7.* Chemical data on the soil profile of a lowland slope illustrated in Fig. 15 (after Smith 1976c).

| Depth (cm) | pH | Loss at 105 °C (%) | Org. C (%) | Tot. N (%) | Tot. P (%) | C.E.C. | exch. $Ca^{2+}$ | exch. $Mg^{2+}$ | exch. $Na^{+}$ | exch. $K^{+}$ |
|---|---|---|---|---|---|---|---|---|---|---|
| | | | | | | (milli-equivalents · 100 $g^{-1}$) | | | | |
| 0– 5 | 4.2 | 655 | 37.2 | 2.40 | 0.65 | 41.8 | 6.1 | 13.4 | 2.0 | 4.0 |
| 5–13 | 4.2 | 590 | 33.8 | 2.23 | 1.09 | 25.3 | n.d. | n.d. | n.d. | n.d. |
| 20–40 | 4.3 | 475 | 20.4 | 1.45 | 1.01 | 27.1 | 1.4 | 0.6 | 0.6 | 0.3 |
| 40–50 | 4.5 | 344 | n.d. | 0.77 | 0.82 | 23.1 | 0.9 | 0.6 | 0.4 | 0.1 |
| 75–85 | 5.3 | 426 | n.d. | n.d. | n.d. | 26.5 | 0.6 | 0.4 | 0.3 | tr |

elevations, accumulation of organic matter is small. The soils of these areas are skeletal, gravelly loams which are generally very shallow (Fig. 16, Table 8). The surface of these soils is often covered with a pavement of rocks and gravel. Principal factors involved in the formation of the pavement seem to be wind and water which carry away the fine soil particles. At high elevations frost action may effect some sorting of stones (Zinderen Bakker 1978; cf. Troll 1944). In formerly glaciated areas pavements of grey lava blocks may, in fact, be the remains of ablation moraines.

In some areas (e.g. young black lava flows and scoria deposits) no appreciable soil formation has taken place.

The chemical status of Marion Island soils is described by Smith (1976a,b,c,d, 1978a,b, 1979). Some data are presented in Tables 6 to 9. Similar to those of many polar and subpolar regions, the soils of the island are low in nitrogen and phosphorous (Smith 1976a; Lindeboom 1979). Physical and chemical weathering seem to play a minor role as mechanisms which supply nutrients to the soil (Smith 1977a). Due to the high precipitation leaching of minerals from the soil is rapid. The major source of mineral nutrients is the ocean, directly via salt-spray (e.g. Na, Mg, and K), or indirectly, via the excretion of sea-going animals (N and P; Smith 1978a). Smith (1977a) states that although the lowland coastal areas are most affected, virtually no part of the islands is free of chemical influcence by salt-spray.

Animal excreta are the most important source of nitrogen on the islands. Lindeboom (1979) calculated an annual input of 30–40 metric tons of nitrogen into the vegetated areas of Marion Island from bird excreta directly deposited on the vegetation. Another 40–50 tons of nitrogen originates from ammonia which evaporates from penguin rookeries and is subsequently deposited in other areas. Elephant seals are responsible for the direct deposition of 7–10 tons of nitrogen in vegetated areas and an additional 2–5 tons of nitrogen in the form of ammonia evaporated from their wallows. Nitrogen-fixation by cyanobacteria (notably *Nostoc commune* Vaucher) accounts for the addition of 100–200 kg of nitrogen into the vegetated areas of Marion Island (Lindeboom 1979; cf. Croome 1973). This quantity is small compared to other sources of nitrogen, but may be of significance in areas not influenced by animals (Lindeboom 1979; cf. Smith 1976a,c). The input of nitrogen from the weathering of rocks, which Lindeboom (1979) estimates at less than 6 kg annually on Marion Island, is negligible.

Additional data on the islands' soils are given in the descriptions of the habitats of the plant communities.

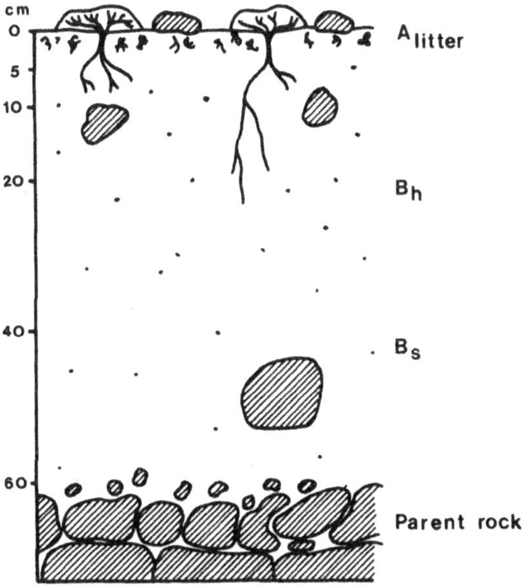


*Fig. 16.* Profile of a fjaeldmark soil (after Smith 1976c).

*Table 8.* Chemical data on the fjaeldmark soil illustrated in Fig. 16 (after Smith 1976c).

| Depth (cm) | pH | Loss at 105 °C (%) | Org. C (%) | Tot. N (%) | Tot. P (%) | C.E.C. | exch. $Ca^{2+}$ | exch. $Mg^{2+}$ | exch. $Na^{+}$ | exch. $K^{+}$ |
|---|---|---|---|---|---|---|---|---|---|---|
| | | | | | | (milli-equivalents · 100 $g^{-1}$) | | | | |
| 3–15 | 5.0 | 198 | 11.9 | 0.73 | n.d. | 20.6 | 0.6 | 1.3 | 1.0 | 0.5 |
| 15–33 | 5.2 | 229 | 6.7 | 0.51 | 0.47 | 21.2 | 0.4 | 0.1 | 0.8 | 0.3 |
| 33–60 | 5.5 | 77 | 2.7 | 0.20 | 0.66 | 19.0 | 0.1 | 0.1 | 0.4 | 0.1 |
| 60–? | 5.6 | 58 | 1.9 | 0.13 | 0.65 | 13.7 | 0.1 | 0.1 | 0.8 | 0.2 |

*Table 9*. Chemical status of the horizons containing living roots in the soils under Marion Island plant communities (after Smith 1976a, 1977d).

| Soil type and vegetation | N | Depth (cm) | pH | Water (%) | Ca (% saturation of C.E.C.) | Mg | Na | K | Total | Org. C (%) | Total N (%) | Total P (%) | $NH_4$-N (mg · 100 $g^{-1}$ soil) | $NO_3$-N |
|---|---|---|---|---|---|---|---|---|---|---|---|---|---|---|
| **Peat soils** | | | | | | | | | | | | | | |
| *Agrostis magellanica* mire | 10 | 0–20 | 4.2 | 1667 ± 199 | 5 ± 1.5 | 13 ± 0.5 | 5 ± 0.7 | 3 ± 1.0 | 26 ± 1.5 | 48 ± 5 | 2.3 ± 0.4 | 0.6 ± 0.5 | 2.4–6.5 | 0.0–0.7 |
| **Slope soils** | | | | | | | | | | | | | | |
| *Blechnum penna-marina* fernbrake | 8 | 0–18 | 4.3 | 715 ± 67 | 21 ± 5.2 | 40 ± 2.2 | 4 ± 0.8 | 7 ± 2.1 | 72 ± 5.4 | 44 ± 1 | 2.2 ± 0.2 | 0.7 ± 0.2 | tr–1.6 | 0.0–tr |
| *Acaena magellanica* drainage line | 6 | 0–45 | 4.2–4.7 | 1002 ± 275 | 26 ± 1.1 | 32 ± 9.0 | 4 ± 1.9 | 2 ± 1.0 | 65 ± 10.8 | 43 ± 11 | 2.5 ± 0.4 | 0.7 ± 0.2 | tr–2.9 | 0.0–tr |
| *Poa cookii* tussock grassland | 8 | 0–50 | 4.0–4.1 | 360 ± 23 | 14 ± 1.2 | 13 ± 3.0 | 5 ± 2.2 | 1 ± 0.5 | 34 ± 4.1 | 25 ± 5 | 1.7 ± 0.3 | 1.6 ± 0.5 | 0.9–8.4 | 0.0–1.0 |
| **Fjaeldmark soils** | | | | | | | | | | | | | | |
| *Azorella selago* fjaeldmark | 4 | 0–10 | 5.2* | 217 ± 168 | 3 ± 2.1 | 9 ± 3.4 | 3 ± 1.9 | 2 ± 3.3 | 17 ± 5.9 | 6.0* | 0.7* | 0.4* | 0.0–0.5 | 0.0–tr |

N = number of determinations;
* = single determination;
tr = trace (<0.5 mg · 100 $g^{-1}$ soil).
Values expressed as ranges or as means ± standard deviations.

## 2.6 Phytogeography

Knowledge of the phytogeography of the Subantarctic zone is incomplete. This is mainly due to the fragmentary nature of data available on the cryptogamic floras of many southern regions and also because no consistent taxonomic assessment of these floras has been made.

Marion and Prince Edward Islands form part of the Antarctic Kingdom (Engler 1964). Good (1974) also places them within the Antarctic Kingdom. This author, however, includes larger parts of South America and New Zealand and leaves out the Antarctic continent and some outlying island groups.

A division of the southern circumpolar regions based on floristic considerations was proposed by Skottsberg (1905) and later revised by himself (Skottsberg 1960). This division is shown in Fig. 17. Skottsberg defined the Subantarctic Zone as the area between 48 and 60° latitude, including Marion and Prince Edward Islands and Crozet, but excluding Bouvet and the South Sandwich Islands. Within the Subantarctic Zone Skottsberg (1960) discerned three phytogeographic provinces: (1) The Kerguelen Province consists of Marion and Prince Edward Islands, Crozet, Kerguelen, Heard and MacDonald. (2) The Magellanian Province comprises South Georgia, Falkland Islands and Subantarctic South America. (3) Macquarie, Auckland, Campbell, Antipodes and Snares Islands together form the Province of the Subantarctic Islands of New Zealand.

A more restricted definition of the Subantarctic Zone is given by Wace (1960, 1965, 1969). His division of southern circumpolar regions into four zones (High Antarctic, Maritime Antarctic, Subantarctic and Cool Temperate; Fig. 18) is based on physiognomic and ecological characteristics of the vegetation of these areas. Wace (loc. cit.) defines the Subantarctic Zone as the region south of the southern limit of growth of trees and woody shrubs, and north of the limit of closed phanerogamic communities. This zonation coincides with a zonation based on the range of mean monthly temperatures (Holdgate 1964; Gressit 1970). The temperature regime of the Subantarctic Zone is characterized by the absence of months warmer than 8.5 °C which means, according to Holdgate (1964), that it is too cold to permit tree growth. The temperature of the coldest month ranges from +3 ° to −3 °C. The Subantarctic Zone comprises six islands and island groups, all situated within a few degrees latitude of the Antarctic Convergence: Marion and Prince Edward Islands, Crozet, Kerguelen, Macquarie, Heard and MacDonald, and South Georgia. The terrestrial vegetation of these islands is characterized by the presence of mire in which the important peat-forming plants are bryophytes (other than *Sphagnum* spp.), closed communities of tussock-forming grasses, cushion-forming flowering plants, and herbfield communities with large-leaved perennial species (Wace 1960, 1965, 1969).

Twenty-two species of vascular plants are indigenous to Marion and Prince Edward Islands. These species and their distribution are listed in Table 10. Data on the distribution of these plants were compiled by Rothuizen (1977). Seven species are restricted to the Kerguelen Province. One of these (*Elaphoglossum randii*) is endemic to Marion and Prince Edward Islands (Alston & Schelpe 1957; Huntley 1971). Seven species are restricted to islands of the Kerguelen Province, e.g. *Polystichum marionense* (Fig. 19), *Ranunculus moseleyi, Poa cookii* (Fig. 35) and *Pringlea antiscorbutica* (Fig. 20), but most of the islands' vascular species show a widespread distribution throughout the temperate regions of the southern hemisphere. Wace (1965) suggests that many of these species, which occur in remote islands as well as on the continents, may have been driven out of Tertiary Antarctica and have achieved circumpolar ranges today by transoceanic dispersal during the Quaternary.

The status of two species on Marion and Prince Edward Islands, *Juncus* cf. *effusus* L. and *Potamogeton* spec., is not conclusively ascertained. Huntley (1971) considers these species, which show a very limited distribution on the islands, as recent additions to the indigenous flora. Only vegetative material of *Potamogeton* was found. *Juncus* cf. *effusus,* however, is known to flower on Marion Island (V. R. Smith, personal communication).

Two subspecies of *Montia fontana* have been found on the islands. Material collected on Marion in 1873 by Moseley during the Challenger expedition is identified by Moore (1963) as subspecies *chondrosperma* (Fenzl.) Walters. This subspecies is considered as almost certainly introduced by man into the Southern Hemisphere (Moore 1963; cf.

Greene & Greene 1963; Greene & Walton 1975). Specimens collected by Huntley on Marion (coll. nr. 999) and Prince Edward Island (coll. nr. 647) belong to the subspecies *fontana*. The latter may have migrated to the Southern Hemisphere via the mountains of either the Old or New Worlds or it may be there as a result of its introduction by man (Moore 1963). The presence of *Montia*-pollen grains in peat deposits on Marion of at least 10 000 years old (Schalke & Zinderen Bakker 1967, 1971) shows the native status of *Montia fontana* on this island, but does not exclude the possibility that one of the subspecies was introduced by man.

A list of moss species known from Marion and Prince Edward Islands and data on their distributions are given in Table 11, compiled from Clifford 1953, Fineran 1971, Greene 1968a,b, Hébard 1970a, Sainsbury & Allison 1962, Zanten 1971,

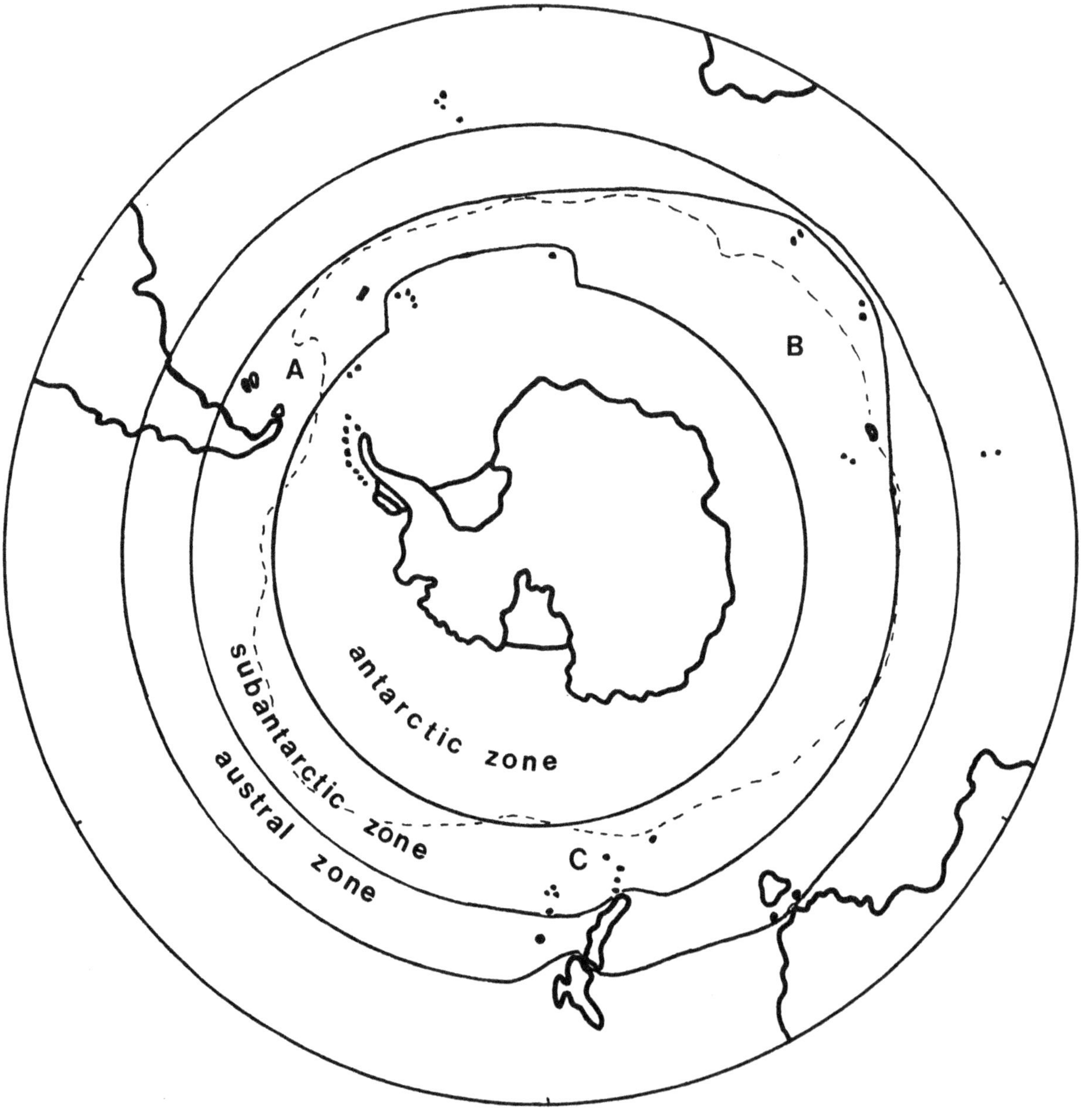


*Fig. 17.* Phytogeographical division of southern circumpolar regions after Skottsberg (1905, 1960). A = Magellanian Province; B = Kerguelen Province; C = Province of the Subantarctic islands of New Zealand.

Seppelt 1977 and my own data. One third of the 72 species of Musci known from the islands are widely spread throughout southern cold and temperate regions. Five species show a bipolar distribution, eight are cosmopolitan. No mosses are endemic to Marion and Prince Edward, but ten species are restricted to two or more islands or island groups of the Kerguelen province. Five species are found in the Kerguelen Province and in South America or its outlying islands, 13 in the Kerguelen Province and in the Australia-New Zealand sector of the Southern Hemisphere.

A high incidence of endemism is shown by the hepatics of Marion and Prince Edward islands (Grolle 1971a). Twenty percent (seven species) of the liverworts are endemic, and a further 20% are restricted to the Kerguelen Province. Five of the 36 species are found in islands of the Kerguelen

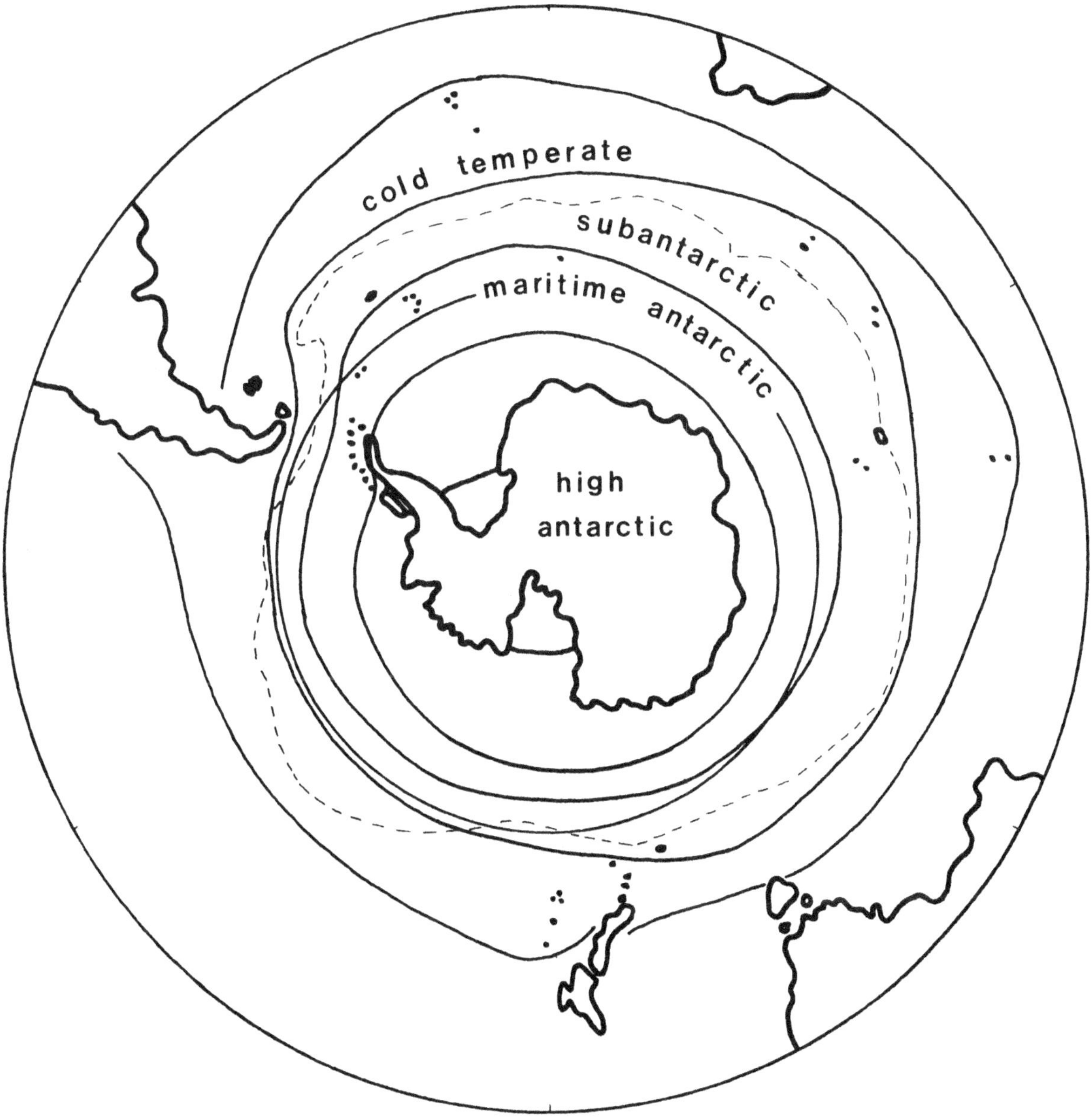


*Fig. 18.* Phytogeographical division of southern circumpolar regions after Wace (1960, 1965, 1969).

*Table 10.* The indigenous vascular plants of Marion and Prince Edward Islands and their distribution.

| | Marion Island | Prince Edward Island | Iles Crozet | Iles Kerguelen | Heard Island | Macdonald Island | New Amsterdam | St. Paul | Macquarie | Campbell Island | Auckland Islands | Snares Islands | Antipodes Islands | New Zealand | Tasmania | Australia | Tristan da Cunha | Gough Island | South Georgia | Falkland Islands | Fuegia | South America | |
|---|---|---|---|---|---|---|---|---|---|---|---|---|---|---|---|---|---|---|---|---|---|---|---|
| *Acaena magellanica* (Lam.) Vahl. | + | + | + | + | + | | | | + | | | | | | | | | | + | + | + | + | |
| *Agrostis magellanica* Lam. | + | + | + | + | | | | | + | + | + | | + | + | | | | | | + | + | + | |
| *Azorella selago* Hook. f. | + | + | + | + | + | + | | | + | | | | | | | | | | | + | + | | |
| *Blechnum penna-marina* (Poir.) Kuhn. | + | + | + | + | | | + | + | + | + | + | | + | + | + | + | + | + | + | + | + | + | |
| *Callitriche antarctica* Engelm. | + | + | + | + | + | | + | | + | + | + | + | + | | + | + | | | + | + | + | + | |
| *Colobanthus kerguelensis* Hook. f. | + | + | + | + | | | | | | | | | | | | | | | | | | | |
| *Cotula plumosa* Hook. f. | + | + | + | + | | | | | + | + | + | | + | | | | | | | | | | |
| *Crassula moschata* Forst. f. | + | + | + | + | | | | | + | + | + | + | + | + | | | | | | + | + | + | Chatham Island |
| *Elaphoglossum randii* Alston & Schelpe | + | + | | | | | | | | | | | | | | | | | | | | | |
| *Grammitis kerguelensis* Tard. | + | + | + | + | | | + | + | | + | + | | + | + | + | + | + | + | + | + | + | + | Ascension Island |
| *Hymenophyllum peltatum* (Poir.) Desv. | + | + | + | + | | | | | + | | | | | + | + | + | + | + | | | | | N-W. Europe |
| *Juncus scheuchzerioides* Gaud. | + | + | + | + | | | | | + | + | | | | + | + | + | | | + | + | + | + | |
| *Limosella australis* R. Br. | + | + | + | + | | | | | | | + | | | + | + | + | | | | | + | + | N. temperate regions; S. Africa |
| *Lycopodium magellanicum* Sw. | + | + | + | + | | | | | | | + | | + | + | + | + | + | + | + | + | + | + | Chatham Island; Juan Fernand |
| *Lycopodium saururus* Lam. | + | + | + | + | | | + | | | | | | | | | | + | + | | | | | * |
| *Montia fontana* L. | + | + | + | + | | | | | + | + | + | | + | + | | + | | | + | + | + | + | ** |
| *Poa cookii* Hook. f. | + | + | + | + | + | + | | | | | | | | | | | | | | | | | |
| *Polystichum marionense* Alston & Schelpe | + | + | + | | | | | | | | | | | | | | | | | | | | |
| *Pringlea antiscorbutica* R. Br. | + | + | + | + | | + | | | | | | | | | | | | | | | | | |
| *Ranunculus biternatus* Sm. | + | + | + | + | | | | | + | | | | | | | | | | + | + | + | + | |
| *Ranunculus moseleyi* Hook. f. | + | | + | + | | | | | | | | | | | | | | | | | | | |
| *Uncinia compacta* R. Br. | + | + | + | + | | | + | | + | + | | | | + | + | + | + | + | | | | | |

*Reunion; Madagascar; Mauritius; Rodrigues; St. Helena; S. Africa
**Northern Hemisphere; New Guinea; E. Afr. mnts.

Province and the South American sector. Four species show a similar distribution pattern, but include islands of the Tristan da Cunha group. The relationship between the liverworth-floras of the islands and the Australia – New Zealand region is much less pronounced than in the mosses. A list of species and their distributions is given in Table 12, compiled from Arnell (1953, 1963), Grolle (1964, 1971a,b, 1972), Hodgson (1962), Kuwahara (1966), Seppelt (1977) and my own data.

*Fig. 19. Polystichum marionense.*

*Fig. 20. Pringlea antiscorbutica.*

Fifty species of lichens are known from Marion and Prince Edward Islands (Lindsay 1976 and my own data, Table 13). Two of these appear to be endemics. Eleven species are restricted to islands of the Kerguelen Province. Remarkable is the high proportion (40%) of species with bipolar distributions.

Distribution patterns of the plant species of the islands indicate a strong relationship between the floras of Marion and Prince Edward and the other islands of the Kerguelen Province. Similar relationships were found in several faunal groups, e.g. Entomostraca (Smith & Sayers 1971), Araneida (Lawrence 1971) and Insecta (Dreux 1971).

Most of the plant species of the islands are widely distributed in southern cold and temperate regions, indicating the ability of these species to cross wide expanses of ocean. Several of the flowering plants have fruits adapted to transport by animals, e.g. *Acaena magellanica* and *Uncinia compacta*. The barbed fruits of these species were often observed attached to the feathers of birds (cf. Kok 1975; Fig. 21). On South Georgia Walton (1977) made similar observations on the fruits of *Acaena*. The ability of oceanic birds in subantarctic regions to cover large distances within a short time is well documented (e.g. Falla 1960; Berutti et al. 1975). The most spectacular example cited by Falla (1960) is of a giant petrel (*Macronectes giganteus*) which was picked up in Western Australia five weeks after its probable date of departure from Signy Island, South Orkney Islands, 11 000 km away. Driftwood of *Nothofagus pumilis* found on Marion (Zinderen Bakker 1971a) indicates the possibility of transport by sea currents from South America to the islands. Similar observations were made on Tasmania and

*Table 11.* **The Musci of Marion and Prince Edward Islands and their distribution.**

| | Marion Island | Prince Edward Island | Iles Crozet | Iles Kerguelen | Heard and Macdonald Islands | Macquarie | Campbell Island | Auckland Islands | New Zealand | Tasmania | Australia | Tristan de Cunha + Gough Island | Bouvet | South Georgia | Falkland Islands | Fuegia | South America | Antarctic zone | Southern Africa | |
|---|---|---|---|---|---|---|---|---|---|---|---|---|---|---|---|---|---|---|---|---|
| *Acrocladium auriculatum* (Mont.) Mitt. | + | | | | | | + | + | + | + | + | | | | | + | + | | | |
| *Andreaea acuminata* Mitt. | + | + | + | + | | | | + | + | + | | + | | | | + | | | | |
| *Andreaea acutifolia* Hook. f. & Wils. | + | | + | + | | + | + | + | + | | ? | | | | + | + | | | | |
| *Andreaea gainii* Card. | + | + | | | | | | | | | | | | + | | | | + | | |
| *Andreaea regularis* C. Muell. | + | + | | | | | | | | | | + | + | + | + | + | + | + | | |
| *Anisothecium hookeri* (C. Muell.) Broth. | + | | + | + | + | | | | | | | + | | + | | + | | + | | |
| *Bartramia patens* Brid. | | | | | | | | | | | | | | | | | | | | |
| var. *papillata* (Hook. f. & Wils.) Zant. | + | + | | | + | + | + | + | | | | | | | | | | | | |
| *Blindia magellanica* Schimp. | + | | | | | | | | + | + | ? | | | | | + | + | | | |
| *Brachythecium austro-salebrosum* (C. Muell.) Par. | + | | | + | | | | | | | | | + | + | + | + | | | | |
| *Brachythecium paradoxum* (Hook. f. & Wils.) Jaeg. | + | | + | + | | | | | + | + | + | | | | | + | + | | | |
| *Brachythecium rutabulum* (Hedw.) B.S.G. | + | | | | | + | + | + | + | + | | | | | | + | + | | | Northern Hemisphere |
| *Brachythecium subplicatum* (Hamp.) Jaeg. | + | | + | + | | | | | + | | | | | + | + | + | + | | | |
| *Breutelia integrifolia* (Tayl.) Jaeg. | + | | + | + | | | | | | | | + | | + | | + | + | | | |
| *Breutelia pendula* (Sm.) Mitt. | + | | + | + | | + | + | + | + | + | + | + | | + | + | + | | | | |
| *Bryum amblyolepsis* Card. | + | | | + | | | | | | | | | | | | | + | | | |
| *Bryum argenteum* Hedw. | + | | | + | | | | | | | | | | + | | | + | | | cosmopolitan |
| *Bryum dichotomum* Hedw. | + | | + | | | | | | + | | | | | | | + | | | | |
| *Bryum eatonii* Mitt. | + | | | + | | | | | | | | | | | | | | | | |
| *Bryum kerguelense* Mitt. | + | | | + | | | | | | | | | | | | | | | | |
| *Bryum laevigatum* Hook. f. & Wils. | + | + | | + | | + | + | | + | + | + | | | | + | + | + | | + | |
| *Campylopus arboricola* Card. & Dix. | + | + | | | | | | + | + | + | | | | | | | | | | |
| *Campylopus cavifolius* Mitt. | + | | | + | | | | | | | | | + | | + | + | + | | | |
| *Campylopus clavatus* (R. Brown) Wils. | + | | | | | + | + | + | + | + | + | | | | | | + | | | |
| *Campylopus introflexus* (Hedw.) Brid. | + | + | + | + | | + | + | + | + | + | + | + | | + | + | + | + | | + | introduced in W. Euro |
| *Campylopus polytrichoides* De Not. | + | | | | | | | | | | | + | | | | | | | + | S-W. Europe; Ceylon |
| *Campylopus subnitens* Kaal. | + | + | + | | | | | | | | | | | | | | | | + | |
| *Catagonium politum* (Hook. f. & Wils.) Broth. | + | + | | + | | | | | + | + | + | | | | | + | + | | + | |
| *Ceratodon cf. purpureus* (Hedw.) Brid. | + | | | | + | + | | + | | | | | | | | | | | | cosmopolitan |
| *Cheilotheca chilensis* (Mont.) Broth. | + | | | | | | | | + | | | | | | | | + | | | |
| *cf. Cratoneuron filicinum* (Hedw.) Spruc. | + | | | | | | | | + | | | | | | | | + | | + | |
| *Dicranoloma billardieri* (Brid.) Par. | + | | + | + | + | | + | + | + | + | + | + | | | + | + | + | | + | |
| *Dicranoweisia brevipes* (C. Muell.) Card. | | + | | | | | | | | | | | | | + | | | | | |
| *Dicranoweisia breviseta* Card. | + | + | | | | | | | | | | | | | | | + | | | |
| *Distichium capillaceum* (Hedw.) B.S.G. | + | | + | | | + | | | | | + | | | | | + | + | | | Northern Hemisphere |
| *Distichophyllum fasciculatum* Mitt. | + | + | | | | | | | | | | + | | | | | | | | |
| *Distichophyllum imbricatum* Mitt. | + | | + | | | | | | | | | | | | | | | | | |
| *Ditrichum conicum* (Mont.) Mitt. | + | + | | + | | | | | | | | + | | | | + | + | | + | |
| *Ditrichum immersum* Zant. | + | | + | | | | | | | | | | | | | | | | | |
| *Ditrichum strictum* (Hook. f. & Wils.) Mitt. | + | + | + | + | | + | + | + | + | + | + | | | | + | + | + | | + | |
| *Drepanocladus aduncus* (Hedw.) Warnst. | + | + | | | | + | | + | + | | | | | | | + | + | | | Northern Hemisphere |
| *Drepanocladus uncinatus* (Hedw.) Warnst. | + | + | + | + | + | | | | + | + | + | + | | + | + | + | + | + | + | cosmopolitan |
| *Eriopus apiculatus* (Hook. f. & Wils.) Mitt. | + | | | | | | | + | + | + | + | | | | + | + | + | | | |
| *Eustichia longirostris* (Brid.) Brid. | + | | | | | | | | | | | + | | | | | + | | + | |
| *Funaria laxa* (Hook. f. & Wils.) Broth. | + | | + | + | | | | | | + | | | | | | | | | | |
| *Grimmia kerguelensis* Card. | + | + | + | + | | | | | | | | + | | | | | | | | |
| *Holodontium pumilum* (Mitt.) Broth. | + | + | | | | | | + | | | | | | | | + | + | | | |

*able 11.* (Continued).

| Species | Marion Island | Prince Edward Island | Iles Crozet | Iles Kerguelen | Heard and Macdonald Islands | Macquarie | Campbell Island | Auckland Islands | New Zealand | Tasmania | Australia | Tristan de Cunha + Gough Island | Bouvet | South Georgia | Falkland Islands | Fuegia | South America | Antarctic zone | Southern Africa | |
|---|---|---|---|---|---|---|---|---|---|---|---|---|---|---|---|---|---|---|---|---|
| *ygroamblystegium filum* (C. Muell.) Reim. | + | | + | + | | | | | | | | | | | | + | | | | |
| *ypnum cupressiforme* Hedw. | + | + | + | + | | + | + | + | + | + | + | + | | | | + | + | | + | cosmopolitan |
| *opterygium pulchellum* (Hedw.) Jaeg. | | | | | | | | | | | | | | | | | | | | |
| var. *antarcticum* (Mitt.) Zant. | + | + | | + | | | | | | | | | | | | + | | | | |
| *eptodontium proliferum* Herz. | + | + | | | | | | | | | | | | | | | + | | | |
| *ielichoferia campylocarpa* (Hook. & Arnott) Mitt. | + | | + | + | | | | | + | + | + | | | + | | | + | + | + | |
| var. *australis* (Hamp.) Zant. | + | | | | | | | | + | | + | | | | | | | | | |
| *rthodontium lineare* Schwaegr. | + | | | | | | | | + | + | + | | | | + | + | + | | + | introduced in Europe |
| *rthotricum crassifolium* Hook. f. & Wils. | + | | + | + | | + | + | + | + | | | | | + | + | + | + | | | |
| *hilonotis cf. augustifolia* Kaal. | + | | + | | | | | | | | | | | | | | | | | |
| *ilonotis scabrifolia* (Hook. f. & Wils.) Broth. | + | + | | + | | | | + | + | + | + | + | | + | + | + | + | | + | |
| *hilonotis tenuis* (Tayl.) Jaeg. | + | + | | | | | | + | + | + | + | + | | | | | | | | |
| *agiothecium platyphyllum* Moenk. | + | + | | | | | | | | | | | | | | | | | | Northern Hemisphere |
| *ohlia albicans* (Wahlenb.) Lindb. | + | | | | | | + | | | | | | | | | | | | | cosmopolitan |
| *ohlia nutans* (Hedw.) Lindb. | + | | | | | | | | | | | | | + | | | | | | cosmopolitan |
| *olytrichum juniperinum* Hedw. | + | | + | | | + | + | | + | + | + | + | | + | | | + | + | | cosmopolitan |
| *olytrichum piliferum* Hedw. | + | | | | | | | | | | + | | | + | + | + | + | + | + | Northern Hemisphere |
| *ilopilum cf. tristaniense* Dix. | + | | | | | | | | | | | + | | | | | | | | |
| *ychomnion ringianum* Broth. & Kaal. | + | | + | | | | | | | | | | | | | | | | | |
| *acomitrium crispulum* (Hook. f. & Wils.) Hook. f. & Wils. | + | + | + | + | + | + | + | + | + | + | + | | + | + | | + | + | + | + | |
| *acomitrium lanuginosum* (Hedw.) Brid. | + | | + | + | | + | + | | | | | | | + | | | | + | | cosmopolitan |
| *chistidium falcatum* (Hook. f. & Wils.) Zant. | + | | | + | | | | | | | | | | | | | | | | |
| *histidium stylostegium* (C. Muell.) Zant. | + | + | | + | | | | | | | | | | | | | | | | |
| *ortula cf. geheebiaeopsis* (C. Muell.) Broth. | + | | + | + | | | | | | | | | | | | | | | | |
| *ortula cf. rubra* Mitt. | + | | | | | + | | | + | | + | | ? | + | + | + | | | | |
| *ematodon flexipes* Mitt. | + | | | | | + | | | + | + | + | | | | | | | | | |
| *illia autro-leucophaea* (Besch.) Broth. | + | + | | | | | | | | | | | | + | | + | | | | |
| *errucidens microcarpus* (Mitt.) Zant. | + | + | | + | | | | | | | | | | | | | | | | |

Macquarie Island (Barber et al. 1959; Costin 1965). It is not known, however, whether or not any of the islands' terrestrial plant species possesses propagules which may survive prolonged immersion in sea water.

Aerial long distance transport is probably the most important mechanism by which cryptogamic species reached the islands (cf. Grolle 1969; Zanten 1976, 1978). Zanten (loc. cit.) has shown that spores of a number of widely distributed moss species are able to survive desiccation and freezing which may occur during aerial transport. The presence of pollen grains from South American sources in peat cores from the islands further demonstrates the possibility of aerial long distance transport. From the occurrence of these polen grains Schalke & Zinderen Bakker (1971) conclude that the wind system around Antarctica has not changed during the last 16 000 years. Western Subantarctic (Kerguelen Province and South American sector) and circumsubantarctic distribution patterns may be readily explained by wind dispersal. For Eastern Subantarctic distribution patterns (Kerguelen Province and Australia – New Zealand sector) aerial transport is not such an obvious explanation.

Some striking differences between the various plant groups appear from the distribution data, e.g. the large proportion of endemics in the Hepaticae

*Table 12.* The hepatics of Marion and Prince Edward Islands and their distribution.

| | Marion Island | Prince Edward Island | Iles Crozet | Iles Kerguelen | Heard Island | Macquarie | Auckland Islands | Campbell Island | Antipodes Islands | New Zealand | Tasmania | Australia | Tristan de Cunha + Gough Island | South Georgia | Falkland Islands | Fuegia | South America | Africa | Maritime Antarctic | |
|---|---|---|---|---|---|---|---|---|---|---|---|---|---|---|---|---|---|---|---|---|
| *Acrobolbus ochrophyllus* (H. f. & T.) Schust. | + | + | + | + | | | + | | | + | | | + | | | + | + | + | | |
| *Anastrophyllum auritum* (Lehm.) Steph. | + | | + | | | | | | | | | | | | | | | + | | |
| *Andrewsianthus carinatus* Grolle | + | | | | | | | | | | | | | | | | | | | |
| *Andrewsianthus lancistipus* Grolle | + | + | | | | | | | | | | | | | | | | | | |
| *Andrewsianthus marionensis* (S. Arn.) Grolle | + | | | | | | | | | | | | | | | | | | | |
| *Blepharidophyllum densifolium* (Hook.) Angstr. | + | + | + | + | | | | | | | | | + | | | + | + | | | |
| *Cephalozia randii* S. Arn. | + | | + | | | | | | | | | | | | | | | | | |
| *Clasmatocolea humilis* (H. f. & T.) Grolle | + | + | + | + | | | | | | | | | + | | | | + | | | |
| *Clasmatocolea vermicularis* (Lehm.) Grolle | + | + | | | | | | | | | | | + | + | + | + | + | + | | |
| *Colura calyptrifolia* (Hook.) Dum. | + | + | | | | | | | | | | | + | | | | + | + | | S-W. Europe |
| *Diplophyllum marionense* S. Arn. | + | | | | | | | | | | | | | | | | | | | |
| *Fossombronia australis* Mitt. | + | + | | + | + | | | | | + | | | | | | | | | | |
| *Herzogobryum atrocapillum* (H. f. & T.) Grolle | + | | + | + | | | | | | | | | | | | | | | | |
| *Herzogobryum teres* (Carringt. & Pears.) Grolle | + | | | | | | | | | + | + | | + | | | + | | | | |
| *Herzogobryum vermiculare* (Schiffn.) Grolle | + | | | + | | | | | | | | | | + | + | | + | | | |
| *Hygrolembidium ventrosum* (Mitt.) Grolle | + | | | + | | | | | | | | | | | | | | | | |
| *Jamesoniella colorata* (Lehm.) Schiffn. | + | + | + | | | + | + | + | | | + | | + | | | | + | + | | |
| *Jamesoniella grandiflora* (Lindenb. & Gottsche) Steph. | + | + | | | | | | | | | + | | + | | | | + | + | | |
| *Jensenia pisicolor* (H. f. & T.) Grolle | + | + | + | + | | | | | | | | | + | | + | + | + | | | |
| *Jungermannia coniflora* Schiffn. | + | + | | + | | | | | | | | | | | | | | | | |
| *Lepidozia laevifolia* (H. f. & T.) Gottsche et al. | + | + | + | | | + | + | + | | | | | | | | | | | | |
| *Leptoscyphus expansus* (Lehm.) Grolle | + | + | + | + | | | | | | | | | + | + | + | + | + | + | | |
| *Lethocolea radicosa* (L. & L.) Grolle | + | + | + | | | | | | | | | | | | + | + | + | | | |
| *Lophocolea randii* S. Arn. | + | + | + | | | | | | | | | | | | | | | | | |
| *Lophozia cylindriformis* (Mitt.) Steph. | + | + | | + | | | | | | | | | | | | | | | | |
| *Marchantia berteroana* L. & L. | + | + | + | | | + | + | + | + | + | + | + | + | + | | | + | + | + | |
| *Metzgeria decipiens* (Mass.) Schiffn. & Gottsche | + | + | | | | | | | + | + | + | + | | | + | | + | | | Java; Japan; Hawaii |
| *Metzgeria grollei* Kuwah. | + | | | | | | | | | | | | | | | | | | | |
| *Metzgeria marionensis* S. Arn. | | + | | | | | | | | | | | | | | | | | | |
| *Pachyglossa fissa* (Mitt.) Herz. & Grolle | + | + | | + | | | | | | | | | | | | + | + | | | |
| *Plagiochila crozetensis* Kaal. | + | + | + | | | | | | | | | | | | | | | | | |
| *Plagiochila heterodonta* (H. f. & T.) Gottsche et al. | + | + | + | + | | | | | | | | | | | | + | + | | | |
| *Schistochila carnosa* (Mitt.) Steph. | + | + | | | | | | | | | | | | + | | + | + | | | |
| *Symphyogyna marionensis* S. Arn. | + | + | | | | | | | | | | | | | | | | | | |
| *Temnona quadripartitum* (Hook.) Mitt. | + | + | + | + | | | + | + | | + | | | | | | + | | | | |
| *Tylimanthus viridis* Mitt. | + | | + | + | | | | | | | | | + | | | + | + | | | |

compared to mosses and lichens, and the strong bipolar element in the lichens. It seems premature to draw any phytogeographic conclusions from this. These patterns may well be caused by the paucity of the data (cf. Zanten 1971, 1978) or be due to the different taxonomic concepts of the specialists working on each group.

## 2.7 History

The islands were discovered on 13 and 14 January 1772 by the French sailor Marion Dufresne, who, thinking that he had found the outposts of the not yet discovered southern continent, called the larger one 'Terre de l'Esperance'. The other one he named

'Ile de la Caverne', after the large cave he sighted near the coast of this island (Marsh 1948, Zinderen Bakker 1971a, Brossard 1972). A few years later in 1776, the islands were sighted by Captain Cook, who gave them the name 'Prince Edward Islands'. After hearing about the discovery by Marion Dufresne, he renamed the larger one Marion Island. The smaller one kept the name Prince Edward Island (Hydrographic Department 1948, Marsh 1948, Zinderen Bakker 1971a).

Possibly Marion Dufresne was not the first one to sight these islands. In 1663 Barent Barentsz. Ham, captain of the Dutch East India Company vessel 'Maerseveen' discovered two islands, which he named 'Dina' (on some maps called 'Denia') and 'Maerseveen' (Leupe 1868). Zinderen Bakker (1971a) suggests that, although the positions given by first mate Michiel Gerritsz Boos in the logbook are very different from those of Marion and Prince Edward, these are the same islands.

Sealers were probably the first people to land on the islands. By 1802 several sealers' camps were established both on Marion and Prince Edward (Tizard et al. 1885). These camps were not permanently inhabited. Parties of sealers were put ashore on the islands at the end of the winter when the seals started to arrive on the islands. At the end of the season they were picked up with their capture. During most of the 19th and early 20th century sealers had settlements on the islands and visits by sealing expeditions from South Africa, England, Norway and the United States are recorded (Marsh 1948, Rand 1956, Roberts 1958). The last sealing expedition to the islands took place in 1930, when the Cape Town-based sealer Kildalkey landed a party on Marion Island. Unfavourable weather curtailed this expedition and forced the Kildalkey to return. No further attempt to resume sealing on the islands has been made (Marsh 1948, Rand 1955b, 1956).

Three shipwrecks are known to have taken place on Prince Edward Island. On 19 June 1849 the British emigrant ship Richard Dart, on passage from England to New Zealand, was wrecked on this island. Eleven survivors out of a complement of 63 were rescued on 2 September. The South African sealing vessel Maria was wrecked on 17 May 1857. The survivors spent the winter on the island and were rescued on the 6th of December. On 22 October 1912 the Cape Town-based sealer Sea Bird was wrecked on Prince Edward Island. All but two of the crew survived a stay of nearly six months on the island and were rescued on the 14th of April of the following year. The Norwegian sealing vessel Solglimt was wrecked at Ship's Cove on Marion Island on 16 October 1908. The crew was rescued a month later (Marsh 1948; Roberts 1958).

Marion and Prince Edward Islands were annexed by South Africa on 29 December 1947 and 4 January 1948 respectively. A meteorological station was established at Transvaal Cove on Marion Island in January 1948 and has been occupied continuously since then (Marsh 1948; La Grange 1954; Schulze 1971).

Before the establishment of the weather station, the only scientific exploration of the islands took place during the short visits of two expeditions: the Challenger-expedition in 1873 (cf. Tizard et al. 1885; Moseley 1874, 1877a,b, 1892) and the Bougainville-expedition in 1939 (Jeannel 1940). Between 1948 and 1965 some biological and geological observations were made (e.g. Rand 1954, 1955a,b, 1956; Trusswell, 1965; La Grange 1962), but it was not until 1965–66 that a systematic exploration of the islands took place by a team of biologists and geologists (cf. Zinderen Bakker 1967; Zinderen Bakker et al. 1967, 1971). Since then several biological expeditions to the islands have been made.

**2.8 Human influences on the islands' ecosystems**

As early as 1802, thirty years after their discovery by the French sailor Marion Dufresne, sealers established several camps on both Marion and Prince Edward Islands. Sealing expeditions from South Africa, England, Norway and the United States are known to have visited the islands during the 19th and early 20th century (Tizard et al. 1885; Marsh 1948; Roberts 1958). Rand (1956) found evidence of the occupation of a sealers' camp at Mixed Pickle Cove during 1909–1911 and 1920–1921. The last sealing expedition to the islands took place in 1930 by the Cape Town-based sealer Kildalkey. Over 1000 Elephant Seals were killed before bad weather forced the ship to return prematurely to Cape Town (Marsh 1948; Rand 1955b).

The impact of sealing activities on the population

*Table 13.* The lichens of Marion and Prince Edward Islands and their distribution (after Lindsay 1976 and own data).

| | Marion Island | Prince Edward Island | Iles Crozet | Iles Kerguelen | Heard and Macdonald Islands | Campbell Island | Auckland Islands | New Zealand | Australia | Tristan da Cunha + Gough Island | South Georgia | Falkland Islands | Fuegia | South America | Maritime Antarctic | |
|---|---|---|---|---|---|---|---|---|---|---|---|---|---|---|---|---|
| *Aspiciliopsis macrophthalma* (Hook. f. & Tayl.) Dodge | + | + | + | + | + | | | | | | | | | | | |
| *Arthroraphis citrinella* (Ach.) Poelt | + | | | | | | | | | | | | | | | bipolar |
| *Buellia coniops* (Wahlenb. ex Ach.) Th. Fr. | + | | | | | | | | | | + | | | | + | bipolar |
| *Caloplaca cirrochrooides* (Vain.) Zahlbr. | + | | | | | | | | | | | | | | + | |
| *Caloplaca depauperata* (Müll. Arg.) Zahlbr. | + | | | + | + | | | | | | | | | | | |
| *Caloplaca vitellinella* (Nyl. ex Cromb.) Zahlbr. | + | | | + | + | | | | | | | | | | | |
| *Catillaria impolita* D. Linds. | + | | | | | | | | | | | | | | | |
| *Cladonia balfourii* Cromb. *f. chlorophaeoides* (Vain.) Evans | + | + | | | | | | + | | | + | | | | + | bipolar |
| *Cladonia cf. coccifera* (L.) Willd. | + | | | | | | | + | + | | + | | | | | |
| *Cladonia macilenta* (L.) Hoffm. | + | + | | | | | | + | + | + | | | | | | bipolar |
| *Cladonia phyllophora* Hoffm. | + | | | | | | | | | | | | | | | cosmopolitan (x) |
| *Cladonia pyxidata* (L.) Hoffm. | + | | | | | + | | + | + | | + | | | | + | bipolar |
| *Cladonia cf. scabriuscula* (Det. ex Duby) Leight | | + | | | | | | | + | | | | | | | Europe; N. America (x) |
| *Cladonia subsquamosa* (Nyl.) Vain. | + | | | | | | | | | | | | | | | cosmopolitan (x) |
| *Cladonia cf. vulcanica* Zoll | + | | | | | | | | | | | | | | | (x) |
| *Haematomma erythromma* (Nyl.) Zahlbr. | + | | | | | | | | | | + | | | | + | |
| *Lecanora polytropa* (Hoffm.) Robenh. | + | | | | | | | + | | | | | | | + | |
| *Lecidea atrata* (Ach.) Wahlenb. | + | | | | | | | | | | + | | | | + | bipolar |
| *Lecidea aubertii* B. de Lesd. | + | | | + | + | | | | | | | | | | | |
| *Lecidea homalotera* Nyl. ex Cromb. | + | | + | + | + | | | | | | | | | | | |
| *Lecidea cf. interrupta* Darb. | + | | | | | | | | | | | + | | | | |
| *Lecidea lapicida* (Ach.) Ach. | + | | | | | | | + | | | | | | | | bipolar |
| *Lecidea lygomma* Nyl. ex Cromb. | + | | | + | | | | | | | | | | | | |
| *Lecidea urbanskyana* Zahlbr. | + | | | + | | | | | | | | | | | | |
| *Microglena austrocinerascens* D. Linds. | + | | | | | | | | | | | | | | | |
| *Orceolina kerguelensis* (Tuck.) H. Hertel | + | + | | + | | | | | | | | | | | | |
| *Pannaria dichroa* (Hook. f. & Tayl.) Cromb. | + | | + | + | + | | | | | | | | | | | |
| *Pannaria hookeri* (Borr. ex Sm.) Nyl. | + | | | | | | | | | | + | | | | | bipolar |
| *Parmelia saxatilis* (L.) Ach. | + | + | | | | | | + | + | | + | | | | + | bipolar |
| *Parmelia sulcata* Tayl. | + | + | | | | | | + | | | + | + | | | | bipolar |
| *Parmotrema critinum* (Ach.) Choisy | | + | | | | | | | | | | | | | | cosmopolitan |
| *Peltigera canina* (L.) Willd. | + | | | | | | | + | + | | + | | | | | bipolar |
| *Peltigera polydactyla* (Neck.) Hoffm. | + | + | | | | | | + | + | | + | | | | | bipolar |
| *Physcia cf. phaea* (Tuck.) Thoms. | + | | | | | | | | | | | | | | | bipolar |
| *Placopsis bicolor* (Tuck.) B. de Lesd. | + | | + | + | + | | | | | | | | + | | | |
| *Placopsis cf. cribellans* (Nyl.) Räs. | + | | | | | | | + | | + | + | | + | | | |
| *Pseudocyphellaria crocata* (L.) Vain. | | + | | | | | | + | + | | | | | + | | bipolar |
| *Ramalina cf. subfarinacea* (Nyl. ex Cromb.) Nyl. | | + | | | | | | | | | | | | | | bipolar |
| *Rhizocarpon geographicum* (L.) DC. | + | | | + | + | | | | + | | + | | | | + | |
| *Sporastatia cf. testudinea* (Ach.) Massal. | + | | | | | | | | + | | + | | | | | bipolar |
| *Stereocaulon cymosum* Cromb. | + | | + | + | | | | | | | | | | | | |
| *Usnea insularis* (M. Lamb) Dodge | + | ? | | + | | | | | | | | | | | | |
| *Verrucaria ceuthocarpa* Wahlenb. ex Ach. | + | | | | | | | | + | | + | | | | + | |

*Table 13.* (Continued).

| | Marion Island | Prince Edward Island | Iles Crozet | Iles Kerguelen | Heard and Macdonald Islands | Campbell Island | Auckland Islands | New Zealand | Australia | Tristan da Cunha + Gough Island | South Georgia | Falkland Islands | Fuegia | South America | Maritime Antarctic | |
|---|---|---|---|---|---|---|---|---|---|---|---|---|---|---|---|---|
| *Verrucaria durietzii* M. Lamb | + | | | | | | + | + | | | + | + | + | | | |
| *Verrucaria maura* Wahlenb. ex Ach. | + | | | | | | | | | | | | | | | bipolar |
| *Verrucaria mucosa* Wahlenb. ex Ach. | + | | | | | + | + | | + | | + | | | | + | bipolar |
| *Verrucaria tesselatula* Nyl. | + | | | + | | | | | | | + | + | | | + | bipolar |
| *Xanthoparmelia mougeotii* (Schaer.) Hale | + | | | | | | | + | + | | | | | | | |
| *Xanthoria candelaria* (L.) Th. Fr. | | + | | | | | | | | | | | | | + | continental Antarctica cosmopolitan (x) |
| *Xanthoria elegans* (Link) Th. Fr. | + | | | | | | | | + | | + | | | | + | continental Antarctica |

(x) = Data in the table incomplete

*Fig. 21.* Fruit of *Acaena magellanica* (left) attached to the feathers of a skua (*Stercorarius skua*).

sizes of Elephant Seal (*Mirounga leonina* L.) and Fur Seal (*Arctocephalus tropicalis* Gray) must have been devastating. The latter species had nearly disappeared from the islands: in 1952 the Fur Seal population on Marion Island counted only a few hundred individuals (Rand 1956), but the population has increased considerably during the last few decades. In 1975 there were about 7000 Fur Seals on Marion and 2000 on Prince Edward Island (Condy 1978). Penguins were also taken by sealers (Moseley 1892), but probably no large part of the population was killed. Sealing influenced the oceanic ecosystem in the vicinity of the islands more than the terrestrial ecosystem, though to some extent the nutrient flow from the ocean to the terrestrial environment must have been affected. Furthermore the size of breeding and moulting populations of seals also influences the distribution of certain plant communities. No data are available to judge the quantitative effects of these changes.

Three species of vertebrates have been introduced on Marion Island. Mice (*Mus musculus* L.) reached the island probably with sealers' ships during the 19th century. They are now very abundant all over the island and seem to be the predominant herbivores (Smith 1977a). They are also very important carnivores: in their stomach contents large amounts of insect larvae have been found. (V. R. Smith, personal communication). Domestic cats (*Felis catus* L.) were introduced at the weather station in the early 1950's (Anderson & Condy 1974). The cats, too, (Fig. 22) are important carnivores in the Marion Island terrestrial ecosystem (Smith 1977a). In 1975 the feral population of cats on Marion exceeded 2000. It extensively exploits the population of small petrels (Procellariidae), possibly killing as many as 600 000 birds/year (Van Aarde 1977). A campaign aimed at controlling the cat population is in progress (Williams et al. 1979).

*Fig. 22.* Feral cat on Marion Island.

In the 1960's trout were introduced in the Van den Boogaard River, one of the very few perennial streams on Marion Island (Grobbelaar 1974). The population is small and is not flourishing.

After the annexation of the islands by South Africa in 1948 and the building of the weather station at Transvaal Cove, Marion Island, a supply of sheep and domestic chickens was imported annually from 1950 to 1969 to provide food for the occupants of the station. No feral populations of these animals established themselves. The sheep, numbering from 10 to 60, mainly lived on *Poa cookii*. This selective grazing and the accompanying trampling resulted in the destruction of tussock grassland in some areas (Huntley 1971), but since the disappearance of these animals the grassland seems to have recovered from the damage.

In 1950 and 1951 trees were planted near the meteorological station, some of them in soil imported from South Africa. Of these *Pinus pinaster*, *Chaemaecyparis lawsoniana* and *Pittosporum crassifolium* survived for some time (La Grange 1954), but none longer than a few years. Soil was also imported for a small greenhouse. Soil invertebrates and other organisms certainly have been introduced in this way.

Invertebrates which apparently have been introduced by man are the spider *Myro kerguelensis* Cambr. (Lawrence 1971) and the flies *Fannia canicularis* (L.) and *Limnophyes pusillus* Eaton (Dreux 1971; Séguy 1971).

No alien animals are known from Prince Edward Island.

Fourteen species of alien vascular plants have been recorded from Marion Island and one from Prince Edward (Huntley 1971; Gremmen 1975; Gremmen & Smith 1980). Four of these, *Avena sativa* L., *Holcus lanatus* L., *Hypochoeris radicata* L., and *Plantago lanceolata* L., are transient aliens, recorded from the vicinity of the meteorological

*Fig. 23. Poa annua* (centre) growing between the remains of a sealers' camp at Ship's Cove, Marion Island.

station, and presumably introduced with livestock, fodder or soil from South Africa. By 1974 these species had disappeared from the island (Gremmen 1975). *Rumex acetosella* L. has been recorded from two localities. This species apparently is unable to spread into the surrounding vegetation and is classed as a persistent alien. Nine species are naturalized aliens. Of these *Alopecurus australis* Nees., *Agropyron repens* (L.) P. Beauv., *Poa pratensis* L. and *Festuca rubra* L. show a restricted distribution. The distribution-patterns of these species indicate an introduction at one or more sealers' camps (cf. Fig. 23). *Agrostis stolonifera* L., *Cerastium fontanum* Baumg., *Poa annua* L., *Sagina apetala* Ard. and *Stellaria media* (L.) Vill. have a widespread distribution. These species are able to spread by seed and, with the exception of *Agrostis stolonifera*, may rapidly colonise sites on the island where the plant cover is completely or partially destroyed, e.g. by trampling animals or humans. *Sagina apetala* was presumably introduced at the meteorological station. Comparison of data on the distribution of this species in 1966 and 1975 shows a considerable extension of the area in which it occurs (Gremmen 1975). The population of *Sagina* has increased even more since 1975 (V. R. Smith, personal communication).

Marion and Prince Edward Islands were declared nature reserves by the South African government in 1948. For a number of years now authorities have been trying to avoid further introductions of alien organisms.

CHAPTER THREE

# Methods

## 3.1 Phytosociological methods

In the present survey the methods of the Braun-Blanquet School have been applied. Concepts and methods of this floristic-sociological approach have been treated in detail by e.g. Braun-Blanquet (1928, 1932, 1951, 1964), Ellenberg (1956), Becking (1957), Shimwell (1971), Werger (1973, 1974), Westhoff & Maarel (1973, 1978) and Mueller-Dombois & Ellenberg (1974), and only a very brief outline of the methods will be given here.

Westhoff & Maarel (1973, 1978) state the essence of the Braun-Blanquet approach to classification and interpretation of communities in the following three ideas:

1. Plant communities are conceived as types of vegetation, recognised by their floristic composition. The full species compositions of communities better express their relationships to one another and their environments than any other characteristic.
2. Amongst the species that make up the floristic composition of a community, some are more sensitive expressions of a given relationship than others. For practical classification (and indication of the environment) the approach seeks to use those species whose ecological relationships makes them most effective indicators; these are diagnostic species (character-species, differential species and constant companions).
3. Diagnostic species are used to organise communities into a hierarchical classification of which the association is the basic unit. The vast information with which phytosociologists deal must, of necessity, be thus organised; and the hierarchy is not merely necessary but invaluable for the understanding and communication of community relationships that it makes possible.

*Sampling procedure*

A total of 510 relevés were made, 472 on Marion Island and 38 on Prince Edward Island. For logistic reasons a large proportion of the field-work was carried out within a few kilometres from the meteorological station, but during a number of field trips most other parts of Marion Island were visited and sampled. On Prince Edward Island field work was restricted to the eastern coastal plain and the eastern slopes of the highland.

To determine a suitable quadrat-size species/area curves were made for the most important vegetation types (Braun-Blanquet 1964). Although the minimal area concept is much critisized (cf. Werger 1972, 1973; Westhoff & Maarel 1973, 1978), it is commonly used to determine a suitable size for sample plots. In the extremely species-poor communities dominated by *Crassula moschata, Cotula plumosa, Montia fontana* or *Callitriche antarctica* a minimal area of less than 1 $m^2$ was found satisfactory. In *Poa cookii* tussock grassland 4–10 $m^2$ appeared to be representative, depending on the structure of the vegetation, especially on the density of *Poa* tussocks. In communities dominated by *Acaena magellanica* 4 $m^2$ was sufficient, while in most mire communities a sample of 2 $m^2$ was representative. In some communities dominated by bryophytes $1/10^2$ proved to be a representative sample of the stand. In *Blechnum penna-marina* vegetation 4 $m^2$ was sufficient and in fjaeldmark vegetation $4–10^2$ was found as minimal area. These sizes are of the same order as empirical values for minimal plot sizes found in temperate vegetation-types of similar structure (cf. Westhoff & Maarel 1973, 1978; Mueller-Dombois & Ellenberg 1974). When the uniformity of site and vegetation allowed for it, the quadrat-size was usually taken larger than the minimal values given here.

For each sample plot the following habitat characteristics were noted:

a) location of the quadrat as determined on the topographical map by Langenegger & Verwoerd (1971);
b) elevation above sea level as estimated using the same map;
c) slope angle as measured and slope aspect as determined using the topographical map;
d) notes on the exposure of the site to the wind;
e) exposure to salt-spray, rated on a 6-point scale ranging from 0 (no or very little influence) to 5 (very strong exposure to salt-spray). In estimating the exposure to salt-spray the exposure to the wind, direction and distance to the sea and altitude was taken into account;
f) influence of animals. Trampling and influence of excreta was estimated on a 6-point scale similar to the one used for salt-spray. Notes were made on other activities of animals, e.g. destruction of bryophyte carpets by fouraging birds or the deposition of crop pellets;
g) general notes on the geology and topography of the quadrat area;
h) a general description of the soil. Soil depth was measured down to a depth of 40 cm. A sample of the upper 5–10 cm of the soil was taken for subsequent analysis of water content, loss-on-ignition, pH and electric conductivity; and
i) depth of the groundwater level. To measure the groundwater level a c. 40 cm deep hole was made (5 cm diameter), in which after about one hour the depth of the water level was measured. Depths of over 40 cm were not noted.

Before analysing the vegetation of the sample plot the size and form of the stand was noted and a general description of the surrounding vegetation and its habitats was made. The structure of the vegetation was described, noting height and cover percentage of each stratum as well as cover percentage of bare soil. Subsequently a list of all species of vascular plants, mosses and hepatics in the quadrat was made. Of the lichens the crustose species were not listed separately. All secies of bryophytes and non-crustose lichens were collected from each sample plot for subsequent checking of the field-identifications. In many of the collections of the genus *Cladonia* the material did not allow for identification at the species level. In the tables all *Cladonias* are therefore listed collectively.

At each plot each species was rated on the cover-abundance scale of Braun-Blanquet (1928, 1951) with a subdivision of scale unit 2, following proposals by Barkman et al. (1964). The cover-abundance scale is given in Table 14. The cover values in this study represent external cover. In most cases the difference between external and internal cover is small. With some flowering plants (e.g. *Acaena magellanica* and *Agrostis magellanica*), however, the difference may be appreciable. The use of external cover minimizes to a large extent the effects of seasonal differences due to leaf-development of the plants.

The sociability of each species was noted, using the 5-point scale of Braun-Blanquet (1928, 1932, 1951, 1964; Table 15), as well as vitality and phenology, but little use is made of these data in this study.

From a number of communities structure diagrams were drawn in the field.

*Table 14.* The cover-abundance scale of Braun-Blanquet (1928, 1951), with a subdivision of scale unit 2 following proposals by Barkman et al. (1964).

| | |
|---|---|
| r | One or few individuals, covering less than 5% of total plot area |
| + | Occasional and covering less than 5% of total plot area |
| 1 | Abundant and covering less than 5% of total plot area |
| 2m | Very abundant, covering less than 5% of total plot area |
| 2a | Covering 5–12.5% of total plot area, irrespective of number of individuals |
| 2b | Covering 12.5–25% of total plot area, irrespective of number of individuals |
| 3 | Covering 25–50% of total plot area, irrespective of number of individuals |
| 4 | Covering 50–75% of total plot area, irrespective of number of individuals |
| 5 | Covering 75–100% of total plot area, irrespective of number of individuals |

*Table 15.* The sociability scale (Braun-Blanquet 1928, 1932, 1951, 1964).

| | |
|---|---|
| 1 | Growing singly |
| 2 | Growing in small groups of a few individuals or forming small tussocks |
| 3 | Growing in small patches, cushions or large tussocks |
| 4 | Growing in extensive patches, in carpets or broken mats |
| 5 | Growing in great crowds or extensive mats completely covering the whole plot area |

*Table 16.* Presence scale (Braun-Blanquet 1928, 1932, 1951, 1964).

| | |
|---|---|
| I | Species present in 1– 20% of the relevés of a community |
| II | Species present in 21– 40% of the relevés of a community |
| III | Species present in 41– 60% of the relevés of a community |
| IV | Species present in 61– 80% of the relevés of a community |
| V | Species present in 81–100% of the relevés of a community |

## *Tabulation of the data*

The relevés were arranged into phytosociological tables. Tabulation procedures are described in detail by e.g. Ellenberg (1956), Shimwell (1971), Westhoff & Maarel (1973, 1978) and Mueller-Dombois & Ellenberg (1974). Tabulation was for a large part executed manually, but also classification and ordination procedures with computer-programs were used, notably the programs TABORD (Maarel et al. 1978) and ORDINA (Roskam 1971). Also ordination using the reciprical averaging procedure described by Hill (1977) was used.

In the heads of the tables a large number of environmental data are given, as well as data on the structure of the vegetation (e.g. height and cover of the different layers). The bottom part of the tables gives the floristic data of the sample plots.

In principle all relevés were entered into the tables. In a number of cases, however, the identifications of the bryophytes could not be checked, as the samples had been consumed for a large part by beetles (most probably *Ectemnorrhinus similis* Waterhouse; cf. Smith 1977c) inadvertently included with the collections. These relevés were excluded from the tables. Transitions between communities not influenced by animals and biotically heavily influenced communities, have, in general, not been included in the tables. Such transitions were numerous, and inclusion of them in the tables would greatly confuse the table pattern. For these various reasons, about 15% of the relevés were not reproduced in the tables given in the following chapters.

A summary of the floristic composition of the communities is given in a synoptic table (Table 32). In this table the presence of each species is given, using a 5-point scale (Braun-Blanquet 1928, 1932, 1951, 1964; Table 16), together with the mean cover-abundance value of each species in each community. The latter value allows a better comparison of the different communities than the range of cover-abundance values, which is usually given in synoptic tables, since in a number of the communities described here this range is rather large. In the compilation of the synoptic table, fragments, transitions and a-typical facies or variants were excluded. Computations were made using the computerprogram SYNOP (O. van Tongeren 1980, unpublished).

## *Syntaxonomy*

Based on the phytosociological tables a classification of the plant communities was devised. The fundamental unit in the syntaxonomic hierarchy is the association, defined by its characteristic species combination, including character and differential species as well as companions with high presence values (cf. e.g. Westhoff & Maarel 1973, 1978). As a general rule in this study, the rank of association is assigned to phytocoena possessing at least one character-species (cf. Dierschke 1971; Westhoff & Maarel 1973, 1978); but some exceptions have been made. Communities showing a characteristic species combination, including at least one differential species, and which are clearly differentiated structurally and or ecologically from related phytocoena with the rank of association, were also given association-rank (cf. Beeftink 1965). Some phytocoena which each possess character-species, were united into a single association and ranked as subassociations because of their close floristic, structural and ecological similarity (e.g. Lycopodio magellanici – Jamesonielletum coloratae ranunculetosum biternati and catagonietosum politi). This is in accordance with the ideas of Braun-Blanquet (e.g. 1928).

Communities lacking character and differential species are described as sociations when a clear dominant was present, or as 'communities' without receiving any definite phytosociological rank when such a dominant species was lacking (cf. Westhoff & Maarel 1973, 1978). In the naming and typification of associations and subassociations the code of phytosociological nomenclature as proposed by Barkman et al. (1976) is followed. All communities discussed in this study are described here for the first time.

Marion and Prince Edward Islands represent only a small part of the distribution area of many of the species met with (cf. Tables 10 to 13). Although not much detailed phytosociological data are available from the other Subantarctic islands, it seems very probable that Marion and Prince Edward Islands represent only part of the range of a number of the communities described here. It seems premature, therefore, to design a formal hierarchical classification system for units with a rank above the association, based solely on the data presented in this study. Some way of arranging the plant communities of Marion and Prince Edward Islands into groups seems, however, useful and logical. The communities are therefore grouped into community-complexes. Each complex comprises a number of floristically related phytocoena, and may be differentiated from other complexes by one or more characteristic taxa. The name of the complex is derived from one or more of these taxa. The complexes are ecologically significant in that they comprise communities of related environmental situations.

*Vegetation mapping*

During the second season of field work a detailed vegetation mapping of three areas was executed on a field-scale of 1:50. The legend for these maps was made by tabulation of the relevant relevés available at that time and suplemented with a number of relevés made within the mapping areas. After the final classification of the islands' plant communities was made, the legend units were revised when necessary and renamed to fit the syntaxonomic system. Detailed topographical maps of the mapping areas were made using a dumpy level following ordinary levelling procedures as described by White (1969). A map showing the depth of the groundwater level was made of each area. These maps are based on measurements at the cross-over points of a grid with lines 2 m apart, which was laid out over the mapping area. In each area the depth of the groundwater level was measured under comparable circumstances: on the second day of a dry spell, after a period with no exceptionally high precipitation. At the same points soil samples were taken for the determination of the soil water content. These samples were also taken under similar conditions. Only the maps of the mapping area in Nellie Humps, near the meteorological station, are published here. Those of the other study areas are available at the Institute for Environmental Sciences, University of the Orange Free State, Bloemfontein, South Africa, at the Department of Geobotany, Katholieke Universiteit, Nijmegen, the Netherlands, and with the author.

### 3.2 Soil analyses and groundwater measurements

In each quadrat a sample was taken of the upper 5–10 cm of soil, after removal of the plant litter. In some cases, notably from areas at great distance from the meteorological station, the effort in carrying all samples back to the laboratory proved to be prohibitive.

In the laboratory, after crumbling and thouroughly mixing of the soil a subsample was taken for measurement of the soil water content. This was done by drying fresh samples at 105 °C to constant weight and determining the water loss by weighing (Gardner 1965) and is expressed as percentages of the oven-dry weight. The remainder of the soil sample was allowed to dry in a heated room at a temperature of 20–25 °C. All other analyses were performed on these dried samples, after they were crushed and sieved through a 2 mm-mesh sieve.

Soil pH was measured in a suspension of one part air-dry soil and five parts distilled water, since in many of the highly organic soils the usual soil:water ratio of 1:2 (Peech 1965) did not result in a sufficiently wet soil paste. The suspension was stirred several times during 30 min and then allowed to stand for one hour. pH was measured using a glass electrode and reference calomel electrode.

Organic matter content of the soil was determined by the loss-on-ignition method (Ball 1964). Loss-on-ignition is expressed as a percentage of the oven-dry weight.

Electric conductivity was measured in a 1:5 soil:distilled water suspension, using a Philips PW 9501 electric conductivity meter. The results of these measurements are not given in detail.

Weather conditions show an appreciable influence on some of the factors measured here. Electrical conductivity of soils near the sea is influenced strongly by the occurrence of storms bringing much salt-spray to the land, and by leaching by heavy rains (cf. Huntley 1971). Soil water content and depth of the groundwater level is affected by the amount of rainfall. Repeated measurements of groundwater levels at a number of points in the Nellie Humps mapping area showed an appreciable variation of the depth of the water level within a short period of time, which was correlated to the amount of rainfall in the period immediately preceeding the measurements (Fig. 24).

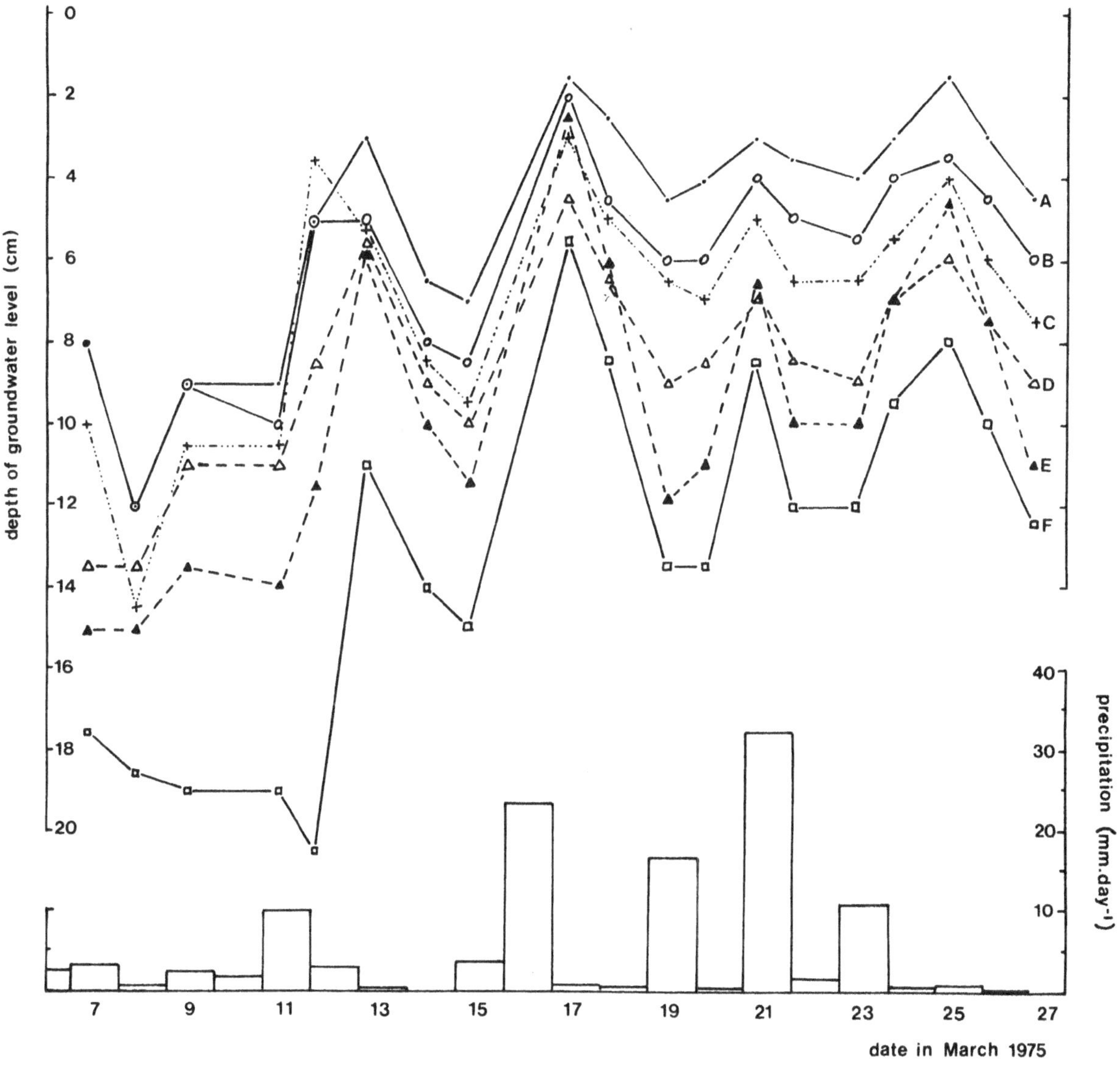


*Fig. 24.* Variation in the depth of the groundwater level at a number of sites in mire and slope vegetation, and the daily precipitation during three weeks. The measuring points were situated in the following communities: A and B: Blepharidophyllo densifolii – Clasmatocoleetum humilis ranunculetosum biternati; C: Blepharidophyllo – Clasmatocoleetum blechnetosum penna-marinae; D and E: Lycopodio magellanici – Jamesonielletum coloratae catagonietosum politi; F: Isopterygio pulchelli – Blechnetum penna-marinae typicum.

CHAPTER FOUR

# The plant communities

The absence of trees, shrubs or other tall-growing plants gives the islands Marion and Prince Edward a rather bleak appearance and from some distance the islands' plant cover looks very monotonous. In the lowland areas two different types of vegetation can be discerned from a distance: most areas have a light brown-green colour, caused by the graminoid species dominating the plant cover here: *Agrostis magellanica* and *Uncinia compacta*. Within these light-brownish areas dark green regions are seen, where the fern *Blechnum penna-marina* is dominant. In the higher-elevated areas, as well as in those parts of the lowlands which are strongly exposed to the wind, the colour of the lava predominates, being grey or black, often with a greenish hue caused by the cushions of *Azorella selago* which dominate the vegetation of these areas. Near places where animals gather, bright green is the predominant colour in the vegetation, the colour of *Poa cookii, Callitriche antarctica* and *Cotula plumosa,* while along the coast locally reddish patches occur where *Crassula moschata* dominates.

On a closer inspection the vegetation does not appear to be as uniform as it looks at first sight. After tabulation of the data collected in this study, 41 plant communities were distinguished (Table 17) which are described in the following chapters. The synoptic table (Table 32) gives a summary of the floristic composition of these communities. Based mainly on their species-composition, although in a few cases also considering structural and ecological factors, the communities were grouped into six community-complexes:

The first group, the *Crassula moschata* complex, is characterized by the presence of *Crassula moschata,* a species which is restricted to areas strongly influenced by salt-spray. The communities of this complex form the plant cover of coastal areas where the influx of minerals via sea water and salt-spray is strong.

The communities of the second group, the *Callitriche antarctica-Poa cookii* complex, are characterized by the importance of *Callitriche antarctica, Montia fontana* and *Poa cookii* in the vegetation. They form the plant cover of areas strongly influenced by trampling and manuring by animals.

The first two community-complexes together are differentiated from all other communities on the islands by *Cotula plumosa* (Table 32).

The dominance of *Acaena magellanica* characterizes the communities of the *Acaena magellanica-Brachythecium* complex (except for one subassociation). The characteristic species in the bryophyte stratum are *Brachythecium subplicatum* or *Brachythecium rutabulum.* The communities of this group occur on sites with a more or less strong lateral water movement through the soil or at the soil surface, like in springs, flushes and drainage lines.

The fourth group of plant communities, the *Juncus scheuchzerioides-Blepharidophyllum densifolium* complex, is characterized by a large number of species, amongst others *Blepharidophyllum densifolium, Clasmatocolea humilis, Distichophyllum fasciculatum, Campylopus arboricola, Andrewsianthus lancistipus* and *Uncinia compacta.*

*Table 17.* List of the plant communities of Marion and Prince Edward Islands.

| | |
|---|---|
| 1. | *Crassula moschata* complex |
| 1.1. | *Cotulo plumosae – Crassuletum moschatae* |
| 1.1. a. | *Crassula – variant* |
| b. | *intermediate variant* |
| c. | *Cotula – variant* |
| 1.2. | *Crassulo moschatae – Clasmatocoleetum vermicularis* |
| 1.3. | *Crassulo moschatae – Azorelletum selaginis* |
| | (*Crassula – Racomitrium crispulum* mosaic) |
| 2. | *Callitriche antarctica – Poa cookii* complex |
| 2.1. | *Montio fontanae – Callitrichetum antarcticae* |
| 2.1.1. | *agrostietosum magellanicae* |
| 2.1.2. | *typicum* |
| 2.1.2.a. | *Montia fontana variant* |
| b. | *intermediate variant* |
| c. | *Callitriche antarctica variant* |
| 2.2. | *Callitricho antarcticae – Poetum annuae* |
| 2.3. | *Poo cookii – Cotuletum plumosae* |
| 2.4. | *Montio fontanae – Clasmatocoleetum vermicularis* |
| 2.5. | *Leptodontio proliferi – Poetum cookii* |
| 2.5.1. | *inops* |
| 2.5.2. | *brachythecietosum rutabuli* |
| | (*Poa cookii – Callitriche antarctica* mosaic) |
| 3. | *Acaena magellanica – Brachythecium* complex |
| 3.1. | *Brachythecietum subplicati* |
| 3.1.1. | *philonotidetosum* |
| 3.1.2. | *inops* |
| 3.1.3. | *breutelietosum integrifoliae* |
| 3.2. | *Acaeno magellanicae – Drepanocladetum uncinati* |
| 3.2.1. | *breutelietosum integrifoliae* |
| 3.2.2. | *brachythecietosum rutabuli* |
| 3.3. | *Acaeno magellanicae – Brachythecietum rutabuli* |
| 3.4. | *Acaeno magellanicae – Agrostietum stoloniferae* |
| 4. | *Juncus scheuchzerioides – Blepharidophyllum densifolium* complex |
| 4.1. | *Community of Juncus scheuchzerioides* |
| 4.1. a. | *Campylopus clavatus variant* |
| b. | *Ranunculus biternatus variant* |
| 4.2. | *Junco scheuchzerioidis – Drepanocladetum uncinati* |
| 4.2.1. | *juncetosum scheuchzerioidis* |
| 4.2.2. | *montietosum fontanae* |
| 4.2.3. | *clasmatocoleetosum humilis* |
| 4.3. | *Distichophylletum fasciculati* |
| 4.4. | *Blepharidophyllo densifolii – Clasmatocoleetum humilis* |
| 4.4.1. | *ranunculetosum biternati* |
| 4.4.2. | *clasmatocoleetosum humilis* |
| 4.4.3. | *blechnetosum penna-marinae* |
| 4.5. | *Uncinio compactae – Ptychomnietum ringiani* |
| 4.6. | *Lycopodio magellanici – Jamesonielletum coloratae* |
| 4.6.1. | *ranunculetosum biternati* |
| 4.6.2. | *catagonietosum politi* |
| | (degradation stages of *Lycopodio – Jamesonielletum* and other mire communities) |
| 4.7. | *Jamesonielletum grandiflorae* |
| 4.8. | *Bryo laevigati – Breutelietum integrifoliae* |
| 4.8.1. | *uncinietosum compactae* |
| 4.8.1.a. | *Breutelia – variant* |
| b. | *Riccardia – variant* |
| 4.8.2. | *acaenetosum magellanicae* |
| 5. | *Blechnum penna-marina* complex |
| 5.1. | *Isopterygio pulchelli – Blechnetum penna-marinae* |
| 5.1.1. | *uncinietosum compactae* |
| 5.1.2. | *brachythecietosum rutabuli* |
| 5.1.3. | *typicum* |
| 6. | *Andreaea – Racomitrium crispulum* complex |
| 6.1. | *Jungermannio coniflorae – Racomitrietum crispuli* |
| 6.2. | *Andreaeo acutifoliae – Racomitrietum crispuli* |
| 6.2.1. | *hymenophylletosum peltati* |
| 6.2.2. | *hypnetosum cupressiformis* |
| 7. | other communities |
| 7.1. | *sociation of Limosella australis* |
| 7.2. | *sociation of Potamogeton spec.* |
| 7.3. | *sociation of Azorella selago* |
| 7.4. | *community of Azorella selago and Acaena magellanica* |

Most of these species are also found in communities of the *Blechnum penna-marina* complex, but usually only occur there in minor quantities. The communities of the *Juncus scheuchzerioides-Blepharidophyllum densifolium* complex form the vegetation of the islands' mires.

Characteristic for the communities of the *Blechnum penna-marina* complex is the dominance of *Blechnum penna-marina.* A number of bryophytes are more or less restricted to these communities, e.g. *Isopterygium pulchellum,* cf. *Amblystegium, Campylopus polytrichoides,* and the lichen *Peltigera canina.* These communities form the vegetation of most of the islands well-drained lowland slopes.

The sixth group contains the communities of areas strongly exposed to the wind. The most conspicuous species in these communities is *Azorella selago* which forms hard cushions in these areas. A large number of species are characteristic for the communities of this group, called the *Andreaea-Racomitrium crispulum* complex, e.g. *Andreaea* spp., *Ditrichum strictum, Herzogobryum vermiculare, Grammitis kerguelensis* (Table 32).

Finally four communities remain, which could not be fitted into any of the groups mentioned sofar. Two of these are waterplant-communities, each containing only one species. These species were not found in any other community on the islands, and consequently these communities do not have any floristic relation at all to any of the other communities. The other two are communities dominated by *Azorella selago*; this species does not form small, separate cushions here, but comparatively loose mats (Fig. 63). These communities also could not satisfactorily be assigned to any of the aforementioned complexes, but they are not sufficiently differentiated from e.g. the *Blechnum penna-marina* complex or the *Andreaea-Racomitrium crispulum* complex to warrant the formation of a separate group. Therefore they are put into this rest-group.

## 4.1 Communities of areas strongly influenced by salt-spray: the Crassula moschata complex

The importance of salt-spray as a source of plant nutrients in Subantarctic and Antarctic areas has been recognised by many authors (e.g. Allen & Northover 1967; Ashton 1965; Grobbelaar 1974; Holdgate 1967b; Holdgate et al. 1967; Johnson 1975; Laing 1909; Smith 1976c, 1977a). Influx of minerals from the ocean via air-borne spray is probably important for plant growth in all parts of Marion and Prince Edward Islands (Grobbelaar 1974; Smith 1976c), but seems to have a noticeable effect on the floristic composition of the vegetation only in a narrow belt along the coast.

Comparison of data on salt-deposition on Marion Island (Huntley 1971) with data from other areas shows that the amount of minerals entering the islands' terrestrial ecosystems through salt-spray is comparatively high (Table 18). The values given for Marion Island are from a transect on the eastern side of the island. Along the western coast, which is exposed to the predominant winds, salt deposition probably is still higher. Analysis of the mineral content of rainwater in transects perpendicular to the coast by Grobbelaar (1974) showed that the influence of the ocean reached much further inland along the westcoast than on the north-eastern side of Marion (Table 19).

Kelp and shells deposited on top of coastal cliffs witness the direct impact of high waves. Wave action may exert a mechanical, erosive influence upon soil and vegetation of areas adjacent to the shore. The impact of the waves seems to be of considerable age, since smoothed pebbles are found at depths of 20–60 cm in coastal soils (Fig. 25),

*Table 18.* Deposition of Na in coastal areas of Marion Island, United Kingdom and Japan.
Marion: Archway Bay (after Huntley 1971); U.K.: Lizard Peninsula, Cornwall (from Malloch 1972); Hokkaido, Japan (Fujiwara & Umejima 1962).

| Distance from the sea (m) | Na deposition in kg $ha^{-1}$ $day^{-1}$ | | |
|---|---|---|---|
| | Marion | U.K. | Hokkaido |
| c. 1 | | 4.0 | 52 |
| c. 20 | 35 | 6.7 | |
| c. 50 | 30 | 4.9 | |
| c. 100 | 18 | 1.2 | |
| 200 | | | 6.5 |
| 225 | 10 | | |
| 435 | 4 | | |
| 500 | | 0.5 | 1.7 |
| 645 | 4 | | |
| 850 | | | 1.4 |

suggesting that some of the soil material was deposited by the sea (Smith 1976c). Some chemical data on the soil profile given in Fig. 25 are reproduced in Table 20.

The most prominent species in the vegetation of the areas strongly influenced by salt-spray are *Crassula moschata* and *Cotula plumosa*. *Crassula moschata*, a small chamaephyte with succulent leaves, is restricted to more or less saline habitats. *Cotula plumosa* is abundant in areas influenced by salt-spray as well as on sites strongly influenced by animal excreta. In saline environments *Cotula* plants generally are only a few cm high, forming small-leaved rosettes, densely covered with long,

*Table 19.* The mineral content of rainwater in transects perpendicular to the coast (after Grobbelaar 1974).

**Transect 'Northwest':**
**SE of Cape Davis; 4/1-6/3 1973**

| Distance from the sea (m) | Mineral content of rainwater in mg $l^{-1}$ | | | | | |
|---|---|---|---|---|---|---|
| | $Na^+$ | $K^+$ | $Ca^{2+}$ | $Mg^{2+}$ | $Cl^-$ | $SO_4^{2-}$ |
| 25 | 144.4 | 5.0 | 1.8 | 16.9 | 227.0 | 34.0 |
| 50 | 40.8 | 1.6 | 1.0 | 4.6 | 71.6 | 9.0 |
| 100 | 22.5 | 0.9 | 0.6 | 2.6 | 39.5 | 4.8 |
| 200 | 13.2 | 0.5 | 0.6 | 1.6 | 22.7 | 3.9 |
| 400 | 11.1 | 0.4 | 0.3 | 1.4 | 18.0 | 2.2 |
| 600 | 7.5 | 0.7 | 0.4 | 0.8 | 14.0 | 2.5 |
| 800 | 7.1 | 0.3 | 0.4 | 0.9 | 11.4 | 2.4 |
| 1000 | 7.2 | 0.2 | 0.4 | 0.9 | 11.2 | 1.9 |

**Transect 'East':**
**Gentoo Lake to Junior's Kop; 4/1-7/3 1973**

| Distance from the sea (m) | Mineral content of rainwater in mg $l^{-1}$ | | | | | |
|---|---|---|---|---|---|---|
| | $Na^+$ | $K^+$ | $Ca^{2+}$ | $Mg^{2+}$ | $Cl^-$ | $SO_4^{2-}$ |
| 25 | 230.0 | 8.5 | 9.5 | 27.2 | 370.0 | 54.0 |
| 50 | 20.1 | 0.8 | 0.9 | 2.2 | 35.5 | 4.2 |
| 100 | 10.1 | 0.5 | 0.4 | 1.1 | 18.6 | 2.4 |
| 200 | 7.8 | 0.4 | 0.3 | 0.7 | 14.0 | 1.8 |
| 400 | 7.0 | 0.3 | 0.3 | 0.9 | 12.7 | 1.8 |
| 600 | 5.0 | 0.2 | 0.3 | 0.6 | 8.8 | 0.6 |
| 800 | 5.4 | 0.2 | 0.3 | 0.7 | 8.9 | 1.1 |
| 1000 | 5.3 | 0.2 | 0.3 | 0.6 | 8.7 | 1.1 |
| 1500 | 4.6 | 0.2 | 0.2 | 0.6 | 7.7 | 1.1 |
| 2000 | 4.7 | 0.2 | 0.2 | 0.6 | 7.6 | 1.8 |

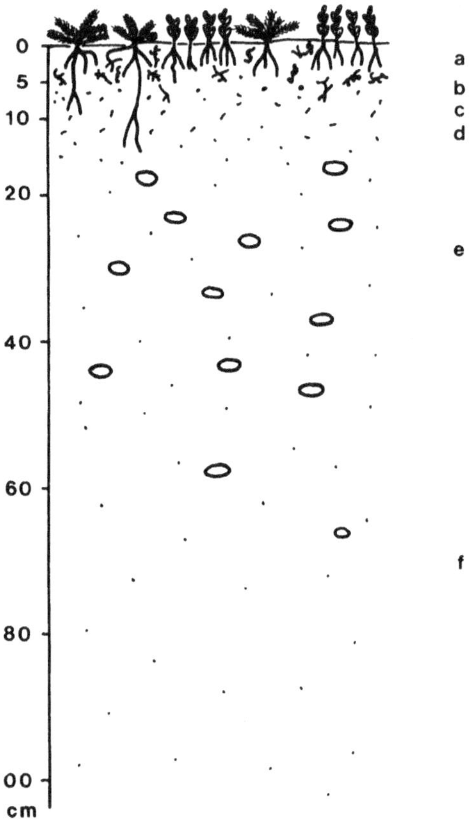


*Fig. 25.* Profile of the soil of a coastal area strongly influenced by salt-spray (after Smith 1976c). a, b and c: light to dark brown fibrous peat; d: black – dark brown sticky clay; e: dark brown clay with many smoothed pebbles; f: dark brown sticky clay with only few pebbles.

*Table 20.* Chemical data on the profile of a coastal soil illustrated in Fig. 25 (after Smith 1976c).

| Depth (cm) | pH | Loss at 105 °C (%) | Org. C (%) | Tot. N (%) | Tot. P (%) | Milli-equivalents · 100 $g^{-1}$ C.E.C. | exch. $Ca^{2+}$ | exch. $Mg^{2+}$ | exch. $Na^+$ | exch. $K^+$ |
|---|---|---|---|---|---|---|---|---|---|---|
| 0– 6 | 5.2 | 590 | 44.1 | 2.17 | 0.54 | 83.8 | 13.5 | 41.6 | 16.3 | 1.8 |
| 6–11 | 5.2 | 681 | 39.8 | 2.43 | 0.48 | 73.9 | 14.8 | 45.2 | n.d. | 1.3 |
| 11–14 | 5.3 | 804 | 38.4 | 2.36 | 0.49 | 69.9 | 10.0 | 28.0 | 20.7 | 1.1 |
| 14–50 | n.d. | 807 | 31.8 | 2.48 | 0.76 | n.d. | 4.3 | 24.7 | 17.4 | 0.8 |
| 50–80 | 4.1 | 533 | 21.5 | 1.39 | 0.95 | 32.1 | 4.6 | 16.8 | 11.3 | 0.6 |

white hairs. This is in strong contrast to the very lush *Cotula* plants which are up to 25 cm high and grow on nitrogenous soils. Similar differences in growth form of this species have been described from Kerguelen by Werth (1906) and Chastain (1958), and from Macquarie by Gilham (1961).

The belt of salt-spray vegetation is widest along the westcoast of the islands, locally over 200 m wide. In other coastal areas only small stands of this vegetation occur, except on sites strongly exposed to the influence of waves and salt-spray, such as East Cape, Duiker's Point and Storm Petrel Bay.

Three associations are distinguished in the vegetation of areas strongly influenced by salt-spray. Characteristic for all three is the presence of *Crassula moschata,* a species which only very rarely occurs in any other type of vegetation on the islands.

Association 1.1 Cotulo plumosae-Crassuletum moschatae
(type relevé 42)

Wherever soil material has accumulated in crevices and depressions between the coastal rocks above the tidal zone, *Crassula moschata* may establish itself. On sites less strongly exposed to the erosive action of the waves larger patches of Cotulo plumosae-Crassuletum are found (Fig. 26). This low-growing vegetation, often only 2–4 cm high, is very poor in species and besides *Crassula moschata* and *Cotula plumosa* only *Ranunculus biternatus* reaches appreciable cover (Table 21, Figs. 27, 28). The association does not possess any character-species. Differential species is *Cotula plumosa.*

Stands of this community are found on generally shallow soils (5–40 cm deep) consisting of fibrous

*Fig. 26.* Black lava coastal rocks with Cotulo plumosae – Crassuletum moschatae (Mixed Pickle Cove, Marion).

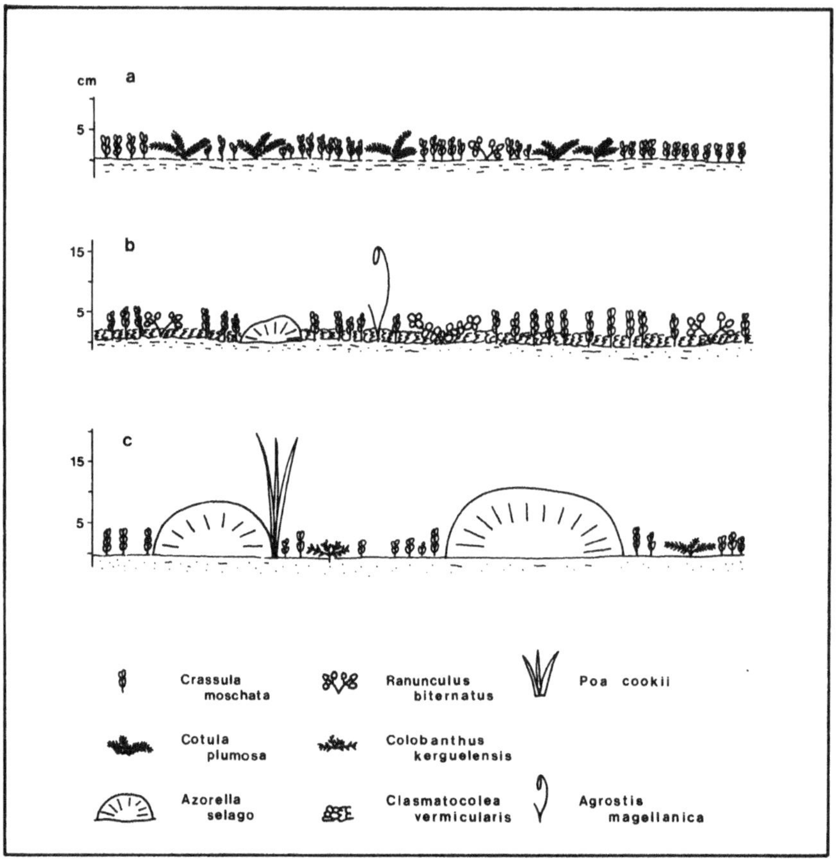


*Fig. 27.* Structure-diagram of the communities of the *Crassula moschata* complex. a = Cotulo plumosae - Crassuletum moschatae; b = Crassulo moschatae - Clasmatocoleetum vermicularis; c = Crassulo moschatae - Azorelletum selaginis.

peat and clay. The soils are wet but well-drained and the high electric conductivity of the soil extracts indicates saline conditions.

Three variants of the Cotulo plumosae-Crassuletum moschatae are discerned, differentiated by the relative abundances of *Cotula* and *Crassula* (Table 21). On sites within reach of large waves the *Crassula*-variant is found. *Cotula* is present here only in very small numbers. On sites less affected by waves and salt-spray, on very well-drained soils, often at a somewhat greater distance from the shore, *Cotula plumosa* becomes dominant and *Crassula* occurs only in small numbers. On intermediate sites a variant is found in which both *Crassula* and *Cotula* reach high cover values.

Stands of Cotulo plumosae-Crassuletum moschatae form a nearly continuous belt along the westcoasts of the islands. Along the other coasts extensive stands only occur in areas strongly exposed to the influence of the sea, e.g. East Cape, Duiker's Point, Paddy Rocks, Storm Petrel Bay and Whale Bird Point.

Association 1.2 Crassulo moschatae-Clasmatocoleetum vermicularis
(type relevé 273)

On sites within the salt-spray zone where impeded drainage has resulted in the presence of a groundwater level near the soil surface, stands of this association occur (Fig. 29). The structure of the vegetation is more intricate than in the first association of this complex: a well-developed bryophyte stratum as well as a herb layer is present, the latter usually not more than 5 cm high (Fig. 27).

Character-species of the Crassulo moschatae-Clasmatocoleetum vermicularis is the moss *Eriopus*

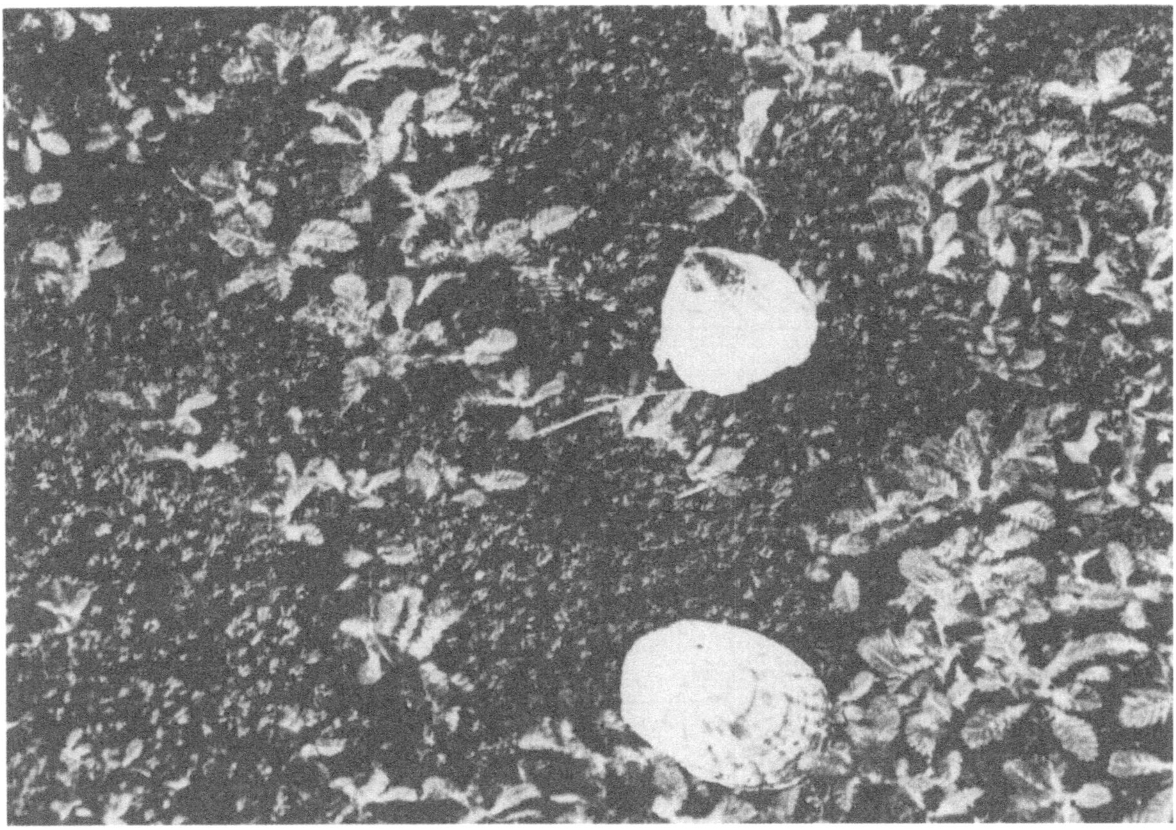

*Fig. 28.* Cotulo plumosae – Crassuletum moschatae, intermediate variant.

*apiculatus*. Differential species are *Clasmatocolea vermicularis*, the dominant bryophyte, *Callitriche antarctica* and *Hygroamblystegium filum* (Table 21). This community is found on soils consisting of fibrous peat, mainly formed by *Crassula moschata*, overlying clay deposits. Stands of this association were met with at only a few localities along the north and westcoast of Marion Island.

Association 1.3 Crassulo moschatae-Azorelletum selaginis
(type relevé 279)

This community, dominated by *Azorella selago*, is found on sites within the salt-spray zone which are strongly exposed to the wind. The prominence of cushion-plants (*Azorella selago* and *Colobanthus kerguelensis*, both differential species of this association) and the large proportion of bare soil (Figs 27 and 30) reflect the extreme environmental conditions. Stands of Crassulo moschatae-Azorelletum selaginis are found on soils with a much lower content of organic matter than those on which the other two associations of the *Crassula* complex occur. Related to this is a comparatively low water content of the soil (Table 21).

Stands of this association are found all around the islands in areas strongly influenced by salt-spray, but are most extensive along the west-coast, which is directly exposed to the predominant, westerly winds.

Extensive parts of black lava flows along the coast are covered by *Azorella*-dominated vegetation on a substratum of lava rocks interspersed with shallow deposits of clay and fibrous peat (Figs. 29 and 31). This vegetation is a mosaic of two communities: the rocks are occupied by a fragmentarily developed fjaeldmark community (cf. Par. 4.6, the *Andreaea-Racomitrium crispulum* complex).

*Table 21*. The *Crassula moschata* complex.

COTULO PLUMOSAE - CRASSULETUM MOSCHATAE — Crassula - variant

| | Crassula - variant | | | | | | | | | | | |
|---|---|---|---|---|---|---|---|---|---|---|---|---|
| RELEVE NUMBER | 55 | 272 | 337 | 54 | 61 | 507 | 52 | 53 | 62 | 63 | 71 | 274 |
| LOCATION: | | | | | | | | | | | | |
| SOUTHERN LATITUDE (46°) MIN. | 52 | 49 | 38 | 52 | 52 | 52 | 52 | 52 | 52 | 52 | 52 | 49 |
| SEC. | 45 | 25 | 30 | 45 | 45 | 14 | 45 | 45 | 45 | 45 | 45 | 25 |
| EASTERN LONGITUDE (37°) MIN. | 51 | 42 | 00 | 51 | 51 | 38 | 51 | 51 | 51 | 51 | 51 | 42 |
| SEC. | 45 | 10 | 30 | 45 | 45 | 10 | 45 | 45 | 45 | 45 | 45 | 10 |
| ELEVATION (M A.S.L.) | 10 | 20 | 10 | 10 | 8 | 30 | 12 | 12 | 10 | 12 | 10 | 20 |
| INCLINATION (°) | 3 | | 30 | | | | 3 | 3 | | 3 | | |
| ASPECT | SE | | E | | | | SSE | SSE | | ENE | | |
| SOIL DEPTH (CM) | 10 | | 20 | 10 | 20 | | 35 | 45 | 20 | 27 | 30 | |
| DEPTH OF GROUNDWATER LEVEL (CM) | --- | - | --- | --- | --- | -30 | --- | --- | --- | --- | --- | --- |
| SOIL WATER CONTENT (X10 %) | 50 | 65 | 70 | 60 | 43 | | 80 | 83 | 94 | 80 | 73 | 87 |
| SOIL PH | 5,4 | | | | 5,6 | | 4,9 | 4,8 | 5,3 | 5,2 | 5,2 | |
| SOIL LOSS ON IGNITION (%) | 78 | | 87 | 72 | 61 | | 91 | 90 | 88 | 88 | 79 | |
| INFLUENCE OF SALTSPRAY | 5 | 5 | 5 | 5 | 5 | 5 | 5 | 5 | 5 | 4 | 4 | 3 |
| INFLUENCE OF TRAMPLING | | | | | | | | | | | | |
| INFLUENCE OF MANURING | | | | | 1 | | | | 1 | | | |
| HEIGHT OF HERB LAYER (CM) | 2 | 2 | 2 | 2 | 2 | 3 | 4 | 3 | 2 | 2 | 2 | 2 |
| COVER OF HERB LAYER (%) | 97 | 60 | 80 | 100 | 99 | 80 | 95 | 97 | 85 | 97 | 100 | 95 |
| HEIGHT OF BRYOPHYTE LAYER (CM) | 1 | | | | | | | 1 | | 1 | 1 | 1 |
| COVER OF BRYOPHYTE LAYER (%) | <1 | | | | | | | 3 | | 3 | 3 | 3 |
| COVER OF ALGAE (%) | | | | | | | | | | | | |
| COVER OF BARE SOIL (%) | 3 | 40 | 20 | | 1 | 20 | 5 | 3 | 15 | 3 | | 5 |
| COVER OF ROCKS (%) | | | | | | | | | | | | |
| QUADRAT SIZE (M2) | 1 | 9 | 2 | 1 | 1 | 36 | 1 | 1 | 1 | 1 | 1 | 9 |
| TOTAL NUMBER OF SPECIES | 3 | 3 | 3 | 2 | 2 | 4 | 4 | 4 | 5 | 6 | 8 | 11 |
| CHARACTER-SPECIES OF THE COMPLEX | | | | | | | | | | | | |
| CRASSULA MOSCHATA | 5 | 4 | 5 | 5 | 5 | 5 | 5 | 5 | 5 | 5 | 5 | 5 |
| DIAGNOSTIC TAXA OF ASSOCIATIONS AND LOWER UNITS | | | | | | | | | | | | |
| COTULA PLUMOSA | | | | R | + | 1 | + | | + | + | R | |
| ERIOPUS APICULATUS | | | | | | | | | | | | |
| CLASMATOCOLEA VERMICULARIS | | | | | | | | | | | | |
| CALLITRICHE ANTARCTICA | | | | | | | | | | | | 1 |
| HYGROAMBLYSTEGIUM FILUM | | | | | | | | | | | | |
| AZORELLA SELAGO | | + | | | | | | | | | | |
| COLOBANTHUS KERGUELENSIS | | | | | | | 1 | 1 | | | + | |
| CAMPYLOPUS CAVIFOLIUS | | | | | | | | | | | | |
| RACOMITRIUM CRISPULUM | | | | | | | | | | | | |
| HERZOGOBRYUM COLL. | | | | | | | | | | | | |
| PLAGIOCHILA HETERODONTA | | | | | | | | | | | | |
| CRUSTOSE LICHENS | | | | | | | | | | | | |
| COMPANION SPECIES | | | | | | | | | | | | |
| RANUNCULUS BITERNATUS | | | + | | | + | 2A | 2A | 2A | 2B | 2A | 1 |
| MONTIA FONTANA | | | | | | | | | 1 | + | + | |
| AGROSTIS MAGELLANICA | | | | | | | | | 1 | R | R | + |
| POA COOKII | | + | R | | | R | | | | | | + |
| JUNCUS SCHEUCHZERIOIDES | | | | | | | | | | | + | |
| FOSSOMBRONIA AUSTRALIS | | | | | | | | | | | 1 | R |
| LOPHOCOLEA RANDII | + | | | | | | | | | | | |

COTULO PLUMOSAE - CRASSULETUM MOSCHATAE — intermediate variant

| | intermediate variant | | | | | | | | | |
|---|---|---|---|---|---|---|---|---|---|---|
| RELEVE NUMBER | 44 | 41 | 43 | 51 | 42 | 45 | 48 | 56 | 46 | 50 |
| LOCATION: | | | | | | | | | | |
| SOUTHERN LATITUDE (46°) MIN. | 52 | 52 | 52 | 52 | 52 | 52 | 52 | 52 | 52 | 52 |
| SEC. | 45 | 45 | 45 | 45 | 45 | 45 | 45 | 45 | 45 | 45 |
| EASTERN LONGITUDE (37°) MIN. | 51 | 51 | 51 | 51 | 51 | 51 | 51 | 51 | 51 | 51 |
| SEC. | 45 | 45 | 45 | 45 | 45 | 45 | 45 | 45 | 45 | 45 |
| ELEVATION (M A.S.L.) | 10 | 10 | 10 | 13 | 10 | 10 | 10 | 10 | 10 | 12 |
| INCLINATION (°) | 3 | | | 3 | | 3 | | 5 | 3 | 20 |
| ASPECT | NE | | | SSE | | NE | | ESE | NE | SSE |
| SOIL DEPTH (CM) | 50 | 20 | 50 | 45 | 25 | 35 | 25 | 25 | 30 | 20 |
| DEPTH OF GROUNDWATER LEVEL (CM) | --- | --- | --- | --- | --- | --- | --- | --- | --- | --- |
| SOIL WATER CONTENT (X10 %) | 68 | | 62 | 83 | 60 | 67 | 72 | 63 | 52 | 49 |
| SOIL PH | 4,7 | | 4,5 | 4,9 | 4,8 | 4,9 | 4,7 | 5,1 | 5,1 | |
| SOIL LOSS ON IGNITION (%) | 93 | 89 | 85 | 91 | 94 | 83 | 92 | 91 | 80 | 80 |
| INFLUENCE OF SALTSPRAY | 5 | 5 | 5 | 5 | 5 | 5 | 5 | 5 | 5 | 5 |
| INFLUENCE OF TRAMPLING | | | | | | | | | | |
| INFLUENCE OF MANURING | | | | | | | | | | |
| HEIGHT OF HERB LAYER (CM) | 2 | 3 | 2 | 4 | 2 | 2 | 2 | 3 | 3 | 3 |
| COVER OF HERB LAYER (%) | 80 | 85 | 85 | 70 | 75 | 95 | 75 | 99 | 90 | 100 |
| HEIGHT OF BRYOPHYTE LAYER (CM) | 1 | 1 | | | | | | 1 | 1 | |
| COVER OF BRYOPHYTE LAYER (%) | <1 | 3 | | | | | | 1 | <1 | |
| COVER OF ALGAE (%) | | | | | | | | | | |
| COVER OF BARE SOIL (%) | 20 | 15 | 15 | 30 | 25 | 5 | 25 | 1 | 10 | |
| COVER OF ROCKS (%) | | | | | | | | | | |
| QUADRAT SIZE (M2) | 1 | 4 | 1 | 1 | 1 | 1 | 1 | 1 | 1 | 1 |
| TOTAL NUMBER OF SPECIES | 4 | 7 | 4 | 4 | 3 | 4 | 3 | 4 | 6 | 3 |
| CHARACTER-SPECIES OF THE COMPLEX | | | | | | | | | | |
| CRASSULA MOSCHATA | 5 | 5 | 5 | 4 | 4 | 5 | 4 | 5 | 5 | 4 |
| DIAGNOSTIC TAXA OF ASSOCIATIONS AND LOWER UNITS | | | | | | | | | | |
| COTULA PLUMOSA | 2A | 2A | 2B | 2B | 2B | 3 | 3 | 3 | 3 | 4 |
| ERIOPUS APICULATUS | | | | | | | | | | |
| CLASMATOCOLEA VERMICULARIS | | | | | | | | | | |
| CALLITRICHE ANTARCTICA | | | | | | | | | | |
| HYGROAMBLYSTEGIUM FILUM | | | | | | | | | | |
| AZORELLA SELAGO | | + | | | | | | | | |
| COLOBANTHUS KERGUELENSIS | | | | + | | | | | | |
| CAMPYLOPUS CAVIFOLIUS | | | | | | | | | | |
| RACOMITRIUM CRISPULUM | | | | | | | | | | |
| HERZOGOBRYUM COLL. | | | | | | | | | | |
| PLAGIOCHILA HETERODONTA | | | | | | | | | | |
| CRUSTOSE LICHENS | | | | | | | | | | |
| COMPANION SPECIES | | | | | | | | | | |
| RANUNCULUS BITERNATUS | 1 | + | 1 | 1 | 2A | 1 | 2A | + | 1 | 1 |
| MONTIA FONTANA | | | | | | | | | | |
| AGROSTIS MAGELLANICA | | + | | | | | | | + | |
| POA COOKII | | R | R | | | | | | | |
| JUNCUS SCHEUCHZERIOIDES | | | | | | | | | | |
| FOSSOMBRONIA AUSTRALIS | | | | | | | | | + | |
| LOPHOCOLEA RANDII | | | | | | | | | + | |

COTULO PLUMOSAE - CRASSULETUM MOSCHATAE — Cotula - variant

| | Cotula - variant | | | | | | | | | | |
|---|---|---|---|---|---|---|---|---|---|---|---|
| RELEVE NUMBER | 47 | 57 | 338 | 58 | 49 | 70 | 59 | 60 | 69 | 67 | 68 |
| LOCATION: | | | | | | | | | | | |
| SOUTHERN LATITUDE (46°) MIN. | 52 | 52 | 38 | 52 | 52 | 52 | 52 | 52 | 52 | 52 | 52 |
| SEC. | 45 | 45 | 35 | 45 | 45 | 43 | 45 | 45 | 43 | 45 | 45 |
| EASTERN LONGITUDE (37°) MIN. | 51 | 51 | 00 | 51 | 51 | 51 | 51 | 51 | 51 | 51 | 51 |
| SEC. | 45 | 45 | 25 | 45 | 45 | 47 | 45 | 45 | 47 | 45 | 45 |
| ELEVATION (M A.S.L.) | 10 | 10 | 7 | 11 | 10 | 10 | 12 | 10 | 12 | 12 | 12 |
| INCLINATION (°) | 3 | 25 | | 3 | 3 | 30 | 10 | 5 | 10 | 15 | |
| ASPECT | S | E | | E | SE | SW | ENE | ENE | N | E | |
| SOIL DEPTH (CM) | 20 | 10 | | 25 | | 20 | 17 | 10 | | | |
| DEPTH OF GROUNDWATER LEVEL (CM) | --- | --- | --- | --- | - | --- | --- | --- | -60 | - | -99 |
| SOIL WATER CONTENT (X10 %) | 50 | 67 | 54 | | 83 | 55 | 59 | 60 | 75 | 74 | 69 |
| SOIL PH | 5,0 | 5,1 | | | 4,6 | 5,5 | 5,7 | 5,5 | 4,7 | 5,0 | 5,4 |
| SOIL LOSS ON IGNITION (%) | | 87 | | 88 | 93 | 81 | 81 | 86 | 90 | 94 | 93 |
| INFLUENCE OF SALTSPRAY | 5 | 5 | 5 | 5 | 4 | 5 | 5 | 5 | 3 | 3 | 3 |
| INFLUENCE OF TRAMPLING | | | | | | | | | | | |
| INFLUENCE OF MANURING | | | | | | | | | | | |
| HEIGHT OF HERB LAYER (CM) | 3 | 5 | 2 | 7 | 3 | 10 | 6 | 3 | 5 | 3 | 4 |
| COVER OF HERB LAYER (%) | 97 | 100 | 99 | 100 | 97 | 100 | 100 | 97 | 97 | 99 | 90 |
| HEIGHT OF BRYOPHYTE LAYER (CM) | | | | | 1 | | | | 1 | 1 | 1 |
| COVER OF BRYOPHYTE LAYER (%) | | | | | 7 | | | | <1 | 70 | 30 |
| COVER OF ALGAE (%) | | | | | | | | | | | |
| COVER OF BARE SOIL (%) | 3 | | 1 | | 3 | | | 3 | 3 | | 5 |
| COVER OF ROCKS (%) | | | | | | | | | | | |
| QUADRAT SIZE (M2) | 1 | 1 | 4 | 1 | 1 | 1 | 1 | 1 | 1 | 1 | 1 |
| TOTAL NUMBER OF SPECIES | 4 | 5 | 3 | 4 | 8 | 4 | 4 | 5 | 5 | 9 | 10 |
| CHARACTER-SPECIES OF THE COMPLEX | | | | | | | | | | | |
| CRASSULA MOSCHATA | 2B | 2B | 2A | 2A | 2A | + | + | 1 | + | + | R |
| DIAGNOSTIC TAXA OF ASSOCIATIONS AND LOWER UNITS | | | | | | | | | | | |
| COTULA PLUMOSA | 4 | 5 | 5 | 5 | 5 | 5 | 5 | 5 | 5 | 5 | 5 |
| ERIOPUS APICULATUS | | 4 | | | | | | | | | |
| CLASMATOCOLEA VERMICULARIS | | | | | | | | | | | |
| CALLITRICHE ANTARCTICA | | | | | | | | | | | |
| HYGROAMBLYSTEGIUM FILUM | | | | | | | | | | | |
| AZORELLA SELAGO | | | | | 1 | | | | | | + |
| COLOBANTHUS KERGUELENSIS | | | | | + | | | | | | + |
| CAMPYLOPUS CAVIFOLIUS | | | | | | | | | | | |
| RACOMITRIUM CRISPULUM | | | | | | | | | | | |
| HERZOGOBRYUM COLL. | | | | | | | | | | | |
| PLAGIOCHILA HETERODONTA | | | | | | | | | | | |
| CRUSTOSE LICHENS | | | | | | | | | | | |
| COMPANION SPECIES | | | | | | | | | | | |
| RANUNCULUS BITERNATUS | + | 1 | | 1 | 2B | + | + | 1 | 2A | 3 | 2A |
| MONTIA FONTANA | | R | | 1 | + | | + | + | | + | + |
| AGROSTIS MAGELLANICA | | + | | | | + | | R | | + | |
| POA COOKII | R | | + | | | | | | | + | + |
| JUNCUS SCHEUCHZERIOIDES | | | | | | | | | + | | 1 |
| FOSSOMBRONIA AUSTRALIS | | | | | | | | | | | |
| LOPHOCOLEA RANDII | | | | | + | | | | | + | |

CRASSULO MOSCHATAE - CLASMATOCOLEETUM VERMICULARIS; CRASSULO MOSCHATAE - AZORELLETUM SELAGINIS; CRASSULA - RACOMITRIUM mosaic

| | CRASSULO MOSCHATAE - CLASMATOCOLEETUM VERMICULARIS | | | | | CRASSULO MOSCHATAE - AZORELLETUM SELAGINIS | | | | | | | CRASSULA - RACOMITRIUM mosaic | | |
|---|---|---|---|---|---|---|---|---|---|---|---|---|---|---|---|
| RELEVE NUMBER | 275 | 273 | 271 | 270 | 276 | 277 | 505 | 570 | 506 | 600 | 261 | 262 | 578 | 280 | 279 |
| LOCATION: | | | | | | | | | | | | | | | |
| SOUTHERN LATITUDE (46°) MIN. | 49 | 49 | 50 | 50 | 49 | 49 | 52 | 53 | 52 | 52 | 53 | 52 | 52 | 53 | 52 |
| SEC. | 25 | 25 | 05 | 05 | 25 | 25 | 14 | 15 | 14 | 38 | 00 | 45 | 57 | 25 | 25 |
| EASTERN LONGITUDE (37°) MIN. | 42 | 42 | 40 | 40 | 42 | 42 | 38 | 52 | 39 | 37 | 52 | 52 | 52 | 42 | 42 |
| SEC. | 10 | 10 | 05 | 05 | 10 | 10 | 10 | 10 | 10 | 57 | 00 | 05 | 05 | 10 | 10 |
| ELEVATION (M A.S.L.) | 25 | 20 | 20 | 20 | 30 | 30 | 25 | 10 | 30 | 30 | 15 | 15 | 20 | 25 | 25 |
| INCLINATION (°) | | | | | | 7 | | | | | | | 3 | 40 | 15 |
| ASPECT | | | | | | E | | | | | | | W | N | S |
| SOIL DEPTH (CM) | | | | | | 25 | | 10 | | | | | 12 | 10 | 10 |
| DEPTH OF GROUNDWATER LEVEL (CM) | - | -9 | -12 | -12 | -8 | --- | -35 | --- | -25 | - | - | | --- | --- | --- |
| SOIL WATER CONTENT (X10 %) | 89 | 139 | 90 | 104 | 161 | 28 | | 34 | | | 44 | 57 | 33 | 22 | 34 |
| SOIL PH | | | | 4,5 | | 5,2 | | | | | 5,1 | 4,8 | 5,6 | 5,5 | 5,5 |
| SOIL LOSS ON IGNITION (%) | 90 | | 82 | 82 | | 46 | | | | 43 | 52 | 60 | 51 | 41 | 55 |
| INFLUENCE OF SALTSPRAY | 4 | 5 | 4 | 3 | 4 | 4 | 4 | 5 | 4 | 3 | 2 | 2 | 2 | 4 | 4 |
| INFLUENCE OF TRAMPLING | | | | | | | | | | | | | | | |
| INFLUENCE OF MANURING | | | | | | | | | | | | | | | |
| HEIGHT OF HERB LAYER (CM) | 2 | 3 | 4 | 4 | 2 | 20 | 15 | 20 | 10 | 25 | 20 | 17 | 15 | 15 | 15 |
| COVER OF HERB LAYER (%) | 95 | 90 | 40 | 15 | 25 | 30 | 30 | 70 | 80 | 85 | 55 | 65 | 20 | 15 | 20 |
| HEIGHT OF BRYOPHYTE LAYER (CM) | 1 | 1 | 1 | 1 | 1 | | 1 | | 1 | 1 | 1 | 1 | 2 | 2 | 1 |
| COVER OF BRYOPHYTE LAYER (%) | 60 | 70 | 95 | 100 | 90 | | <1 | | <1 | <1 | <1 | <1 | <1 | <1 | <1 |
| COVER OF ALGAE (%) | | | | | | | | | | | | | | | |
| COVER OF BARE SOIL (%) | 5 | 10 | | | 5 | 70 | 70 | | 20 | 15 | 45 | 35 | 70 | 20 | 10 |
| COVER OF ROCKS (%) | | | | | | | | 1 | | | | | 15 | 65 | 70 |
| QUADRAT SIZE (M2) | 9 | 9 | 4 | 9 | 6 | 25 | 100 | 25 | 100 | 36 | 25 | 16 | 25 | 16 | 25 |
| TOTAL NUMBER OF SPECIES | 8 | 10 | 5 | 7 | 8 | 4 | 9 | 6 | 8 | 8 | 12 | 14 | 7 | 12 | 12 |
| CHARACTER-SPECIES OF THE COMPLEX | | | | | | | | | | | | | | | |
| CRASSULA MOSCHATA | 5 | 4 | 3 | 2B | 2B | 1 | 1 | 3 | 2A | + | + | + | + | 1 | 1 |
| DIAGNOSTIC TAXA OF ASSOCIATIONS AND LOWER UNITS | | | | | | | | | | | | | | | |
| COTULA PLUMOSA | | | | | | | + | + | + | + | R | + | | | |
| ERIOPUS APICULATUS | + | 2B | | | | | | | | | | | | | |
| CLASMATOCOLEA VERMICULARIS | 4 | 4 | 5 | 5 | 5 | | | | | | | | | | |
| CALLITRICHE ANTARCTICA | + | 1 | + | + | 1 | | | | 1 | + | | | | | |
| HYGROAMBLYSTEGIUM FILUM | | + | | 1 | + | | | | | | | | | | |
| AZORELLA SELAGO | R | + | + | | | 3 | 3 | 3 | 5 | 5 | 3 | 3 | 2B | 2B | 2B |
| COLOBANTHUS KERGUELENSIS | | | | | | + | 1 | 1 | + | | 1 | + | + | + | 1 |
| CAMPYLOPUS CAVIFOLIUS | | | | | | | + | | + | | | | | + | + |
| RACOMITRIUM CRISPULUM | | | | | | | | | | | + | | 1 | + | 1 |
| HERZOGOBRYUM COLL. | | | | | | | | | | | | | | + | + |
| PLAGIOCHILA HETERODONTA | | | | | | | | | | | | + | | + | + |
| CRUSTOSE LICHENS | | | | | | | | + | | | | | 1 | + | 1 |
| COMPANION SPECIES | | | | | | | | | | | | | | | |
| RANUNCULUS BITERNATUS | 1 | 1 | | 1 | 1 | | | | | + | + | + | 1 | + | + |
| MONTIA FONTANA | | | | | | | | | | | | + | | | |
| AGROSTIS MAGELLANICA | | 3 | | 1 | R | R | + | | | | 1 | 3 | 1 | | |
| POA COOKII | | | R | | | | + | R | + | + | 2A | 1 | | | |
| JUNCUS SCHEUCHZERIOIDES | + | | | | + | | | | | | + | 1 | | | |
| FOSSOMBRONIA AUSTRALIS | + | | | R | + | | | | | | + | + | | + | |
| LOPHOCOLEA RANDII | | 2B | | | | | | | | + | | | | | |

*Fig. 29.* Black lava coastal area along the westcoast of Marion Island. The vegetation in the depression is Crassulo moschatae – Clasmatocoleetum vermicularis, in the other areas Crassulo moschatae – Azorelletum selaginis and Crassula – Racomitrium crispulum mosaic.

*Fig. 30.* Crassulo moschatae – Azorelletum selaginis (Fur Seal Bay, Marion).

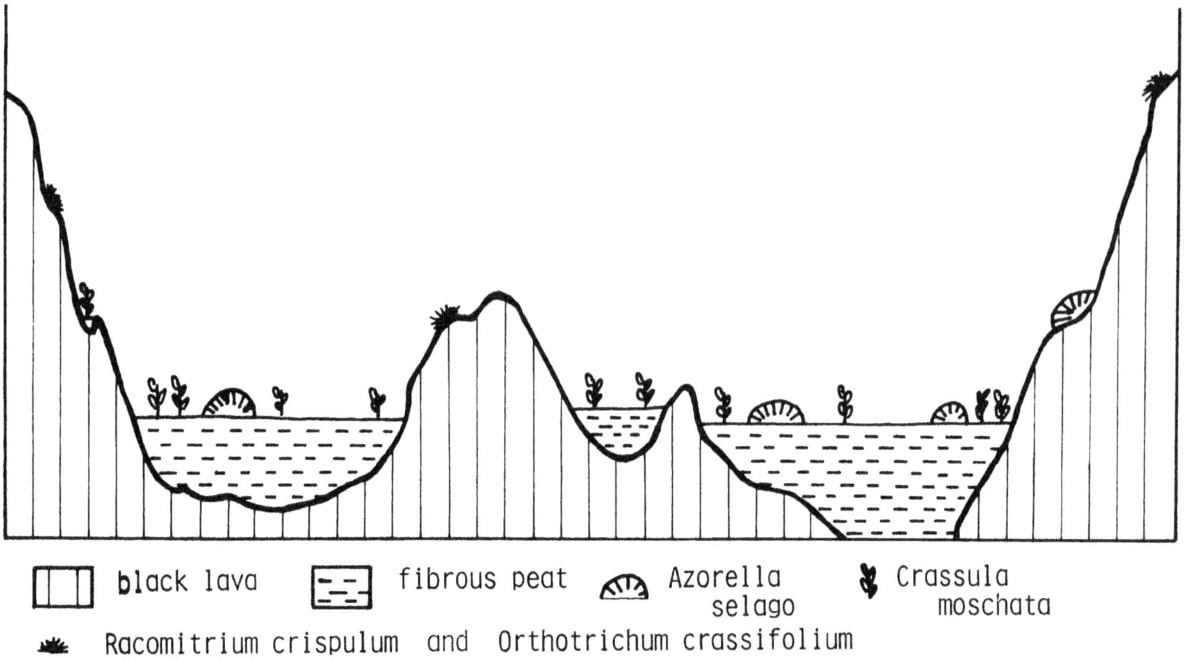


*Fig. 31.* Structure-diagram of Crassula – Racomitrium crispulum mosaic.

The second component of the mosaic is formed by fragments of Crassulo moschatae – Azorelletum selaginis or Cotulo plumosae-Crassuletum moschatae on the clay and peat deposits (cf. Table 21).

*Synchorological remarks*

Communities similar to the Cotulo plumosae-Crassuletum moschatae are widely spread in Subantarctic and southern temperate regions. A community of *Crassula moschata* and *Cotula plumosa,* associated with *Ranunculus pseudotrullifolius* Skottsb., forms extensive stands in the coastal zone of Iles Kerguelen (Chastain 1958; Cour 1959; Aubert de la Rue 1964) and Iles Crozet (Werth 1906). *Ranunculus pseudotrullifolius* is endemic to these island groups. A detailed floristic description of this vegetation is not available, but it seems very similar to the Cotulo-Crassuletum and is probably a vicariant community.

*Crassula moschata* is the dominant species in communities found within the salt-spray zone of most of the islands south of New Zealand: Auckland Islands, Antipodes Islands (Cockayne 1909), Campbell Island (Laing 1909, Oliver & Sorensen 1951) and Macquarie Island (Taylor 1955a). The descriptions in the literature are not sufficient for any detailed comparison, but these communities seem to be similar in structure and ecology to the Cotulo-Crassuletum of Marion and Prince Edward Islands.

The *Crassula*-association described by Skottsberg (1913) from the Falkland Islands, in which the most prominent species are *Crassula moschata, Colobanthus subulatus* and *Plantago barbata,* differs appreciably in floristic composition, but seems to resemble the Cotulo-Crassuletum in ecology.

No communities comparable to the Crassulo moschatae-Clasmatocoleetum vermicularis or the Crassulo moschatae-Azorelletum selaginis have been reported from other Subantarctic regions. Very recently Cleef (1981) described a number of very wet *Tillaea* (= *Crassula*) communities from the Colombian Andes and summarized these in the order Tillaeetalia. It is possible that the Cotulo-Crassuletum described here belongs to that order.

## 4.2 Communities of areas strongly influenced by animals: the Callitriche antarctica-Poa cookii complex

Animals have a distinct influence on the vegetation of Subantarctic and southern temperate regions (Moseley 1892; Taylor 1955a; Wace 1960, 1961; Gilham 1961; Aubert de la Rue 1964; Greene 1964; Follmann 1965; Huntley 1971; Smith & Walton 1975; Smith 1977a, 1978a, 1979). Marion and Prince Edward Islands are the breeding grounds of vast numbers of seabirds (La Grange 1962; Rand 1954, 1955a; Zinderen Bakker Jr. 1967, 1971; Williams et al. 1979) and seals (Rand 1955b, 1956; Zinderen Bakker Jr. 1967; Condy 1978, 1979). These provide an important pathway through which large amounts of minerals are transported from the ocean to the terrestrial environment (Zinderen Bakker Sr. 1971a, 1978; Smith 1976b, 1977a, 1978a; Lindeboom 1979; Williams et al. 1978).

Animal excreta are an important source of phosphorus and nitrogen. Comparison of the chemical status of soils under *Poa cookii* tussock grassland, which is influenced by burrowing species of Procellariidae, with uninfluenced soils show a two- to three-fold increase in phosphorous content, as well as a considerable increase of the ammonium concentration in the soil of the influenced area (Smith 1976b,c, 1979; Table 9). The effect of manuring by animals is generally not reflected by a large increase of the total nitrogen content of the soil, as can be seen from the data given in Table 22. Smith (1978a) states that the total nitrogen content of these soils depends largely on the soil organic matter level, and to a much smaller extent on the intensity of manuring. A striking difference in $NH_4^+$-nitrogen and $NO_3^-$-nitrogen content exists between soils of unaffected and of manured sites. A ten- to thirty-fold increase as a result from manuring is usual. In mud between *Poa cookii* tussocks in a King Penguin rookerie concentrations were found to be 1000 times as high as in unaffected soils (Smith 1978a; Table 22).

Not only excreta, but also feathers, carcasses, excretion from nasal glands and other matter are deposited by animals. The influence of such deposits on the vegetation has not been studied separately from the effect of manuring, except for the effects of crop pellets regurgitated by Kelp Gulls (*Larus dominicanus*). These pellets, consisting largely of shell fragements of *Gaimardia trapesina*, are a source of Ca to the soil. Smith (1976c) found that exchangeable Ca concentrations in affected soils were about twice as high as in soils not influenced by crop pellets. No corresponding difference in the vegetation was found. Possibly this is due to the strong influence of salt-spray at the sites studied, which resulted in an extremely species-poor vegetation, consisting of only *Cotula plumosa* and *Crassula moschata*.

The second important influence of animals is trampling which may result in partial or complete destruction of the plant cover and in modification of the soil structure. The impact of trampling on the soil depends on its water content and the content of

*Table 22.* The nitrogen status of some Marion Island soils (after Smith 1978a). Wandering Albatross: samples collected from the vicinity of the nests. King Penguin: samples from the periphery of a rookerie. N = number of determinations; n.d. = not determined; tr. = trace ($<0.5$ mg · 100 $g^{-1}$).

| | N | organic N (%) | | $NH_4^+$-N (mg · 100 $g^{-1}$) | | $NO_3^-$-N (mg · 100 $g^{-1}$) | |
|---|---|---|---|---|---|---|---|
| | | mean ± s.d. | range | mean ± s.d. | range | mean ± s.d. | range |
| Uninfluenced | 10 | 2.6 ± 0.72 | 1.8– 3.9 | 1.4 ± 1.26 | 0– 3.5 | 0.2 ± 0.24 | 0– 0.6 |
| Wandering Albatross: | | | | | | | |
| occupied nests | 3 | 3.0 ± 0.32 | 2.8– 3.3 | 54.5 ± 55.72 | 11.6– 117.5 | 0.5 ± 0.11 | tr.– 0.6 |
| unoccupied nests | 15 | 2.4 ± 0.64 | 1.7– 3.7 | 10.2 ± 15.25 | 4.1– 67.2 | 1.9 ± 2.55 | 0– 4.8 |
| King Penguin: | | | | | | | |
| top of peat stools | 3 | 3.7 ± 0.18 | 3.5– 3.9 | 34.4 ± 17.60 | 14.1– 45.4 | 67.1 ± 18.54 | 52.1– 87.8 |
| mud between peat stools | 3 | 11.0 ± 1.07 | 10.0–12.1 | 1344.2 ± 97.02 | 1244.0–1437.7 | 260.3 ± 88.05 | 174.0–350.0 |
| Elephant Seal: | | | | | | | |
| mounds between wallows | 3 | n.d. | n.d. | 6.1 ± 1.81 | 4.1– 7.6 | 37.6 ± 14.1 | 23.3– 51.5 |
| mud in wallows | 3 | n.d. | n.d. | 190.2 ± 101.8 | 119.8– 306.0 | 90.7 ± 38.52 | 47.9–122.6 |

organic matter. In relatively dry, mineral soils trampling generally results in soil compaction. Wet, peaty soil, however, is easily churned up by trampling animals. The effect of trampling on the plant cover not only depends on the strength and growth form of the plants, but also on the soil properties mentioned above. In soils which are easily churned up trampling may lead to considerable damage to the root system of the plants. In mineral soils such damage is less likely to occur. In large penguin rookeries the soil is eroded away locally to a depth of several metres as a result of intensive trampling. Erosion gullies of over 3 m deep were encountered at the edge of the rookerie at Kildalkey Bay, the largest rookerie on the islands. Water is the main erosive agent, as wind plays only a minor role (cf. Gilham 1961).

No large herbivores are present on Marion and Prince Edward Islands. Consequently, grazing is no important factor affecting the vegetation.

The impact of seals and birds on soils and vegetation varies considerably with the size and habits of the animals concerned. Herds of Elephant Seals moving inland from their landing beaches during the moulting season (November–April) have a devastating effect on the plant cover of wet, peaty areas. The seals churn up the soil over large areas and heavily damage the vegetation. As a result of this generally yearly recurring disturbance the vegetation of these areas has a pioneer-character, consisting largely of therophytes, notably *Callitriche antarctica* and *Montia fontana*. The latter species is generally classified as hemicryptophyte, but behaves here as a therophyte. By depressing the peat locally in their moulting areas Elephant Seals form wallows: hollows of a few square metres in area, up to 1 m deep, filled with strongly nitrogenous mud. The mud of unoccupied wallows is often colonized by *Callitriche antarctica* and *Montia fontana*.

Yearly recurring destruction of the plant cover and disturbance of the soil also takes place in the breeding colonies of King and Macaroni penguins (Fig. 32). Within the rookeries heavy trampling and

*Fig. 32.* King Penguin colony at Kildalkey Bay, Marion.

manuring is prohibitive for plant growth and the soil remains bare. Most penguins are confined to the coast, but large groups of non-breeding King Penguins may wander up to 2 km inland during the moulting season, locally exerting a strong influence upon soil and vegetation. Rockhopper Penguins breed nearly everywhere along the coasts of the islands. They do not concentrate into large rookeries. Consequently the impact of these more dispersed animals is comparatively small.

Large concentrations of burrowing species of small prions and petrels (Procellariidae) are found on coastal and some inland slopes. By trampling and manuring these birds strongly modify the soil and the vegetation of these slopes (cf. Smith 1976b). Because of their small size and weight these animals do not have such a destructive impact as the larger species of penguins and seals. Groups of breeding Sooty, Grey-headed and Yellow-nosed Albatrosses, which are found on inland and coastal slopes, have a similar influence.

The influence of solitary breeding birds is restricted to the small areas around the nest sites. Croome (1972) found that one seasons' occupation of a nest by Wandering Albatrosses or Giant Petrels resulted in a significant modification of the chemical status of the soils. According to this author a five-fold increase in the organic nitrogen content is not uncommon, and he suggests that the influence of the birds on the nitrogen content of the soil may last for 5–6 years after the nest is abandoned.

Trampling and manuring are generally connected on the islands. In some cases, however, minerals from animal excreta are carried by water or wind to areas not occupied by animals. A striking example of this was seen at Kaalkoppie, a tuffaceous hill on the west coast of Marion Island. Part of its top and seaward slope is occupied by a penguin rookerie. On the landward side the top of the hill supports a luxuriant growth of *Poa cookii*. Guano, washed down from the rookerie, apparently allows plant growth on this otherwise bare hillside.

The plant cover of areas influenced by animals differs strongly from all other vegetation on the islands. Characteristic species for all communities of this complex are *Poa cookii*, *Callitriche antarctica*, and *Montia fontana* (Tables 23 and 32). Bryophytes, which are important in many other types of vegetation on the islands, play only a minor role here. In this respect the vegetation of areas influenced by animals resembles the communities of saline habitats. Five associations together form the *Callitriche antarctica-Poa cookii* complex (Table 23). They range from low-growing pioneer communities to tussock grassland of some 40 cm height with a comparatively complex structure (Fig. 33).

Association 2.1 Montio fontanae-Callitrichetum antarcticae
(type relevé 139)

Areas where the original plant cover has been destroyed by trampling, and which are strongly influenced by manuring, are rapidly colonized by *Callitriche antarctica* and *Montia fontana*. These species form the character-combination of this association, which is the pioneer vegetation on wet, peaty and more or less heavily manured soils. The community is very poor in species and has a very simple structure (Fig. 33). *Montia* and *Callitriche* frequently occur in all other communities of the *Callitriche antarctica-Poa cookii* complex, but have their optimum in the present association (Table 23). Two subassociations were found.

Subassociation 2.1.1 agrostietosum magellanici
(type relevé 186)

After destruction of the bryophyte layer in *Agrostis magellanica*-dominated mire vegetation (cf. Par. 4.4) by trampling or foraging birds (notably Lesser Sheatbill, *Chionis minor*, and Kelp Gull, *Larus dominicanus*) *Montia fontana* invades the vegetation. The influence of animals is comparatively light, because of the small numbers and small size of the birds concerned. Thus, enrichment of the soil by excreta is slight, but seems imperative for the development of this community. In the absence of prolonged animal activity the vegetation reverts to one of the mire communities, probably within a few years. Differential species of this subassociation is *Agrostis magellanica*.

Subassociation 2.1.2 typicum
(type relevé 139)

This subassociation, which does not possess any differential taxa, develops when intensive trampling

*Table 23.* The *Callitriche antarctica – Poa cookii* complex.

| | MONTIO FONTANAE - CALLITRICHETUM ANTARCTICAE | | | | | | | | | | | | | | | | | | | | | | | | | CALLITRICHO ANTARCTICAE - POETUM ANNUAE | | | | | | | |
|---|---|---|---|---|---|---|---|---|---|---|---|---|---|---|---|---|---|---|---|---|---|---|---|---|---|---|---|---|---|---|---|---|---|
| | agrostietosum magellanicae | | | | | | typicum | | | | | | | | | | | | | | | | | | | | | | | | | | |
| | | | | | | | Montia - variant | | | | | intermediate variant | | | | | | | Callitriche - variant | | | | | | | | | | | | | | |
| RELEVE NUMBER | 259 | 179 | 182 | 186 | 255 | 75 | 416 | 190 | 348 | 463 | 572 | 385 | 145 | 573 | 139 | 191 | 376 | 141 | 569 | 349 | 181 | 568 | 317 | 326 | 190 | 584 | 183 | 386 | 583 | 582 | 576 | 176 | 320 |
| LOCATION: | | | | | | | | | | | | | | | | | | | | | | | | | | | | | | | | | |
| SOUTHERN LATITUDE (46°) MIN. | 53 | 51 | 51 | 51 | 53 | 52 | 52 | 51 | 37 | 50 | 53 | 57 | 57 | 53 | 57 | 50 | 57 | 57 | 53 | 37 | 51 | 53 | 52 | 38 | 50 | 52 | 51 | 57 | 52 | 52 | 53 | 50 | 38 |
| SEC. | 05 | 00 | 00 | 00 | 05 | 45 | 40 | 00 | 20 | 48 | 17 | 50 | 55 | 17 | 55 | 25 | 30 | 55 | 17 | 20 | 00 | 10 | 10 | 30 | 25 | 30 | 00 | 50 | 30 | 30 | 10 | 40 | 23 |
| EASTERN LONGITUDE (37°) MIN. | 51 | 47 | 47 | 47 | 51 | 52 | 51 | 47 | 59 | 48 | 52 | 50 | 51 | 52 | 51 | 47 | 50 | 51 | 52 | 59 | 47 | 51 | 51 | 00 | 47 | 51 | 47 | 50 | 51 | 51 | 52 | 48 | 00 |
| SEC. | 55 | 30 | 30 | 30 | 55 | 45 | 30 | 30 | 15 | 05 | 07 | 35 | 00 | 07 | 00 | 45 | 20 | 00 | 07 | 20 | 30 | 57 | 28 | 20 | 45 | 32 | 30 | 42 | 32 | 32 | 03 | 10 | 25 |
| ELEVATION (M A.S.L.) | 20 | 55 | 45 | 45 | 15 | 15 | 18 | 55 | 125 | 110 | 20 | 100 | 40 | 15 | 40 | 25 | 140 | 40 | 20 | 120 | 45 | 25 | 10 | 15 | 20 | 25 | 45 | 100 | 25 | 25 | 10 | 40 | 15 |
| INCLINATION (°) | 10 | | 3 | | 20 | 10 | 3 | | | | | | | 5 | 5 | | | 5 | | | | | | | | | | | | | | | |
| ASPECT | SW | | N | | ESE | E | N | | | | | | | SE | E | | | E | | | | | | | | | | | | | | | |
| SOIL DEPTH (CM) | | | | | | 45 | | | | | | | | | | | | | | | | | | | | | | | | | 4 | | |
| DEPTH OF GROUNDWATER LEVEL (CM) | -35 | -35 | -25 | -25 | -25 | --- | -30 | -30 | - 6 | + 1 | -30 | - | -30 | -25 | - | - 2 | -35 | - | - 4 | -15 | -25 | +50 | - 1 | - 1 | -25 | - | -25 | -35 | - | --- | --- | -45 | - |
| SOIL WATER CONTENT (X10 %) | 111 | 49 | 114 | 129 | 100 | 80 | 75 | 119 | 214 | 174 | 115 | 77 | 111 | 155 | 75 | 76 | 80 | 78 | 213 | 107 | 119 | | 167 | 80 | 76 | 76 | 110 | 48 | 82 | 65 | 16 | 47 | 12 |
| SOIL PH | 4,3 | 4,9 | 4,4 | 4,3 | 4,2 | 4,8 | 4,4 | 4,4 | 4,9 | 7,6 | | 4,2 | 4,3 | 4,2 | 4,4 | 4,4 | 4,5 | 4,5 | 4,4 | 4,4 | 4,8 | 4,7 | 5,2 | 4,7 | 3,9 | 4,9 | 4,6 | 4,8 | 5,0 | 4,5 | | 5,2 | |
| SOIL LOSS ON IGNITION (%) | | | 94 | 96 | 96 | 92 | | 93 | 98 | 89 | 96 | 92 | 99 | 96 | 95 | 97 | 81 | 96 | 93 | | 94 | | 92 | | 98 | 59 | 62 | 80 | 79 | 80 | 63 | 50 | |
| INFLUENCE OF SALTSPRAY | | | | | | | | | | | | | | | | | | | | | | | | | | | | | | | | | |
| INFLUENCE OF TRAMPLING | 2 | 1 | 2 | 3 | 2 | 1 | 4 | 1 | 4 | 1 | 5 | 4 | 5 | 5 | 4 | 5 | | 5 | 4 | 5 | 4 | 5 | 5 | 5 | 5 | 5 | 3 | 5 | 5 | 5 | 5 | | 5 |
| INFLUENCE OF MANURING | 2 | 1 | 1 | 3 | 2 | 1 | 4 | 1 | 4 | 3 | 5 | 4 | 5 | 5 | 4 | 5 | | 5 | 5 | 5 | 3 | 5 | 5 | 5 | 5 | 5 | 2 | 5 | 5 | 5 | 5 | | 5 |
| HEIGHT OF HERB LAYER (CM) | 10 | 10 | 5 | 15 | 7 | 15 | 3 | 12 | 4 | 6 | 10 | 3 | 4 | 7 | 4 | 6 | 5 | 25 | 10 | 3 | 3 | 2 | 2 | 7 | 15 | 10 | 15 | 12 | 15 | 10 | 15 | 25 | 15 |
| COVER OF HERB LAYER (%) | 100 | 100 | 100 | 100 | 100 | 99 | 60 | 100 | 100 | 100 | 100 | 100 | 100 | 100 | 100 | 100 | 100 | 100 | 100 | 100 | 95 | 99 | 65 | 100 | 100 | 95 | 100 | 100 | 100 | 99 | 100 | 100 | 99 |
| HEIGHT OF BRYOPHYTE LAYER (CM) | | 1 | | | | 2 | 1 | | | | | | | | | | | | | | | 2 | | | | 3 | | | | | | 3 | |
| COVER OF BRYOPHYTE LAYER (%) | | 1 | | | | 1 | <1 | | | | | | | | | | | | | | | <1 | | | | 1 | | | | | | 3 | |
| COVER OF ALGAE (%) | | | | | | | | | | | | | | | | | | | | | | | | | | | | | | | | | |
| COVER OF BARE SOIL (%) | | | | | | | 40 | | | | | | | | | | | | | | 5 | 1 | 35 | | | 5 | | | | | | | 1 |
| COVER OF ROCKS (%) | | | | | | | | | | | | | | | | | | | | | | | | | | | | | | | | | |
| QUADRAT SIZE (M2) | 4 | 4 | 9 | 3 | 4 | 1 | 2 | 2 | 9 | 5 | 25 | 4 | 6 | 25 | 9 | 4 | 4 | 9 | 5 | 9 | 4 | 2 | 6 | 9 | 4 | 25 | 3 | 3 | 25 | 20 | 9 | 1 | 25 |
| TOTAL NUMBER OF SPECIES | 4 | 7 | 6 | 3 | 3 | 7 | 8 | 2 | 3 | 3 | 5 | 3 | 5 | 4 | 3 | 3 | 4 | 4 | 4 | 3 | 4 | 4 | 3 | 3 | 2 | 9 | 5 | 4 | 4 | 7 | 5 | 6 | 1 |
| CHARACTER-SPECIES OF THE COMPLEX: | | | | | | | | | | | | | | | | | | | | | | | | | | | | | | | | | |
| MONTIA FONTANA | 2B | 4 | 5 | 5 | 5 | 5 | 4 | 5 | 5 | 5 | 5 | 4 | 4 | 3 | 2B | 2B | 2B | 4 | 1 | + | + | | | | | 3 | 5 | 1 | | 1 | + | | |
| CALLITRICHE ANTARCTICA | 3 | | 1 | + | 2A | | 1 | | | | 2B | 4 | 5 | 5 | 5 | 5 | 5 | 5 | 5 | 5 | 5 | 5 | 5 | 5 | 5 | 1 | 1 | 3 | 3 | 1 | 1 | | |
| POA COOKII | | | + | | | | | | 2B | | 1 | | 1 | + | + | + | R | 2A | R | + | | | | | 2A | | | | | + | | + | |
| OTHER DIAGNOSTIC TAXA: | | | | | | | | | | | | | | | | | | | | | | | | | | | | | | | | | |
| POA ANNUA | | 1 | | | | | | | | + | | | | | | | 1 | | | | R | | 1 | 2B | | 5 | 5 | 5 | 5 | 5 | 5 | 5 | 5 |
| COTULA PLUMOSA | | | | | | + | | | | | | | | | | | | | | | | | | | | | | | + | | + | | |
| CLASMATOCOLEA VERMICULARIS | | | | | | | R | | | | | | | | | | | | | | | | | | | | | | | | | | |
| MARCHANTIA BERTEROANA | | | | | | | 1 | | | | | | | | | | | + | | | | | | | | | | | | | | | |
| LOPHOCOLEA PANDII | | | | | | + | | | | | | | | | | | | | | | | | | | | | | | | | | R | |
| LEPTODONTIUM PROLIFERUM | | | | | | | | | | | | | | | | | | | | | | | | | | | | | | | | | |
| BRACHYTHECIUM RUTABULUM | | | | | | | | | | | | | | | | | | | | | | | | | | 1 | | | | | | | |
| AGROSTIS STOLONIFERA | | | | | | | | | | | | | | | | | | | | | | | | | | | | | | | 1 | | |
| LOPHOZIA CYLINDRIFORMIS | | | | | | | | | | | | | | | | | | | | | | | | | | | | | | | | | |
| DREPANOCLADUS UNCINATUS | | 1 | | | | 1 | | | | | | | | | | | | | | | | + | | | | 1 | | | | | | | |
| ACAENA MAGELLANICA | | + | | | | | | | | 1 | | | | | | | | | | | | | | | | | R | | | | | | |
| MIELICHHOFERIA CAMPYLOCARPA | | | | | | | | | | | | | | | | | | | | | | | | | | | | | | | | | |
| AGROSTIS MAGELLANICA | 5 | 4 | 4 | 4 | 4 | 2B | + | | | | + | | 1 | + | | | | | + | | | 1 | R | | | + | + | | | | | | |
| COMPANION SPECIES | | | | | | | | | | | | | | | | | | | | | | | | | | | | | | | | | |
| RANUNCULUS BITERNATUS | + | 3 | 2B | | | + | + | 3 | 2A | | 1 | 2A | + | | | | | | | | 1 | + | | 2A | | 1 | | + | 1 | 1 | | 2B | |
| JUNCUS SCHEUCHZERIOIDES | | + | 1 | | | | + | | | | | | | | | | | | | | | | | | | + | | | | 1 | | | |
| FOSSOMBRONIA AUSTRALIS | | | | | | | + | | | | | | | | | | | | | | | | | | | | | | | | | | |
| AZORELLA SELAGO | | | | | | | | | | | | | | | | | | | | | | | | | | | | | | | | | |

has resulted in the complete destruction of the original plant cover of wet, peaty soils, rich in nitrogen. Three variants are discerned, according to the relative abundances of *Montia fontana* and *Callitriche antarctica*. One variant is dominated by *Montia*, another by *Callitriche*, while still another is intermediate for these species. On sites where manuring has been extremely heavy *Callitriche* seems to reach dominance more often than *Montia*, but no clear environmental gradient fits the pattern of changing dominance of these two species. It is believed that the availability of propagules at the time when the bare soil is available for colonization plays an important role in determining which species reaches dominance.

Stands of this subassociation are found in all areas on the islands which are influenced by animals. Extensive stands occur near large penguin rookeries and colonies of Elephant Seals. Very small stands were met with near nests of Wandering Albatrosses (*Diomedea exulans*) and large species of petrels.

Both *Callitriche antarctica* and *Montia fontana* are able to partake in a perennial vegetation, and without further disturbance other species, e.g. *Cotula plumosa* and *Poa cookii*, invade the vegetation. In most cases, however, succession is prevented by annually recurring destruction of the plant cover.

Association 2.2 Callitricho antarcticae-Poetum annuae
(type relevé 583)

*Poa annua*, which species together with *Callitriche antarctica* forms the character-combination of this association, is a naturalized alien on Marion and Prince Edward Islands (Gremmen 1975). As a general rule in this study, alien species have not

| POO COOKII - COTULETUM PLUMOSAE | | | | | | | | | | | | | | transition | | | MONTIO FONTANAE - CLASMATOCOLEETUM VERMICULARIS | | | | | | | | | LEPTODONTIO PROLIFERI - POETUM COOKII | | | | | | | | | | | | | | | |
|---|---|---|---|---|---|---|---|---|---|---|---|---|---|---|---|---|---|---|---|---|---|---|---|---|---|---|---|---|---|---|---|---|---|---|---|---|---|---|---|---|---|
| | | | | | | | | | | | | | | | | | | | | | | | | | FRAGMENTS | Inops | | | | | | | brachythecietosum rutabuli | | | | | | | | |
| 260 | 194 | 258 | 257 | 193 | 251 | 516 | 515 | 250 | 637 | 154 | 567 | 562 | 571 | 143 | 146 | 563 | 142 | 210 | 76 | 291 | 560 | 197 | 175 | 196 | 199 | 152 | 129 | 350 | 323 | 377 | 322 | 410 | 458 | 457 | 427 | 436 | 437 | 421 | 422 | 425 | 554 |
| 53 | 50 | 53 | 53 | 50 | 53 | 55 | 55 | 53 | 52 | 51 | 53 | 57 | 53 | 57 | 57 | 57 | 57 | 52 | 52 | 53 | 51 | 50 | 40 | 50 | 50 | 51 | 51 | 37 | 38 | 57 | 38 | 52 | 50 | 50 | 53 | 51 | 51 | 53 | 53 | 53 | 52 |
| 05 | 25 | 05 | 05 | 25 | 05 | 20 | 20 | 05 | 35 | 15 | 10 | 11 | 15 | 55 | 55 | 11 | 55 | 10 | 45 | 10 | 15 | 45 | 50 | 45 | 40 | 15 | 15 | 20 | 25 | 30 | 25 | 40 | 37 | 37 | 50 | 15 | 15 | 48 | 48 | 50 | 21 |
| 51 | 47 | 51 | 51 | 47 | 51 | 35 | 35 | 51 | 38 | 57 | 52 | 52 | 52 | 51 | 51 | 52 | 51 | 51 | 52 | 51 | 50 | 48 | 48 | 48 | 48 | 57 | 57 | 59 | 00 | 50 | 00 | 51 | 48 | 48 | 52 | 50 | 50 | 52 | 52 | 52 | 49 |
| 55 | 45 | 55 | 55 | 45 | 55 | 57 | 58 | 55 | 00 | 40 | 01 | 02 | 10 | 00 | 00 | 02 | 00 | 15 | 45 | 15 | 37 | 10 | 15 | 10 | 00 | 40 | 40 | 18 | 25 | 20 | 25 | 30 | 33 | 33 | 55 | 37 | 37 | 54 | 54 | 55 | 55 |
| 20 | 25 | 25 | 20 | 25 | 20 | 40 | 40 | 15 | 35 | 70 | 25 | 30 | 15 | 40 | 45 | 35 | 35 | 30 | 15 | 58 | 25 | 40 | 35 | 30 | 40 | 70 | 70 | 120 | 40 | 140 | 20 | 17 | 30 | 30 | 35 | 35 | 30 | 35 | 30 | 25 | 140 |
| 5 | 20 | 3 | 20 | 15 | 40 | | | 40 | 30 | 3 | 15 | | | 10 | 10 | 1 | 10 | | | 5 | 5 | 5 | | 7 | | 5 | 3 | | 20 | 5 | 20 | 23 | 40 | 50 | 50 | 45 | 50 | 40 | 40 | 57 | 25 |
| SW | W | SE | S | N | SSE | | | SE | NW | SW | SW | | | E | SE | E | E | | | SSE | E | NW | | SE | | SW | SSE | | NE | NE | E | S | NE | N | N | NE | NW | NE | NE | SE | SE |
| | | | | | | | | | | | | | | | | | | | | | | | | | | | | | | 12 | | | 40 | 25 | | | | | | 20 | |
| -30 | - | -30 | - | -40 | - | - | - | --- | - | -45 | - | - | - | -30 | -30 | -25 | -20 | -25 | -25 | -20 | - | -18 | -22 | -20 | -17 | -25 | -22 | -15 | - | --- | - | --- | --- | --- | - | - | - | - | - | --- | - |
| 109 | 114 | 81 | 87 | 86 | 74 | | | 57 | | 49 | 93 | 111 | 116 | 81 | 72 | 106 | 99 | 97 | 92 | 121 | 116 | 112 | 114 | 108 | 114 | 73 | 115 | 133 | 14 | 61 | 69 | 90 | 36 | 25 | 47 | 61 | 72 | 55 | 75 | 59 | 29 |
| 4,1 | 4,0 | 4,0 | 4,2 | 4,0 | 4,0 | | | 4,8 | | | 4,7 | 4,3 | | 4,3 | 3,8 | 4,2 | 4,2 | 4,2 | 5,3 | 4,0 | 4,1 | 4,2 | 4,2 | 4,0 | 4,0 | 4,3 | 4,2 | 4,8 | 4,2 | 4,1 | 4,7 | 4,0 | 4,0 | 4,2 | 4,0 | 4,2 | 4,2 | 4,4 | 4,0 | 4,4 | 4,0 |
| 98 | 97 | 98 | 95 | 97 | 95 | | | 87 | 93 | 94 | 93 | 96 | 95 | 96 | | 95 | 87 | 96 | 96 | 92 | 93 | 97 | 90 | 87 | 92 | 89 | 94 | 97 | | 88 | 73 | 93 | | 57 | 75 | 82 | 78 | 75 | 76 | 73 | 50 |
| | | | | | | 1 | 1 | | | | | | 2 | | | | | | | | | | | | | | | | | | | | | | | | | | | | |
| 3 | 3 | 4 | 3 | 3 | 3 | 3 | 3 | 4 | 4 | 4 | 3 | 3 | 4 | 3 | 4 | 2 | 3 | 2 | 1 | | 2 | 3 | 4 | 2 | 2 | 5 | 3 | 5 | 4 | 3 | 4 | 2 | 3 | 4 | 2 | 3 | 3 | 3 | 3 | 2 | 4 |
| 4 | 4 | 4 | 4 | 4 | 4 | 4 | 4 | 4 | 5 | 4 | 3 | 3 | 4 | 3 | 4 | 2 | 3 | 2 | | | 2 | 3 | 4 | 3 | 2 | 5 | 4 | 5 | 5 | 3 | 5 | 3 | 4 | 5 | 3 | 4 | 3 | 4 | 4 | 3 | 5 |
| 7 | 25 | 5 | 12 | 12 | 20 | 25 | 25 | 25 | 30 | 10 | 40 | 20 | 35 | 25 | 25 | 25 | 10 | 10 | 3 | 20 | 30 | 20 | 30 | 30 | 25 | 30 | 25 | 25 | 40 | 30 | 30 | 35 | 35 | 35 | 35 | 35 | 35 | 35 | 35 | 35 | 35 |
| 100 | 100 | 100 | 100 | 100 | 100 | 100 | 100 | 100 | 100 | 100 | 99 | 100 | 99 | 100 | 99 | 100 | 60 | 95 | 90 | 100 | 30 | 90 | 70 | 30 | 1 | 100 | 99 | 95 | 90 | 100 | 100 | 95 | 97 | 90 | 99 | 99 | 97 | 100 | 100 | 99 | 90 |
| | | | | | | | | | | | 1 | 2 | 1 | 1 | 1 | 2 | 1 | 2 | 2 | 1 | 2 | 1 | 1 | 1 | 1 | 1 | 3 | 1 | 1 | 2 | 1 | 5 | 2 | 1 | 3 | 2 | 2 | 3 | 2 | 3 | 2 |
| | | | | | | | | | | | <1 | 3 | <1 | 30 | 7 | 65 | 95 | 100 | 95 | 100 | 99 | 90 | 99 | 100 | 100 | <1 | 30 | 3 | <1 | 1 | 3 | 10 | 40 | 35 | 17 | 3 | 10 | 7 | 5 | 15 | 5 |
| | | | | | | | | | | | 1 | | 1 | | | | | | 3 | | | | | | | | | 5 | 10 | | | | | | 1 | 1 | 3 | | | | |
| | | | | | | | | | | | | | | | | | | | | | | | | | | | | | | | | | 3 | 5 | | | | | | | |
| 4 | 4 | 4 | 4 | 4 | 4 | 25 | 25 | 4 | 25 | 7 | 25 | 25 | 21 | 9 | 4 | 25 | 9 | 4 | 1 | 3 | 16 | 4 | 4 | 4 | 4 | 9 | 4 | 9 | 25 | 9 | 25 | 6 | 25 | 25 | 16 | 16 | 16 | 16 | 16 | 16 | 15 |
| 4 | 4 | 4 | 3 | 4 | 4 | 3 | 4 | 3 | 2 | 3 | 7 | 4 | 5 | 8 | 10 | 7 | 8 | 9 | 11 | 11 | 13 | 8 | 11 | 4 | 4 | 5 | 9 | 4 | 4 | 8 | 6 | 13 | 11 | 13 | 13 | 12 | 15 | 14 | 10 | 14 | 14 |
| 5 | | | | | | + | + | | | | + | | + | 1 | 1 | 1 | 3 | 2B | 2A | 5 | + | 4 | | | | 5 | 3 | + | + | 1 | | | 1 | 1 | + | 1 | 1 | 1 | 1 | + | + |
| 2A | 2A | 5 | 1 | 1 | | | | | | | 1 | | + | 1 | + | | + | | + | | | 1 | + | | | + | + | 1 | + | + | | + | | 1 | + | 1 | 1 | + | | + | |
| | R | + | | R | 1 | + | + | R | | 2A | 2A | 3 | 3 | 2B | 2B | 3 | 1 | + | | 1 | 2B | 3 | 4 | 3 | + | 5 | 5 | 5 | 5 | 5 | 5 | 5 | 5 | 5 | 5 | 5 | 5 | 5 | 5 | 5 | 5 |
| | | | | | | | | | + | | | | | | | | | | | | | | | | | | | | | | | | | + | | | | | | | + |
| 5 | 5 | 5 | 5 | 5 | 5 | 5 | 5 | 5 | 5 | 5 | 5 | 5 | 5 | 5 | 5 | 5 | | | + | | | + | | + | | | | | | | + | | + | | | | | | | | |
| | | | | | | | | | | | | | | 3 | 2A | 4 | 5 | 4 | 5 | 5 | 5 | 5 | 5 | 5 | 4 | | | | | | | + | 1 | 2B | | | | | | | |
| | | | | | | | | | | | | | | 1 | 1 | + | 2B | | | | 1 | 2A | 1 | 3 | 4 | R | 1 | | + | 1 | + | + | 3 | 2A | 1 | 1 | 1 | + | + | 1 | 1 |
| | | | | | | | | | | | | 1 | + | | + | | | | | | + | | 2A | | | | | | | + | 1 | 1 | | 1 | 1 | + | 1 | + | + | 1 | |
| | | | | | | | | | | | | + | | | | | | | | + | | | + | | | | + | | | + | | + | 2M | 1 | + | 1 | 1 | + | + | + | 1 |
| | | | | | | | | | | | | | | | | | | | | | | | | | | | | | | | | | + | 2B | 2B | 1 | 2A | 2A | 1 | 1 | 1 |
| | | | | | | | | | | | | | | | | | | | | | | | | | | | | | | | | | | | 1 | 3 | 2A | 1 | 2A | 2A | |
| | | | | | + | | | | | | | | | | | | | | | | | | + | | | | | | | | | | | 1 | | | 1 | + | + | + | |
| | | | | | | | | | | | | | | | | | | | + | | | | | | | | | | | | | | | | + | | 1 | 1 | 1 | 2B | + |
| | | | | | + | | | | | | | | | | | | | | | | + | | | | | | R | | | | | | | | | | | + | + | + | 2B |
| | | | | | | | | | | | | | | | | | | | | | | | | | | | | | | | | + | + | | | + | | + | | + | 1 |
| + | + | + | 1 | R | | | | + | | + | + | | | 2B | 1 | 1 | 3 | + | 1 | + | 1 | + | + | | | + | 1 | + | | | | | + | | | | | | | | |
| | | | | | | | | | | | | | | 1 | + | 1 | | 5 | 2B | | 1 | 1 | + | | | | | | | | | | | + | | | + | | | | + |
| | | | | | | | | | | | | | | | | | + | + | + | | | | | | | | | | | | | | | | | | | | | | |
| | | | | | | | | | | + | | | | | + | | + | + | | | | | | | | | + | | | | | + | | | | | | | | | |
| | | | | | | | + | | | | | | | | | | | | | R | | | R | | | | | | | | R | + | | | | | | R | | | |

been used to characterize syntaxa. *Poa annua,* however, has spread to many sites on the islands where it definitely was not introduced and is doing very well there. Dense swards of this grass, which is sometimes the only species present (relevé 320), are found on relatively dry, mineral, strongly compacted soils, generally heavily manured by animals. In this vegetation most plants of *Poa annua,* elsewhere usually a therophyte, are perennial. Similar behaviour of this species has been noted on Kerguelen (Chastain 1958) and South Georgia (Walton & Smith 1973).

Stands of this association were found near some large King Penguin rookeries and Elephant Seal moulting grounds, e.g. Kildalkey Bay, King Penguin Bay, Trypot Beach and Cave Bay (Prince Edward) and in the immediate vicinity of the meteorological station.

Association 2.3 Poo cookii-Cotuletum plumosae (type relevé 193)

This association does not possess any character-species. Differential species is *Cotula plumosa,* the dominant of this species-poor vegetation. This community occurs on well-drained, organic soils, which are strongly influenced by manuring, on sites where trampling is not so heavy as to do very much damage to the plant cover.

Stands of Poo cookii-Cotuletum plumosae are restricted to the coastal lowlands. *Cotula plumosa* is not found above c. 100 m above sea level, which may be a consequence of the alleged dependence of this species on minerals deposited by salt-spray (Huntley 1971).

*Cotula plumosa* has its optimum in two associations, the present one and the Cotulo plumosae-Crassuletum moschatae, which occurs in saline habitats. *Cotula* differentiates these two communi-

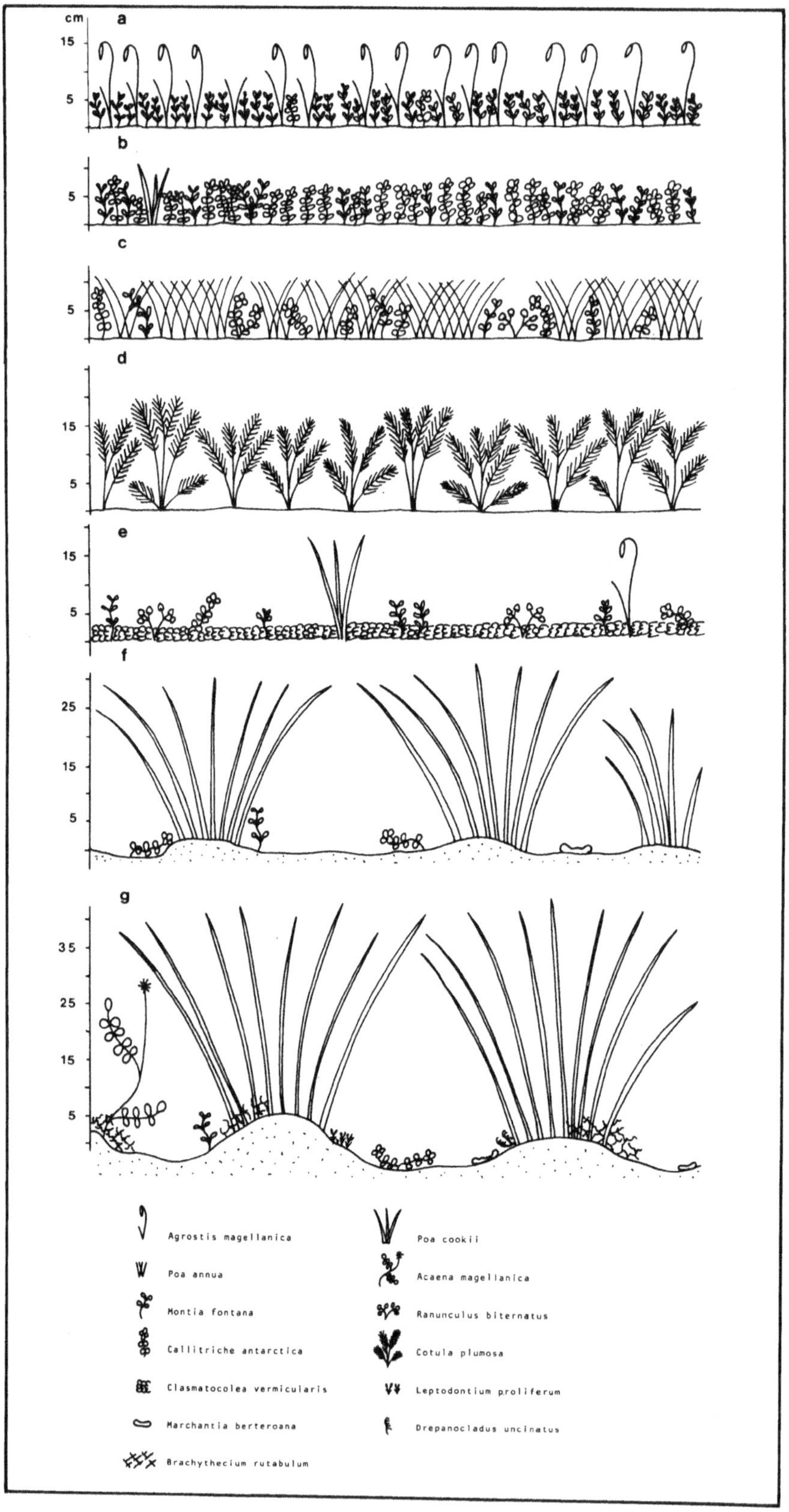


*Fig. 33.* Structure-diagrams of communities of the *Callitriche antarctica – Poa cookii* complex. a = Montio fontanae – Callitrichetum antarcticae agrostietosum magellanicae; b = Montio fontanae – Callitrichetum antarcticae typicum; c = Callitricho antarcticae – Poetum annuae; d = Poo cookii – Cotuletum plumosae; e = Montio fontanae – Clasmatocoleetum vermicularis; f = Leptodontio proliferi – Poetum cookii inops; g = Leptodontio proliferi – Poetum cookii brachythecietosum rutabuli.

ties from all other ones on Marion and Prince Edward Islands. Floristically the Cotulo-Crassuletum and the Poo-Cotuletum are differentiated from each other respectively by *Crassula moschata* and by *Poa cookii* and *Montia fontana* (Table 32). There is a conspicuous difference in growth form between *Cotula* plants of these two communities: in Poo cookii-Cotuletum plumosae stands the *Cotula* plants are usually 10–25 cm high, with long, ascending leaves, which contrast strongly with the small rosettes, densely covered with long, white hairs formed by *Cotula* in the salt-spray zone (Fig. 28, cf. Gilham 1961; Chastain 1958).

Stands of Poo cookii-Cotuletum plumosae cover large parts of the coastal regions in areas influenced by animals.

Association 2.4 Montio fontanae-Clasmatocoleetum vermicularis
(type relevé 76)

This is the only community within the *Callitriche antarctica-Poa cookii* complex in which a bryophyte stratum completely covers the soil. Dominant is the liverwort *Clasmatocolea vermicularis*, the differential species of this association against all other communities of this complex. Its phytosociological position is comparable to that of *Cotula plumosa*, but it does not show morphological differences between the populations of different habitats. *Clasmatocolea vermicularis* is dominant in the association described here and in the Crassulo moschatae-Clasmatocoleetum vermicularis, which occurs on very wet soils within the salt-spray zone (cf. Par. 4.1). It differentiates these two communities from all other ones on Marion and Prince Edward Islands.

The Montio fontanae-Clasmatocoleetum vermicularis has a more complex structure than the communities of this complex described sofar, possessing a well-developed bryophyte stratum as well as a herb layer (Fig. 33e). The small proportion of therophytes in the vegetation and the comparatively large number of species (on average 9 species/relevé) indicate a more mature character of the vegetation compared with the communities of areas influenced by animals described sofar.

Stands of this association are found on wet, organic soils with a groundwater level at about 20 cm below the surface. Trampling is usually light and the influence of manuring moderate, and enrichment through the groundwater from surrounding areas is in some cases more important than direct deposition of excreta by animals. An intermediate vegetation between Montio-Clasmatocoleetum and Poo-Cotuletum was found on slightly drier soils (relevés 143, 146 and 563). Other stands develop towards the Leptodontio proliferi-Poetum cookii (e.g. relevés 560, 197 and 175). Poorly developed, fragmentary stands of this association also occur (e.g. relevés 196 and 199).

Association 2.5 Leptodontio proliferi-Poetum cookii
(type relevé 422)

This association comprises the *Poa cookii* tussock grasslands of the islands. Character-species of the Leptodontio proliferi-Poetum cookii are *Poa cookii, Leptodontium proliferum* and *Marchantia berteroana*. Tussock grassland is found on well-drained soils, generally on slopes, with a moderate to strong influence of animals. The impact of animals is not evenly distributed in this type of vegetation. Trampling is in most cases much stronger in between the tussocks of *Poa cookii* than on the places occupied by individuals of this species, and consequently deposition of excreta also is not homogeneous over all parts of the site (cf. Table 22). In his description of the vegetation of Kerguelen, Chastain (1958) gives the impression that *Poa cookii* tussock grassland there is restricted to areas influenced by saltspray. On Marion and Prince Edward Islands this is not the case, as stands of this association were found at sites more than 4 km inland, at elevations of up to 300 m above sea level (cf. Huntley 1971).

Subassociation 2.5.1 inops
(type relevé 323)

In areas strongly influenced by seals and penguins tussock grassland is generally poor in species and the bryophyte stratum is usually very poorly developed. Disturbance of the soil and vegetation is stronger here than in the other subassociation, and the animals influencing the vegetation here are generally much larger in size. Only when small species of burrowing prions and petrels are present in very high densities, such as is the case on some slopes near Cave Bay (Prince Edward), the subassociation inops is found. When their density is smaller the other subassociation occurs.

Extensive stands of this community occur in all areas strongly influenced by animals, notably by seals and penguins.

Subassociation 2.5.2 brachythecietosum rutabuli (type relevé 422)

This subassociation is found on well-drained, often steep slopes, where the disturbance of soil and plant cover is less strong than on sites occupied by the subassociation inops, which in most cases is due to the much smaller size of the animals concerned (usually burrowing species of Procellariidae). The influence of these animals is therefore less destructive in this case than in most other communities of this complex. This is reflected in the comparatively high number of species in the vegetation (on average 13 species/relevé, compared to 6 or 7 in the subassociation inops) and in the occurrence of a number of species which have their optimum in communities of areas not influenced by animals. Character-species of this subassociation is *Mielichhoferia campylocarpa*. Differential species are *Brachythecium rutabulum, Acaena magellanica, Agrostis stolonifera, Drepanocladus uncinatus* and *Lophozia cylindriformis* (Table 23). The structure of this community, which is more complex than that of most associations of this complex, is illustrated in Fig. 33g.

Extensive stands of this vegetation cover coastal and inland slopes (Fig. 34) influenced by species of Procellariidae or by albatross species, in some cases also by penguins.

*Poa cookii-Callitriche antarctica mosaic*

Soil erosion associated with trampling by penguins may cause the formation of 30–100 cm high peat stools, crowned by *Poa cookii* tussocks and surrounded by hollows with *Callitriche antarctica.* Usually a large proportion of the soil remains bare (Table 24, Fig. 35). The vegetation of such sites is a mosaic of fragments of the Leptodontio proliferi-Poetum cookii on the hummocks and of Montio fontanae-Callitrichetum antarcticae occupying the hollows, and represents a degradation stage of the Leptodontio-Poetum. Large areas at the periphery of some penguin rookeries are covered with this mosaic.

The distribution of the plant communities of the *Callitriche antarctica-Poa cookii* complex in relation to environmental conditions is represented diagrammatically in Fig. 36. The vertical axis represents a gradient of increasing influence by animals, while on the horizontal axis the communities are arranged in a series from wet to dry. The latter gradient coincides largely with a gradient from organic to mineral soils. Soil compaction is a function of both the intensity of the influence of animals and soil characteristics. In the diagram soil compaction is strongest in the top right-hand corner, which represents sites with a dry mineral soil and a very strong influence by animals. At the bottom of the diagram only a few of the most common vegetation types of areas unaffected by animals are shown.

*Synchorological remarks*

Most of the species dominating the communities of the *Callitriche antarctica-Poa cookii* complex are widely distributed on the islands in the Southern Ocean (cf. Tables 10, 11, 12). A few species have a more restricted distribution, e.g. *Poa cookii,* which only occurs on the islands of the Kerguelen Province. However, other species of *Poa,* more of less similar in habit and in ecology, form tussock grasslands in many other Subantarctic and southern temperate regions (cf. Wace 1965).

Although no detailed floristic data are available, it seems likely that the Montio fontanae-Callitrichetum antarcticae, or closely related communities, are present on most Subantarctic islands. On Macquarie and on Campbell Island the pioneer vegetation in abandoned Elephant Seal wallows is dominated by *Callitriche antarctica,* while in a later stage *Montia fontana* and still later *Cotula plumosa* play an important role in the vegetation of these sites (Taylor 1955a; Gilham 1961; Oliver & Sorensen 1951). A similar luxuriant growth of *Callitriche antarctica, Montia fontana* and *Cotula plumosa* is reported from wet, sheltered coastal cliffs on Auckland Island (Cockayne 1909), but it is not clear whether or not this vegetation has a similar ecology as the Montio-Callitrichetum. In his

*Fig. 34.* Coastal slopes, covered with Poo cookii – Cotuletum plumosae and Leptodontio proliferi – Poetum cookii.

description of the vegetation of Kerguelen Chastain (1958) does not mention a community resembling the Montio-Callitrichetum, but Cour (1959) reports lush mats of *Callitriche antarctica* and lists this species as one of the dominants in the vegetation of inland seal wallows on these islands. No related communities are reported from Crozet, but in view of the resemblance of flora and environment of this archipelago to those of Kerguelen and Marion and Prince Edward Islands, it seems likely that such communities do occur there. On South Georgia extensive swards of *Callitriche antarctica* are found around abandoned albatross nests (Greene 1964).

*Poa annua* has been introduced to all Subantarctic islands (Walton 1975) and communities related to the Callitricho antarcticae-Poetum annuae seem to occur on most of them. Because of its adaptation to heavy grazing pressure this species has been very successful on islands with introduced herbivores (Walton 1975). Thus, on South Georgia extensive 'replacement swards' dominated by *Poa annua* are found in areas where the original vegetation has been destroyed by the introduced reindeer (Walton & Smith 1973; Walton 1975; Kightley & Smith 1976).

*Table 24. Poa cookii–Callitriche antarctica* mosaic.

| Relevé number | 192 | 149 | 388 | 127 | 151 | 321 | 187 | 195 | 126 | 150 |
|---|---|---|---|---|---|---|---|---|---|---|
| Location: | | | | | | | | | | |
| Southern Latitude (46°) min. | 50 | 57 | 57 | 51 | 51 | 38 | 51 | 50 | 57 | 57 |
| sec. | 25 | 55 | 47 | 15 | 15 | 25 | 00 | 25 | 40 | 55 |
| Eastern Longitude (37°) min. | 47 | 51 | 50 | 57 | 57 | 00 | 47 | 47 | 51 | 51 |
| sec. | 45 | 00 | 42 | 40 | 45 | 25 | 30 | 45 | 15 | 00 |
| Elevation (m a.s.l.) | 25 | 35 | 100 | 70 | 30 | 15 | 45 | 20 | 70 | 35 |
| Inclination (°) | | 10 | | 7 | 20 | | | 35 | 7 | 10 |
| Aspect | | E | | SSE | W | | | NNW | SSE | E |
| Soil depth (cm) | -35 | - | - | -25 | - | - | - | - | -25 | - |
| Depth of groundwater level (cm) | -35 | - | - | -25 | - | - | - | - | -25 | - |
| Soil water content (X 10%) | 93 | 60 | 67 | 70 | 46 | 121 | 43 | 30 | 41 | 46 |
| Soil pH | 3.8 | 4.2 | 4.7 | 3.3 | 4.5 | | 3.7 | 3.1 | 4.3 | 3.9 |
| Soil loss on ignition (%) | 97 | 95 | | 92 | 96 | | 81 | 80 | 87 | 88 |
| Influence of saltspray | | | | | | | | | | |
| Influence of trampling | 5 | 5 | 5 | 5 | 5 | 5 | 5 | 5 | 5 | 5 |
| Influence of manuring | 5 | 5 | 5 | 5 | 5 | 5 | 5 | 5 | 5 | 5 |
| Height of herb layer (%) | 25 | 15 | 30 | 30 | 30 | 30 | 25 | 30 | 30 | 30 |
| Cover of herb layer (%) | 100 | 97 | 90 | 65 | 65 | 65 | 60 | 30 | 20 | 30 |
| Cover of bare soil (%) | | 3 | 10 | 35 | 35 | 35 | | 70 | 80 | 70 |
| Cover of rocks (%) | | | | | | | | | | |
| Quadrat size (m²) | 9 | 9 | 9 | 25 | 9 | 100 | 9 | 25 | 25 | 16 |
| Total number of species | 2 | 4 | 4 | 4 | 3 | 6 | 3 | 2 | 2 | 6 |
| Poa cookii | 4.3 | 3.3 | 4.3 | 3.3 | 3.3 | 3.3 | 4.3 | 4.3 | 2b.3 | 2b.3 |
| Callitriche antarctica | 5.4 | 4.4 | 2a.3 | 3.4 | 2b.3 | 3.3 | 1.3 | 1.3 | 1.3 | 2a.3 |
| Montia fontana | | 2a.3 | 1.3 | 1.2 | 1.2 | r.1 | | | | +.2 |
| Poa annua | | | 1.3 | | | +.3 | 2a.3 | | | |
| Cotula plumosa | | r.2 | | | | +.3 | | | | +.3 |
| Ranunculus biternatus | | | | | | +.2 | | | | +.2 |
| Agrostis magellanica | | | | +.2 | | | | | | |
| Marchantia berteroana | | | | | | | | | | r.2 |

*Cotula plumosa* occurs on all islands of the Kerguelen Province and on the islands south of New Zealand (Table 10), but communities dominated by this species are only reported from Kerguelen, where a *Cotula plumosa*-'prairie' occurs on deep, organic soils outside the littoral zone (Chastain 1958). This author does not mention the influence of animals with respect to this community. Moseley (1877a), however, observed that on Kerguelen *Cotula plumosa* 'thrives and luxuriates where the soil is enriched by animal dung'. The available information on this community is insufficient to judge the resemblance of it to the Poo cookii-Cotuletum plumosae.

*Clasmatocolea vermicularis,* dominant and character-species of the Montio fontanae-Clasmatocoleetum vermicularis, is reported from only one other Subantarctic island, South Georgia (cf. Table 12), and no communities resembling this association have been described.

*Poa cookii* tussock grassland is found on Kerguelen (Aubert de la Rue 1932, 1964; Chastain 1958; Cour 1959), Crozet (Dreux 1964), Heard and Macdonald Island (Hemsley 1885; Vanhöffen 1903), but floristic and ecological data are insufficient to judge the relationship of those tussock grasslands to the Leptodontio proliferi-Poetum cookii. Some of the *Poa foliosa* and *Poa flabellata* communities reported from Macquarie (Taylor 1955a), South Georgia (Skottsberg 1912; Greene 1964), Gough and Tristan da Cunha (Wace & Holdgate 1958; Wace 1961; Wace & Dickson 1965) appear to be similar in structure and ecology to the Leptodontio proliferi-Poetum cookii.

*Fig. 35. Poa cookii* on large peat stools at the edge of a Macaroni Penguin rookerie (Kildalkey Bay, Marion).

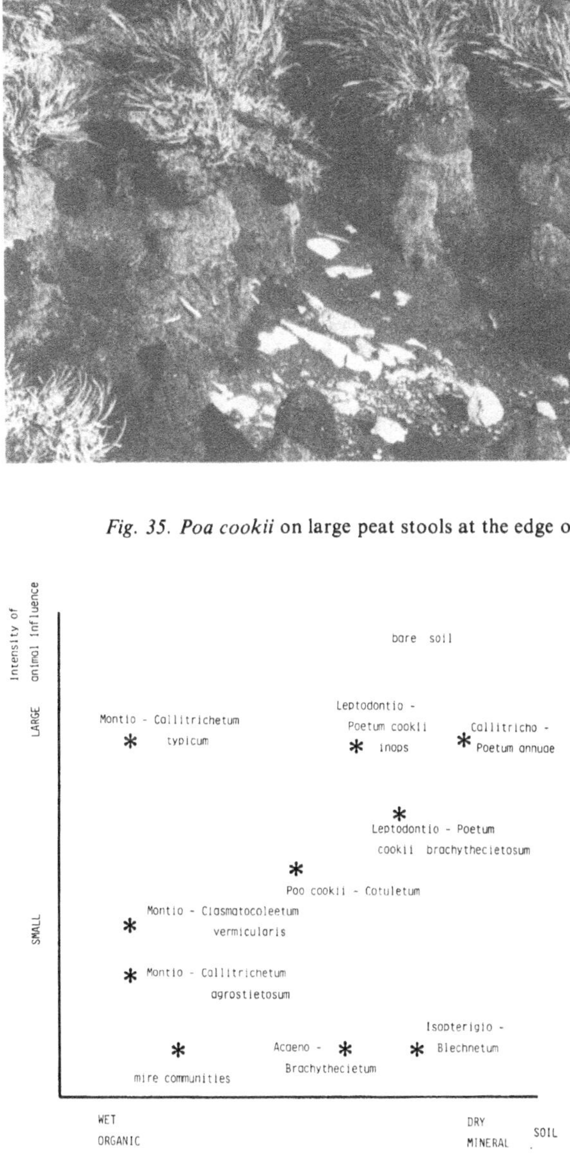


*Fig. 36.* Relations between the plant communities of the *Callitriche antarctica - Poa cookii* complex and the major factors influencing their distribution.

## 4.3 Communities of springs and drainage lines: the Acaena magellanica - Brachythecium complex

Communities dominated by *Acaena magellanica* often form oblong stands along river banks, in springs, flushes, water tracks and drainage lines. Characteristic for the environment in which those communities occur is a more or less rapid lateral water flow through the soil or at the soil surface. The pH of water and soils of springs, flushes, water tracks and drainage lines is higher than in most other habitats on Marion and Prince Edward Islands, indicating minerotrophic conditions (cf. Grobbelaar 1974; Smith 1976c), although the water movement also causes a somewhat higher pH (cf. Sparling 1966). The ionic dominance order of water from springs on Marion is Na > Ca > Mg > K: Cl > $HCO_3$ > $SO_4$, compared with Na > Mg > Ca > K: Cl > $SO_4$ > $HCO_3$ for most other types of fresh water on the island. According to Grobbelaar (1974) this indicates a marked geochemical modification of the mineral content of the drainage water. Smith (1976a, 1977d) found comparatively high concentrations of Ca and Mg in soils of drainage lines (cf. Table 9).

Not only the concentrations of minerals in the soil and drainage water influences the mineral nutrition of the plants, but also the water movement is alleged to have an important effect, resulting in a greater availability of nutrients (Chapman 1965; Ingram 1967). Thus water tracks and springs are sites with an appreciably greater nutrient supply to the vegetation than their surrounding areas.

Associated with drainage lines, water tracks, flushes and springs is a group of plant communities characterized by the abundance of *Acaena magellanica* in the herb layer and of *Brachythecium subplicatum* or *Brachythecium rutabulum* in the bryophyte stratum (Table 25, 32). These communities form a series, ranging in habitat preference from rapidly flowing, shallow, open water, via fast-flowing springs and flushes, generally without open water channel, and water tracks (places with a less abundant water supply than flushes), to drainage lines on slopes, where the groundwater level is usually more than 40 cm below the surface.

Association 3.1 Brachythecietum subplicati
(type relevé 465)

*Brachythecium subplicatum,* the character-species of this association, dominates the bryophyte stratum of the vegetation of springs and flushes and other places with rapidly flowing, shallow water. Usually *Acaena magellanica* dominates the herb layer. Although the water flow, which is related to the amount of rainfall, may vary considerably, the bryophyte layer of the vegetation always remains about saturated with water. Three subassociations are discerned.

Subassociation 3.1.1 philonotidetosum
(type relevé 309)

*Philonotis* cf. *angustifolia* and *Bryum* spec. (no. 9 in Zanten 1971) are the character-species of this subassociation. Only one extensive stand of this community was found, covering an area permanently inundated by Van Den Boogaard River near Tafelberg on Marion Island. A dense mat of bryophytes grows here in 3–10 cm deep running water. The soil consists of 1–3 cm organic matter overlying a bed of pebbles and sandy material (Fig. 37). At the edges of the stand the bryophyte layer is locally poorly developed and the vegetation is dominated by *Montia fontana* (relevé 312). A possible reason for this change in dominance is the stronger impact of fluctuations in the water level along the edges of the inundated area.

Fragmentary stands of this community were found along the borders of some rivers, on sandy deposits within reach of the water. These stands are most common along the upper course of Van Den Boogaard River. The river probably provides an easy dispersal route for *Philonotis* and *Bryum,* which were both found on Marion in vegetative form only. The permanent nature of the river is probably important, assuring a permanent water supply, even during spells without precipitation.

Subassociation 3.1.2 inops
(type relevé 459)

In springs and water tracks with a strong rate of flow dense carpets of *Brachythecium subplicatum* occur. *Acaena magellanica* dominates the herb layer, but in some cases this species is absent and the herb layer is poorly developed (Figs. 37, 38). This subassociation contains only few species, and is characterized by the absence of the differential and character-species of both other subassociations. The vegetation often has the character of a quaking bog. There seems to be a tendency for the cover of *Acaena* in this community to decrease towards the middle of the spring or flush. No explanation for this phenomenon was found.

Subassociation 3.1.3 breutelietosum integrifoliae
(type relevé 465)

This community (Fig. 37) occurs on more solid ground in water tracks, on more mineral soils than the subassociation inops. It is comparatively rich in species (Table 25). Differential species are *Breutelia integrifolia* and *Drepanocladus uncinatus.* Character-species is cf. *Cratoneuron filicinum* (Zanten 1971). The water flow through stands of this community seems to be appreciably less than

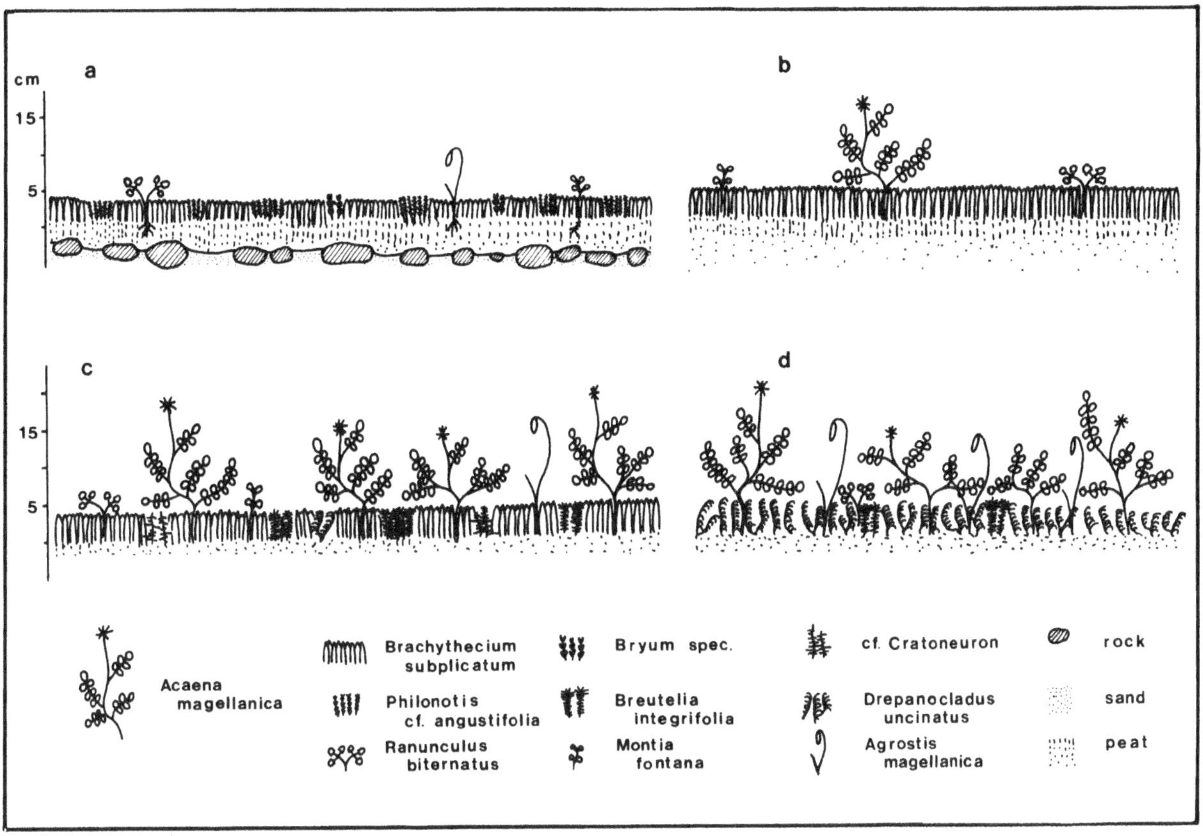


*Fig. 37.* Structure diagrams of communities of the *Acaena magellanica* - *Brachythecium* complex. a = Brachythecietum subplicati philonotidetosum; b = Brachythecietum subplicati inops; c = Brachythecietum subplicati breutelietosum integrifoliae; d = Acaeno magellanicae - Drepanocladetum uncinati breutelietosum integrifoliae.

through the flushes covered with stands of Brachythecietum subplicati inops.

Only one extensive stand of the Brachythecietum subplicati philonotidetosum was found, along Van Den Boogaard River (Marion). Both other subassociations are common on Marion Island. No stands of this association were found on Prince Edward Island, where also none of the characteristic bryophytes of this community, e.g. *Brachythecium subplicatum, Philonotis* cf. *angustifolia, Bryum* spec., *Breutelia integrifolia* and cf. *Cratoneuron filicinum,* have been found. None of these species is known to produce spores on Marion Island, which makes distribution of these species from this island to Prince Edward Island very difficult.

Association 3.2 Acaeno magellanicae-Drepanocladetum uncinati
(type relevé 206)

In drainage lines with a less strong water flow and with a water level at least somewhat below the soil surface, *Drepanocladus uncinatus* is the dominant bryophyte. Together with *Acaena magellanica,* the dominant in the herb layer, this species forms the character-combination of this association. Two subassociations are distinguished.

Subassociation 3.2.1 breutelietosum integrifoliae
(type relevé 206)

This subassociation (Fig. 37) is found in drainage lines with a water level close to the surface (1-13 cm deep). Differential species are *Breutelia integrifolia* and *Agrostis magellanica*.

*Table 25.* The *Acaena magellanica – Brachythecium* complex

| | BRACHYTHECIETUM SUBPLICATI | | | | | | | | | | | | | | | ACAENO MAGELLANICAE - DREPANOCLADETUM UNCINATI | | | | | | | | |
|---|---|---|---|---|---|---|---|---|---|---|---|---|---|---|---|---|---|---|---|---|---|---|---|---|
| | philonotidetosum | | | inops | | | | | | breutelietosum integrifoliae | | | | | | breutelietosum integrifoliae | | | | | brachythecietosum rutabuli | | | |
| RELEVE NUMBER | 312 | 309 | 310 | 380 | 478 | 456 | 459 | 453 | 382 | 479 | 267 | 465 | 263 | 207 | 264 | 208 | 205 | 540 | 462 | 206 | 431 | 557 | 189 | 134 |
| LOCATION | | | | | | | | | | | | | | | | | | | | | | | | |
| SOUTHERN LATITUDE (46°) MIN. | 53 | 53 | 53 | 57 | 50 | 50 | 50 | 50 | 57 | 50 | 51 | 50 | 51 | 51 | 51 | 51 | 51 | 52 | 50 | 51 | 53 | 52 | 51 | 57 |
| SEC. | 05 | 05 | 05 | 05 | 48 | 48 | 48 | 48 | 05 | 48 | 00 | 48 | 00 | 00 | 00 | 00 | 00 | 21 | 48 | 00 | 45 | 21 | 00 | 45 |
| EASTERN LONGITUDE (37°) MIN. | 48 | 48 | 48 | 31 | 48 | 48 | 48 | 48 | 31 | 48 | 48 | 48 | 48 | 48 | 48 | 48 | 48 | 49 | 48 | 48 | 52 | 49 | 47 | 51 |
| SEC. | 40 | 40 | 40 | 15 | 05 | 05 | 05 | 05 | 15 | 05 | 00 | 05 | 00 | 00 | 00 | 00 | 00 | 55 | 05 | 00 | 30 | 55 | 30 | 00 |
| ELEVATION (M A.S.L.) | 230 | 230 | 230 | 75 | 110 | 90 | 110 | 100 | 75 | 110 | 140 | 110 | 120 | 130 | 120 | 130 | 130 | 130 | 110 | 130 | 50 | 135 | 45 | 50 |
| INCLINATION (°) | | | | | | | 4 | 4 | | | 1 | | 5 | | 80 | | | | 10 | | | 35 | | 25 |
| ASPECT | | | | | | | N | N | | | SW | | NE | | NNW | | | | N | | | SSE | | E |
| SOIL DEPTH (CM) | | | | | | | | | | | 4 | | 8 | | 7 | | | | | | | | | 15 |
| DEPTH OF GROUNDWATER LEVEL (CM) | + 5 | + 7 | + 4 | - 7 | + 2 | - 3 | - 3 | - 4 | - 9 | 0 | 0 | + 2 | + 3 | - 2 | - 2 | -30 | - 8 | - 3 | -10 | - 1 | -25 | - | -35 | --- |
| SOIL WATER CONTENT (X10 %) | | | | 317 | | 304 | 204 | 209 | 414 | | 132 | 71 | | 97 | 150 | 50 | 57 | 63 | 156 | 83 | 96 | 76 | 94 | 83 |
| SOIL PH | | | | 5,6 | | 5,3 | 5.4 | 5,8 | 6,0 | | | 5,8 | | 5,4 | | 6,0 | 5,4 | 5,8 | 5,8 | 5,5 | 4,8 | 5,0 | 4,1 | 4,1 |
| SOIL LOSS ON IGNITION (%) | | | | 89 | | 87 | | 78 | 88 | | 64 | 26 | | 40 | | 38 | 33 | 98 | 92 | 34 | 76 | 75 | 91 | 83 |
| INFLUENCE OF SALTSPRAY | | | | | | | | | | | | | | | | | | | | | | | | |
| INFLUENCE OF TRAMPLING | | | | | | | | | | | | | | | | | | | | | | | | |
| INFLUENCE OF MANURING | | | | | | | | | | | | | | | | | | | | | | | | 1 |
| HEIGHT OF HERB LAYER (CM) | 5 | 5 | 5 | 12 | 10 | 15 | 15 | 15 | 30 | 10 | 15 | 20 | 15 | 15 | 15 | 20 | 20 | 25 | 15 | 25 | 25 | 30 | 30 | 20 |
| COVER OF HERB LAYER (%) | 80 | 10 | 3 | 7 | 3 | 3 | 40 | 80 | 99 | 70 | 60 | 40 | 60 | 95 | 60 | 50 | 70 | 70 | 60 | 90 | 100 | 85 | 95 | 99 |
| HEIGHT OF BRYOPHYTE LAYER (CM) | 5 | 5 | 5 | 8 | 5 | 5 | 5 | 5 | 7 | 5 | 5 | 6 | 5 | 5 | 5 | 6 | 5 | 5 | 6 | 5 | 3 | 15 | 4 | 15 |
| COVER OF BRYOPHYTE LAYER (%) | 3 | 100 | 99 | 100 | 100 | 100 | 100 | 100 | 100 | 100 | 100 | 100 | 100 | 100 | 100 | 100 | 100 | 100 | 100 | 100 | 99 | 100 | 90 | 100 |
| COVER OF ALGAE (%) | | | | | | | | | | | | | | | | | | | | | | | | |
| COVER OF BARE SOIL (%) | | | | | | | | | | | | | | | | | | | | | | | | |
| COVER OF ROCKS (%) | 20 | | | | | | | | | | | | | | | | | | | | | | | |
| QUADRAT SIZE (M2) | 4 | 4 | 4 | 4 | 4 | 25 | 15 | 12 | 9 | 4 | 4 | 15 | 3 | 3 | 3 | 3 | 3 | 8 | 9 | 3 | 8 | 20 | 5 | 16 |
| TOTAL NUMBER OF SPECIES | 9 | 9 | 10 | 6 | 4 | 6 | 6 | 6 | 4 | 8 | 10 | 10 | 8 | 7 | 6 | 5 | 9 | 7 | 8 | 7 | 5 | 10 | 6 | 5 |
| CHARACTER-SPECIES OF THE COMPLEX | | | | | | | | | | | | | | | | | | | | | | | | |
| ACAENA MAGELLANICA | | | | | 1 | 1 | 3 | 4 | 5 | 4 | 4 | 3 | 4 | 5 | 4 | 3 | 3 | 3 | 4 | 4 | 5 | 5 | 5 | 5 |
| DIAGNOSTIC TAXA OF ASSOCIATIONS AND LOWER UNITS | | | | | | | | | | | | | | | | | | | | | | | | |
| PHILONOTIS CF. ANGUSTIFOLIA | 1 | 3 | 3 | | | | | | | | | + | | | | | | | | | | | | |
| BRYUM SPEC. | 1 | 1 | 1 | | | | | | | | | | | | | | | | | | | | | |
| BRACHYTHECIUM SUBPLICATUM | + | 5 | 5 | 5 | 5 | 5 | 5 | 5 | 5 | 5 | 5 | 5 | 5 | 5 | 5 | | | | + | | | | | |
| CRATONEURON FILICINUM | | | + | | | | | | | 2M | 1 | 1 | 2A | 2A | 2B | | | | | 1 | | | | |
| BREUTELIA INTEGRIFOLIA | | | | | | | | | | + | 1 | + | 1 | + | 2B | + | 1 | + | 1 | 2A | | | | |
| DREPANOCLADUS UNCINATUS | | | | | | | + | | | + | + | + | 1 | + | | 5 | 5 | 5 | 5 | 5 | 5 | 5 | 4 | 5 |
| BRACHYTHECIUM RUTABULUM | | | | | | | | | | | | | | | | | | 1 | | 1 | | 1 | 2A | 2B |
| BLECHNUM PENNA-MARINA | | | | | | | | | | | | | | | | | R | | + | | 2B | | | 2A |
| AGROSTIS STOLONIFERA | | | | | | | | | | | | | | | | | | | | | | | | |
| PRINGLEA ANTISCORBUTICA | | | | | | | | | | | | | | | | | | | | | | | | |
| AGROSTIS MAGELLANICA | 1 | 1 | + | + | | 1 | + | | | + | 1 | 2A | + | 2B | + | 3 | 3 | 3 | 1 | 2B | | + | | |
| CF. AMBLYSTEGIUM | | | | | | | | | | | | | | | | | | | | | | | | |
| STELLARIA MEDIA | | | | | | | | | | | | | | | | | | | | | | | | |
| CERASTIUM HOLOSTEOIDES | | | | | | | | | | | | | | | | | | | | | | | | |
| COMPANION SPECIES | | | | | | | | | | | | | | | | | | | | | | | | |
| MONTIA FONTANA | 5 | 1 | 1 | + | 1 | 1 | 1 | 2A | | 1 | + | + | + | 1 | 1 | | + | | + | 2A | + | + | + | |
| RANUNCULUS BITERNATUS | 1 | 1 | 1 | 1 | 1 | + | | + | + | 1 | + | 1 | 1 | | | + | + | + | | | + | + | 1 | |
| POA COOKII | | 1 | R | | | + | | + | + | | R | | | | | | | | | | | 1 | | 1 |
| LOPHOCOLEA RANDII | | | | | | | | + | | | 1 | | | | | | | | + | | | 1 | 2B | |
| POA ANNUA | + | 1 | 1 | | | | | | | | | + | | | | | | | | | | | | |
| AZORELLA SELAGO | + | | | | | | | | | | | | | | | | 1 | | | | | 2A | | |
| HYGROAMBLYSTEGIUM FILUM | | | | | | | | | | | | | | | | | R | | | | | | | |
| LEPTODONTIUM PROLIFERUM | | | | | | | | | | | | | | | | | | | | | | | | |

Subassociation 3.2.2 brachythecietosum rutabuli (type relevé 189)

In drainage lines with a groundwater level more than about 25 cm below the surface *Breutelia* and *Agrostis magellanica* disappear and *Brachythecium rutabulum* becomes more frequent (Table 25). This subassociation does not possess any differential species. On average the water content of soils under stands of this community is much higher than in those on which the next association, the Acaeno magellanicae-Brachythecietum rutabuli, occurs. On somewhat drier soils transitional stages between these communities are found (relevés 249, 474 and 425).

Association 3.3 Acaeno magellanicae-Brachythecietum rutabuli (type relevé 425)

In drainage lines and along the banks of streams, on sites with a groundwater level usually more than 40 cm below the surface, and with comparatively dry

| transition | | | ACAENO MAGELLANICAE - BRACHYTHECIETUM RUTABULI | | | | | | | | | | | | | | | | | transition | | ACAENO MAGELLANICAE - AGROSTIETUM STOLONIFERAE | | | | | | | | | fragments | | |
|---|---|---|---|---|---|---|---|---|---|---|---|---|---|---|---|---|---|---|---|---|---|---|---|---|---|---|---|---|---|---|---|---|---|
| | | | | | | | | | | | | | | | | | | | | | | | | | | | fragments | | | | | | |
| 249 | 474 | 425 | 227 | 222 | 371 | 473 | 111 | 375 | 387 | 216 | 226 | 370 | 546 | 136 | 509 | 167 | 543 | 508 | 252 | 424 | 221 | 225 | 241 | 240 | 426 | 219 | 220 | 223 | 430 | 423 | 319 | 324 | 518 |
| 51 | 52 | 53 | 51 | 52 | 58 | 52 | 57 | 57 | 52 | 51 | 57 | 52 | 57 | 52 | 57 | 52 | 52 | 52 | 53 | 53 | 52 | 52 | 52 | 52 | 53 | 52 | 52 | 52 | 53 | 53 | 38 | 38 | 56 |
| 50 | 30 | 45 | 50 | 10 | 00 | 15 | 30 | 50 | 10 | 50 | 50 | 21 | 45 | 52 | 15 | 21 | 21 | 56 | 05 | 45 | 10 | 10 | 00 | 05 | 45 | 10 | 10 | 10 | 45 | 40 | 30 | 32 | 40 |
| 51 | 50 | 52 | 50 | 50 | 51 | 50 | 50 | 50 | 51 | 50 | 50 | 49 | 51 | 37 | 51 | 49 | 49 | 38 | 51 | 52 | 51 | 50 | 51 | 51 | 52 | 51 | 51 | 50 | 52 | 52 | 00 | 00 | 35 |
| 05 | 55 | 45 | 45 | 50 | 00 | 48 | 20 | 42 | 15 | 45 | 55 | 55 | 55 | 00 | 50 | 10 | 55 | 00 | 55 | 45 | 00 | 50 | 10 | 10 | 45 | 15 | 00 | 50 | 30 | 15 | 21 | 19 | 40 |
| 60 | 60 | 50 | 65 | 60 | 70 | 55 | 10 | 150 | 100 | 35 | 70 | 75 | 135 | 70 | 30 | 50 | 130 | 65 | 20 | 30 | 45 | 60 | 45 | 45 | 30 | 35 | 45 | 60 | 50 | 30 | 30 | 35 | 50 |
| 5 | 35 | 30 | | | 10 | 45 | | 3 | | 5 | 45 | 7 | 30 | 25 | | 50 | 35 | | 50 | 45 | | | | | 45 | | | | | 40 | 15 | 10 | |
| SE | WNW | NNE | | | NE | ENE | | NW | | E | S | SE | SE | SSW | | NE | SE | | S | NE | | | | | ENE | | | | | NE | ENE | NE | |
| | | | | | 17 | 15 | 50 | 50 | | | 30 | 15 | | 15 | | | | | | | | | | | 40 | | | | | 20 | | | |
| - | --- | - | - 7 | - | --- | --- | -20 | --- | --- | -35 | --- | --- | - | --- | --- | --- | - | - | - | - | -20 | - | - | - | -20 | - | | - | -25 | --- | - | -13 | -20 |
| 32 | 62 | 47 | 90 | 59 | 64 | 58 | 200 | 76 | 21 | 89 | 70 | 39 | 57 | 52 | | 22 | 66 | | 74 | 55 | 42 | 23 | 23 | 27 | 85 | 22 | 11 | 10 | 98 | 71 | 71 | 80 | |
| 4,9 | 4,9 | 4,8 | 4,9 | 4,6 | 4,3 | 5,4 | 4,8 | 4,7 | 4,9 | 4,5 | 4,7 | 4,6 | 5,1 | 4,4 | | 5,3 | 4,6 | | 4,2 | 4,3 | 4,7 | 5,7 | 5,4 | 5,3 | 5,6 | 9,1 | 5,6 | 5,6 | 5,4 | 4,7 | 5,5 | 4,7 | |
| | 83 | 53 | 74 | 70 | 89 | 75 | 66 | 72 | 35 | 93 | 77 | 76 | 64 | 87 | | 45 | 65 | | 91 | 55 | 45 | 33 | 36 | 37 | 60 | 34 | 12 | 23 | 65 | 63 | 83 | 70 | |
| 1 | | | | | | | | | | | 1 | | | | | | | | 3 | | | | | | | | | | | | | 1 | |
| 2 | | | | | 2 | | | | 1 | | 2 | 1 | | | | | | | 4 | 1 | | | | | 1 | | | | | 1 | | 2 | |
| 35 | 30 | 30 | 25 | 20 | 40 | 35 | 15 | 25 | 25 | 40 | 35 | 30 | 30 | 25 | 30 | 40 | 35 | 30 | 30 | 30 | 30 | 35 | 30 | 30 | 30 | 40 | 30 | 40 | 25 | 30 | 35 | 30 | 20 |
| 100 | 99 | 100 | 100 | 100 | 98 | 95 | 100 | 90 | 99 | 100 | 100 | 99 | 100 | 100 | 100 | 100 | 100 | 99 | 100 | 100 | 100 | 100 | 100 | 100 | 100 | 100 | 95 | 100 | 100 | 100 | 100 | 100 | 100 |
| 5 | 6 | 4 | 5 | 5 | 10 | 15 | 1 | 10 | 7 | 10 | 5 | 10 | 7 | 10 | 10 | 7 | 5 | 1 | 1 | 3 | 3 | 1 | 2 | 2 | 3 | 2 | | | 1 | 3 | 5 | 2 | |
| 95 | 60 | 95 | 100 | 97 | 80 | 100 | 60 | 100 | 100 | 100 | 95 | 50 | 90 | 90 | 100 | 100 | 99 | <1 | 1 | 60 | 60 | <1 | 1 | 3 | 7 | 1 | | | <1 | <1 | 15 | 40 | |
| | | | | | | | | | | | | | | | | | | 1 | | | | | | | | | 5 | | | | | | |
| 4 | 25 | 16 | 4 | 4 | 9 | 36 | 2 | 9 | 9 | 16 | 4 | 9 | 16 | 9 | 25 | 25 | 20 | 18 | 4 | 16 | 4 | 4 | 4 | 4 | 16 | 4 | 4 | 4 | 16 | 16 | 25 | 25 | 25 |
| 11 | 9 | 7 | 6 | 8 | 7 | 6 | 7 | 4 | 6 | 4 | 7 | 9 | 7 | 5 | 10 | 6 | 8 | 12 | 13 | 9 | 9 | 8 | 8 | 9 | 8 | 8 | 2 | 6 | 5 | 5 | 2 | 7 | 4 |
| 5 | 5 | 5 | 5 | 5 | 5 | 5 | 5 | 5 | 5 | 4 | 4 | 5 | 5 | 5 | 5 | 5 | 5 | 5 | 4 | 5 | 5 | 4 | 4 | 4 | 2A | 3 | 3 | 4 | 1 | + | 5 | 5 | 5 |
| 4 | 2B | 3 | 2A | 2A | 1 | 1 | + | | | | 1 | | | | 1 | | + | | | 1 | + | + | + | 1 | 2A | + | | | + | | | | |
| 3 | 3 | 4 | 5 | 5 | 5 | 5 | 4 | 5 | 5 | 4 | 5 | 3 | 5 | 5 | 4 | 5 | 5 | + | 1 | 4 | 4 | 1 | 1 | + | + | 1 | | | | + | | | |
| | 1 | 1 | | + | | 1 | | | | | | | 2A | 2A | 2B | 2B | 3 | 1 | + | | 1 | | | | | | | | | | | | |
| 1 | | + | | | | | | | | | | | | | | | | | 1 | 3 | 2B | 5 | 5 | 5 | 5 | 5 | 4 | 4 | 5 | 5 | | | |
| | | | | | | | | | | | | | | | | | R | | | | | R | R | R | + | | | R | | | | | |
| + | | | 2B | | + | + | + | | | | | | | | + | + | | | 1 | + | | | | | | | | | | | | | |
| | | | | | | | | | 1 | | | 1 | | | 4 | 2A | | + | | | | 1 | | | | | | | | | | | |
| + | + | + | | | | | | R | | | | R | + | | + | | | | | | | | R | | + | | | | | | | | |
| | | | | | | | | | | | | | | | + | | | R | | | | + | | + | | R | | + | | | | | |
| 2A | + | | | 1 | | | 1 | | | | 2A | + | + | | | | + | | 1 | 1 | | R | + | + | | + | | | R | + | | | 3 |
| 1 | + | | + | + | + | | 5 | + | + | 1 | | + | | R | | | | + | | | + | | | + | | | | + | 1 | | | + | |
| 3 | + | 1 | R | 2A | 2B | R | | | | 2A | 4 | | 2A | 1 | + | | 2B | R | 2A | + | + | | | R | + | | | | | | | 2B | 1 |
| | | | | | + | | 1 | | | | + | 1 | | | + | | + | | + | + | | | | | | | | | | + | 2B | 3 | |
| + | | | | | | | | | + | | | | | | | | | | + | | | R | | | | | | + | | | | | |
| 1 | 3 | | | + | | | | | | | | | | | | | | + | | | 2B | R | | | | | | | | | | | + |
| | | | | | | | | | | | | | | | | | | + | | | + | | + | | | + | | | | | | | |
| | | | | | | | | | | | + | | | | | | | | | + | | | | | | | | | | | | | |

soils, a community occurs dominated by *Acaena magellanica* in the herb layer and with a bryophyte stratum dominated by *Brachythecium rutabulum*. Together these two species form the character-combination of this association. The soil here consists of a 20–40 cm thick humus-rich layer overlying a brownish clay of varying depth. Beneath this layer often a soft or hard plinthic layer occurs, usually containing many black Mn nodules (Smith 1976c; Fig. 39). On wet, organic soils *Drepanocladus uncinatus* may reach relatively high cover values (e.g. relevés 227 and 222), while *Blechnum penna-marina* generally indicates somewhat drier conditions (relevés 167 and 543). On comparatively mineral, dry soils *Agrostis stolonifera* invades the vegetation, resulting in transitional stands between the Acaeno magellanicae-Brachythecietum rutabuli and Acaeno magellanicae-Agrostietum stoloniferae (relevés 424 and 221). Some stands occur in which *Brachythecium rutabulum* has a very low cover, and in which a comparatively large number of species are present (relevés 508 and 252). Their syntaxonomic and ecological position is not clear.

*Fig. 38.* Spring with Brachythecietum subplicati (near Diving Petrel Stream, Marion).

The Acaeno magellanicae-Brachythecietum rutabuli is widely distributed on Marion Island but was not found on Prince Edward Island, where *Brachythecium rutabulum* is lacking. On Marion Island *Brachythecium rutabulum* was found only vegetatively, which accounts for it not spreading from this island to Prince Edward. The *Acaena* vegetation of drainage lines on Prince Edward has a very poorly developed bryophyte stratum, since the mosses dominating the vegetation of comparable habitats on Marion are lacking here, and apparently no other species take over this prominent position (cf. relevés 319 and 324).

Association 3.4 Acaeno magellanicae-Agrostietum stoloniferae
(type relevé 240)

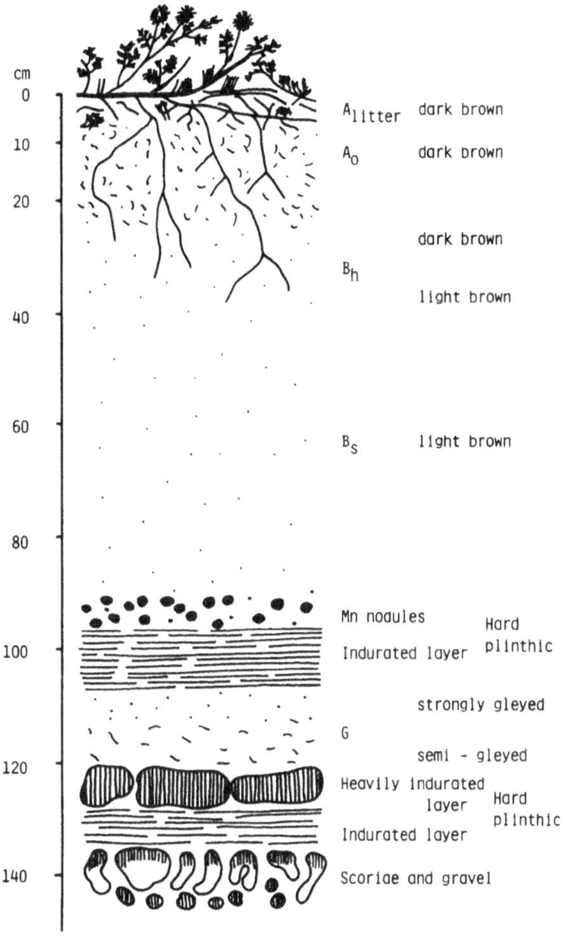


*Fig. 39.* Profile of the soil in a stand of Acaeno magellanicae – Brachythecietum rutabuli (after Smith 1976c).

*Agrostis stolonifera* has been introduced inadvertently by man on Marion Island. It is widely spread along stream-sides and on coastal slopes in the north-eastern part of the island (Gremmen & Smith 1980), where it forms extensive stands on relatively dry, mineral soils on sites with lateral groundwater movement, notably in drainage lines on grey lava slopes and on deposits of gravel and sandy material along rivers (Fig. 41). *Agrostis stolonifera* and *Acaena magellanica* form the character-combination of this association. Differential species is *Pringlea antiscorbutica.*

Stands of Acaeno magellanicae-Agrostietum stoloniferae were only met with in the north-eastern part of Marion Island, where this community seems to replace the Acaeno magellanicae-Brachythecietum rutabuli in part of its habitat-range, notably on comparatively dry, mineral soils.

*Agrostis stolonifera* has not been found on Prince Edward Island.

*Synchorological remarks*

Communities dominated by *Acaena magellanica* are recorded from most Subantarctic islands, but little data on their floristic composition and ecology are available. On Kerguelen and Crozet *Acaena magellanica* dominates the vegetation of large parts of the lowlands (Schenck 1905; Werth 1906; Chastain 1958; Cour 1959; Aubert de la Rue 1964; Dreux 1964, Hébrard 1970), occurring on dry, mineral soils as well as on wet peat. On South Georgia *Acaena magellanica* is one of the pioneers on bare scree or morainic slopes, where it forms extensive stands together with the turf-forming moss *Tortula robusta* Hook. & Grev. (Skottsberg 1912; Greene 1964; Smith & Walton 1975). The *Acaena*-dominated communities on these islands seem to cover a wider range of habitats than the communities of the *Acaena magellanica-Brachythecium* complex on Marion, and their distribution apparently is not related to a lateral water flow through the soil. Insufficient data, however, are available to assess the relationships between the *Acaena*-communities of those islands.

In his description of the vegetation of Macquarie Taylor (1955a) reports dense mats of *Acaena magellanica* growing locally on the gravels of creek fans. No floristic data are available but the habitat description suggests a correspondence to the Acaeno magellanicae-Brachythecietum rutabuli. It should be noted that most of the characteristic species of the communities of the *Acaena-Brachythecium* complex described in this study do not occur on Macquarie (cf. Table 11).

An apparently vicariant community of the Brachythecietum subplicati philonotidetosum is found on South Georgia, where along streamsides and in flushes a vegetation of *Philonotis acicularis* (C. Muell.) Kindb., *Philonotis vagans* (Hook. f. & Wils.) Mitt., *Hygroamblystegium austro-fluviatile* (C. Muell.) and *Montia fontana* occurs (Skottsberg 1912), which resembles this subassociation strongly in structure and ecology. Lindsay (1971) has described a community of a species of *Brachythecium,* which occurs in the South Shetland Islands

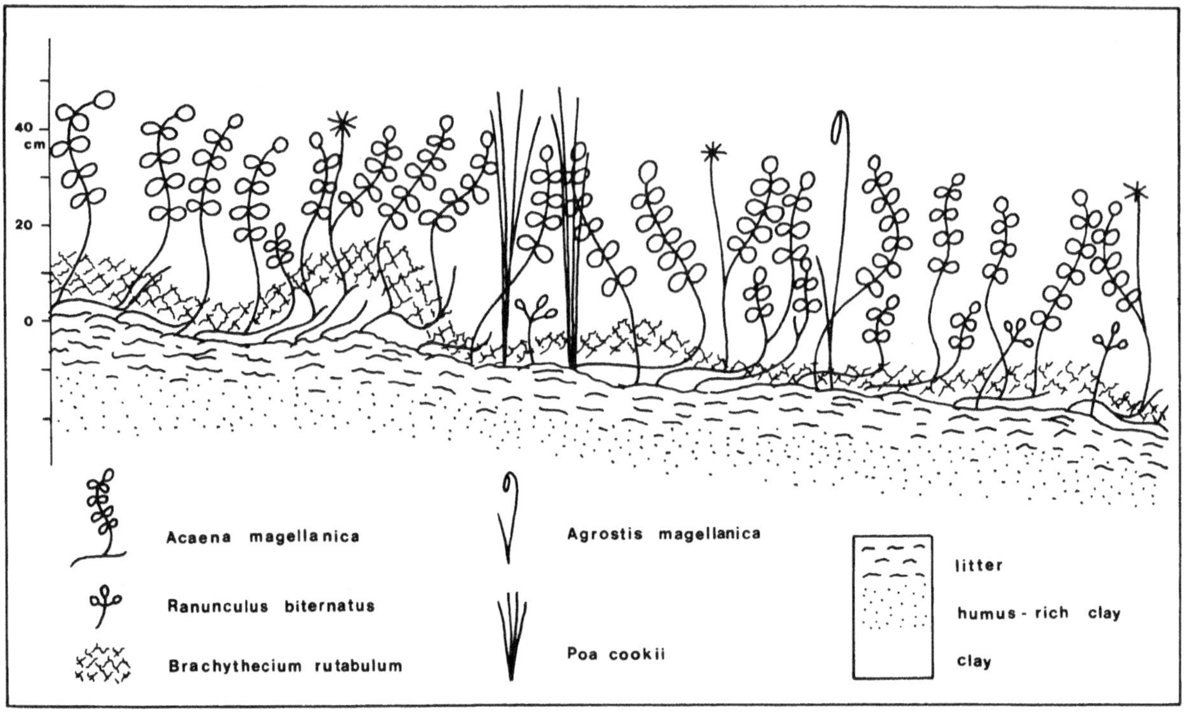


*Fig. 40.* Structure-diagram of Acaeno magellanicae – Brachythecietum rutabuli.

*Fig. 41.* Acaeno magellanicae – Agrostietum stoloniferae (the light band of grassy vegetation) along Van Den Boogaard River.

(Maritime Antarctic), on wet slopes and on flat areas where melt water accumulates during summer. This author states that the more rapid the water flows, the more luxuriant the moss in this cryptogamic community grows. This *Brachythecium* sociation seems to correspond in site ecology to the Brachythecietum subplicati on Marion Island.

### 4.4 Communities of oligotrophic mires: the Juncus scheuchzerioides - Blepharidophyllum densifolium complex

The cool and extremely oceanic climate of Marion and Prince Edward Islands greatly favours peat formation and wherever drainage is impeded peat deposits have accumulated, ranging in depth from a few centimetres to more than 4 m. In the lowland areas of the islands numerous small and large mires, often interconnected by peat deposits on slopes, cover a large part of the surface (Fig. 42). These mires are not ombrotrophic in a strict sense, as most of them receive water which has been in contact with the mineral soil of the surrounding areas. This apparently does not result in a marked influx of minerals into the mires, and about all minerals here are derived from the precipitation (cf. Grobbelaar 1974). As a consequence of the high and nearly continuous rainfall, in most areas movement of plant nutrients in the upper part of the peat deposits may well occur in a downward direction only, thus preventing any influence of geogenous water on the vegetation. Lateral drainage of water through the mire across the surface of the parent rock is indicated by the oxygen content of the soil solution, which first decreases rapidly with depth, but increases again in the lower part of the peat deposit (Smith 1976c). The rate of water flow is very small

*Fig. 42.* Goney Plain, Marion Island, showing an intricate pattern of mires (light coloured areas) in the depressions of the lava flows and Isopterygio pulchelli – Blechnetum penna-marinae on well-drained slopes (dark areas).

compared to the flow of water in the springs and flushes described in the previous chapter. Only in water tracks through the mire expanse does a relatively high water flow occur. Water transport here mostly takes place through the peat soil, and usually no open water channel is found. The vegetation of these water tracks differs strongly from that of other parts of the mire areas, which probably is related to the different plant nutrient regime in the water tracks (cf. Chapman 1965; Sparling 1966; Ingram 1967; see association 4.8). Peat deposits are best developed on grey lava and on black lava flows of the aa type. On block lava and on pahoehoe lava flows of the second volcanic series the very porous nature of the surface does not permit the stagnation of water, thus preventing the formation of peat. The deepest peat deposits occur in the basins in the lava flows. Peat is also found on somewhat sloping ground, but is absent on steep slopes. Between peat and lava usually a 15–70 cm thick layer of loamy clay occurs, originating from the weathering of the lava before the beginning of peat accumulation, or in black lava areas possibly also blown in by the wind from the nearby grey lava areas, which for a large part are covered with thick layers of glacial clay.

The surface of the valley bogs is generally flat or slightly concave. On slopes the surface of the peat deposits follows more closely the topography of the underlying lava flows. Physiognomically the islands' mires show a resemblance to the blanket bogs of western Ireland. The mires have a very even surface, a pattern of hollows and hummocks as in the bogs of the European continent being completely absent. Also no surface structures comparable to the shallow terraces found by Wace (1961) on the surface of the bogs on the Gough Island plateau are present on Marion and Prince Edward Islands.

The most important peat-forming plants in the mires on the islands are bryophytes, notably *Drepanocladus uncinatus, Blepharidophyllum densifolium, Jamesoniella colorata, Racomitrium lanuginosum* and *Breutelia integrifolia*. Graminoid species also play an important role, notably *Agrostis magellanica* and *Uncinia compacta,* and, to a much smaller extent, *Juncus scheuchzerioides*. In relatively dry parts of the mires *Blechnum pennamarina* contributes considerably to the peat formation. As has already been noted in an earlier chapter, *Sphagnum* spp., which are very important in bog vegetation of large parts of the world, do not occur on Marion and Prince Edward Islands, and also do not play a role of any import in the mire vegetation of the other Subantarctic islands.

A large number of species is restricted to communities of the *Juncus scheuchzerioides-Blepharidophyllum densifolium* complex or has its optimum here, e.g. *Juncus scheuchzerioides, Distichophyllum fasciculatum, Peltigera polydactyla, Blepharidophyllum densifolium, Clasmatocolea humilis, Lepidozia laevifolia, Leptoscyphus expansus, Schistochila carnosa, Acrobolbus ochrophyllus, Ptychomnion ringianum, Jamesoniella grandiflora, Jamesoniella colorata, Fossombronia australis, Uncinia compacta, Andrewsianthus lancistipus, Riccardia cf. pinguis, Riccardia cf. multifida, Jensenia pisicolor, Lycopodium magellanicum, Breutelia integrifolia, Bryum laevigatum* and *Symphyogyna marionensis* (Tables 26a,b, 32).

The communities of this complex can be arranged into a series along a gradient from wet to dry. The position of the communities along this gradient, based on more than 1000 measurements of the soil water content in stands of mire vegetation, is shown in Table 33. At the wet end of the gradient inundation or extremely wet soil conditions apparently prevent the occurrence of many species (cf. Table 26a,b). On drier soils not so much the average situation as well as the extreme conditions seem important. During periods of several consecutive days without precipitation, with strong winds and with sunny weather, the water supply to the bryophyte stratum of the vegetation is sometimes insufficient, locally resulting in severe drought-damage. After a period of four days with such weather conditions (a very rare occurrence on Marion and Prince Edward Islands) severe damage was observed to the vegetation along the wetter part of the gradient, notably the Junco scheuchzerioidis-Drepanocladetum uncinati and the Blepharidophyllo densifolii-Clasmatocoleetum humilis. Bryophytes with a mat-forming habit, e.g. *Clasmatocolea humilis, Blepharidophyllum densifolium* and *Drepanocladus uncinatus,* are more susceptible to drought than species which form turves or carpets. These dense types of colony, and turves even more so than carpets, are more resistant to water loss to a desiccating atmosphere than mats of wefts. Furthermore, the densely packed, predominantly vertical structural elements in turves do facilitate the water supply to the upper parts of the plants by providing numerous capillary spaces for water transport (cf. Gimingham & Birse 1957; Gimingham & Smith 1971; Mägdefrau 1935; Gimingham 1967).

The pattern of the plant communities in the mire in a small black lava valley in Nellie Humps, near the meteorological station, is shown in a vegetation map (Fig. 43). This pattern shows a conspicuous correlation with the pattern of the groundwater level (Fig. 45), soil water content (Fig. 46) and elevation (Fig. 44). From the community patterns in this and two other areas of which detailed vegetation maps have been made, which are not reproduced here, conclusions are drawn about the relative position of the mire communities along a wet-dry gradient. These patterns also provide indications on the direction in which the communities develop in a successional series.

Community 4.1 Community of Juncus scheuchzerioides
(type relevé 586)

In this community two types of pioneer vegetation, mostly dominated by *Juncus scheuchzerioides,* are united: one which occurs on bare, wet peat soil, the other is found in 20–80 cm deep water in bog ponds. This community is floristically very poorly characterized: it does not possess any character- or differential species. Conspicuous is the very open character of the vegetation, which is very poor in species, and the relative importance of *Juncus scheuchzerioides.* Bryophytes play a very minor role in the vegetation. Two variants are discerned:

*a. Campylopus clavatus variant*

This is a degradation stage of the Blepharidophyllo densifolii-Clasmatocoleetum humilis or related communities. It is found on wet, oligotrophic peat soils, in areas where the original plant cover has been more or less completely destroyed, often after initial disturbance of the vegetation by fouraging birds (notably Lesser Sheatbill and Kelp Gull) and subsequent erosion by water and wind. *Juncus scheuchzerioides* then colonises the bare peat. Some relics of the original vegetation generally are present, notably tufts of *Agrostis magellanica* and *Uncinia compacta,* and in some cases most of the species present are remains of the community previously occupying the site (e.g. relevé 467).

The *Campylopus clavatus* variant is floristically not well differentiated from the other one, but is clearly different in habitat.

*Juncus scheuchzerioides* shows markedly different growth forms in the two variants of the *Juncus* community. In the *Campylopus* variant *Juncus* plants often are not more than a few centimetres high and have a reddish-brown colour, while plants of this species in the *Ranunculus biternatus* variant grow up to 30 cm high, and are often bright green in colour.

*b. Ranunculus biternatus variant*

*Juncus scheuchzerioides, Agrostis magellanica* and *Ranunculus biternatus* are the most prominent and often the only species in this pioneer community which occurs in 20–80 cm deep water. This community, which occurs in many bog pools on the islands (Fig. 47), is floristically differentiated from the *Campylopus clavatus* variant by *Ranunculus biternatus,* and by the absence of e.g. *Uncinia compacta* and *Campylopus clavatus,* species which are found, although infrequently, in stands of the latter variant. The difference between the growth forms of *Juncus scheuchzerioides* in the two variants of this community has already been remarked upon.

In some stands *Juncus scheuchzerioides* plays only a minor role, and *Agrostis magellanica* or *Ranunculus biternatus* attain dominance. This change in dominance could not be correlated with any difference in environmental conditions.

Association 4.2 Junco scheuchzerioidis-Drepanocladetum uncinati
(type relevé 488)

The dominant bryophyte in this community is *Drepanocladus uncinatus,* which together with *Juncus scheuchzerioides* forms the character-combination of this association. It is found towards the drier side of the *Juncus scheuchzerioides* community, along the edges of bog pools and on extremely wet peat soils in depressions in the mire expanse. Three subassociations are discerned.

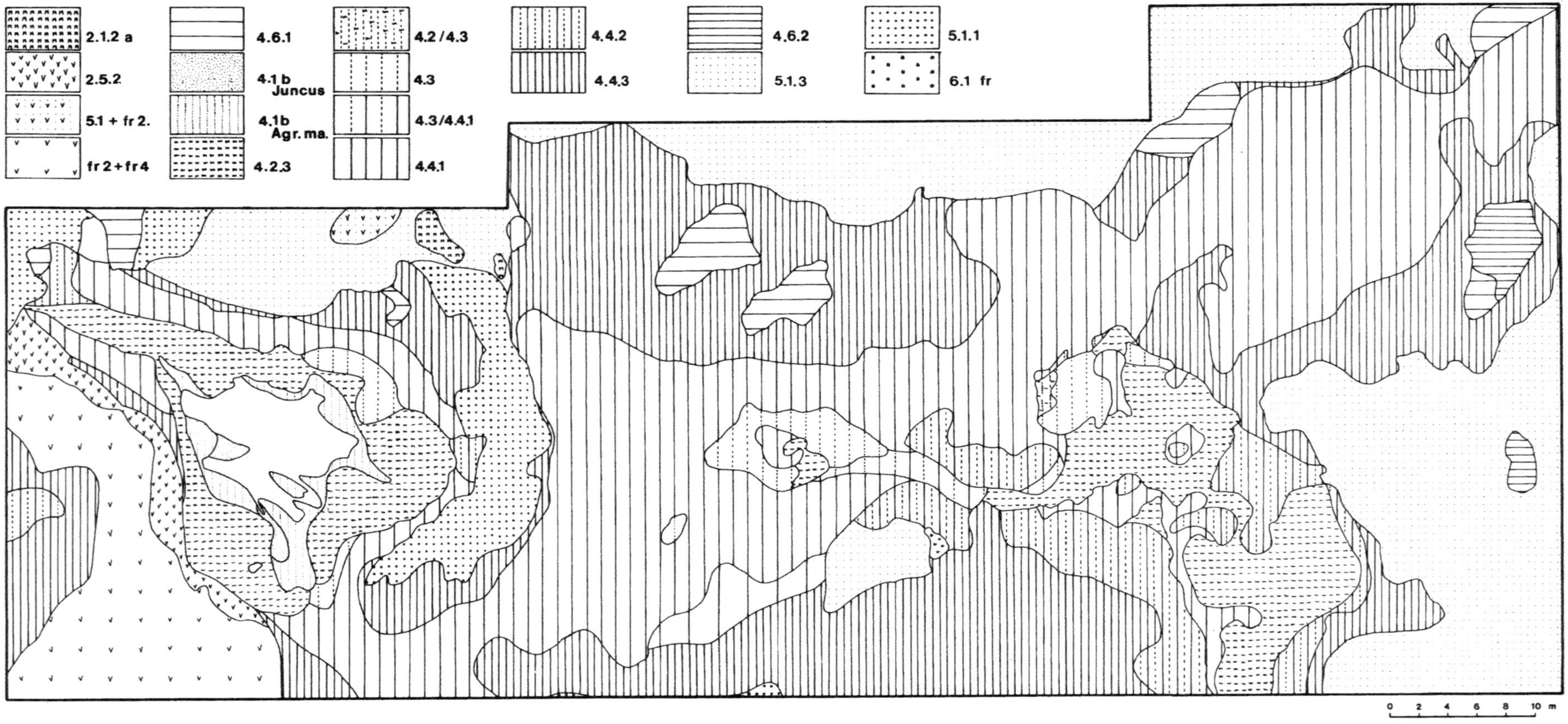


*Fig. 43.* Vegetation map of an area in Nellie Humps, Marion Island (see Table 17 (p. 42) for explanation of the numbers of the legend).

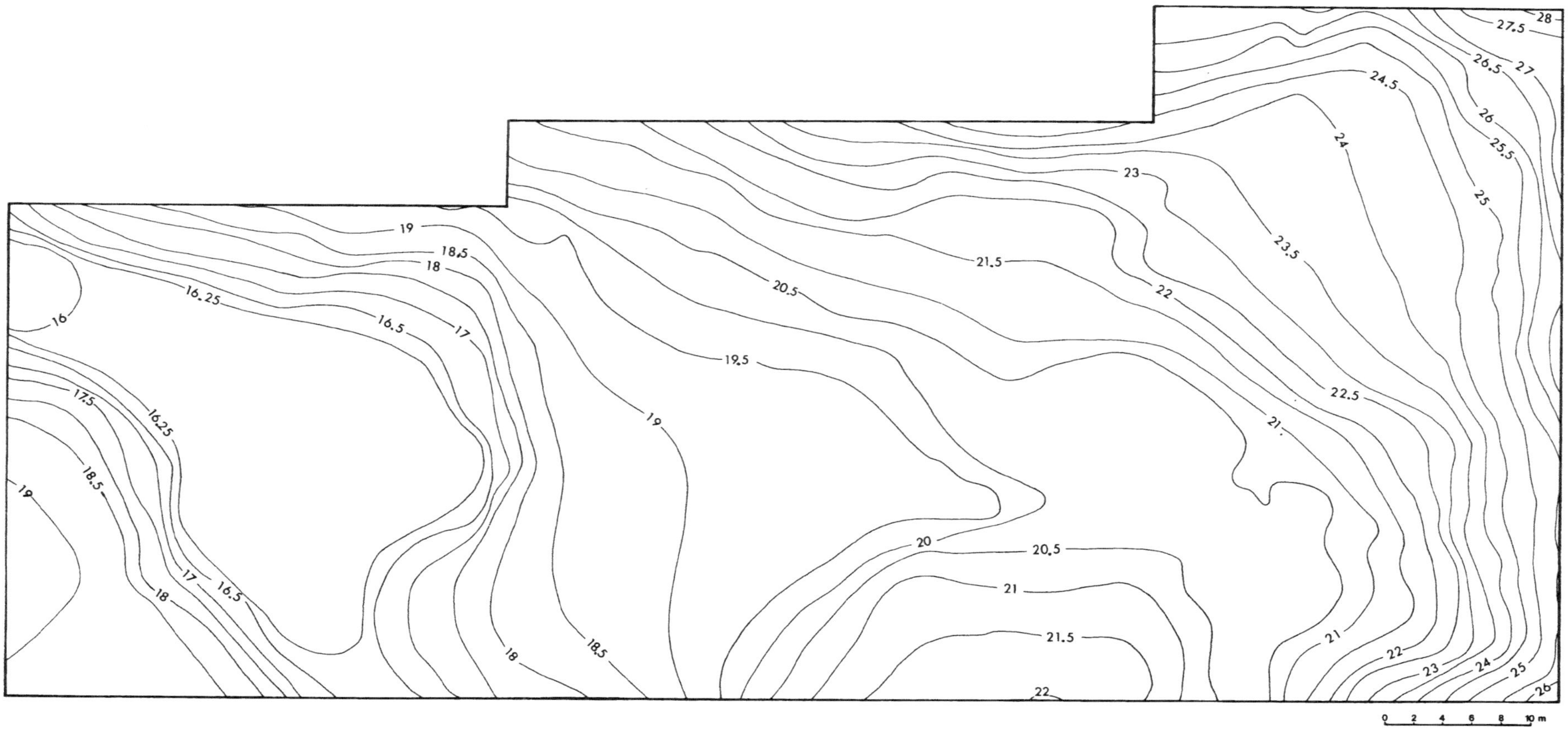


*Fig. 44.* Map of the elevation (m above see level) in the Nellie Humps mapping area (cf. Fig. 43).

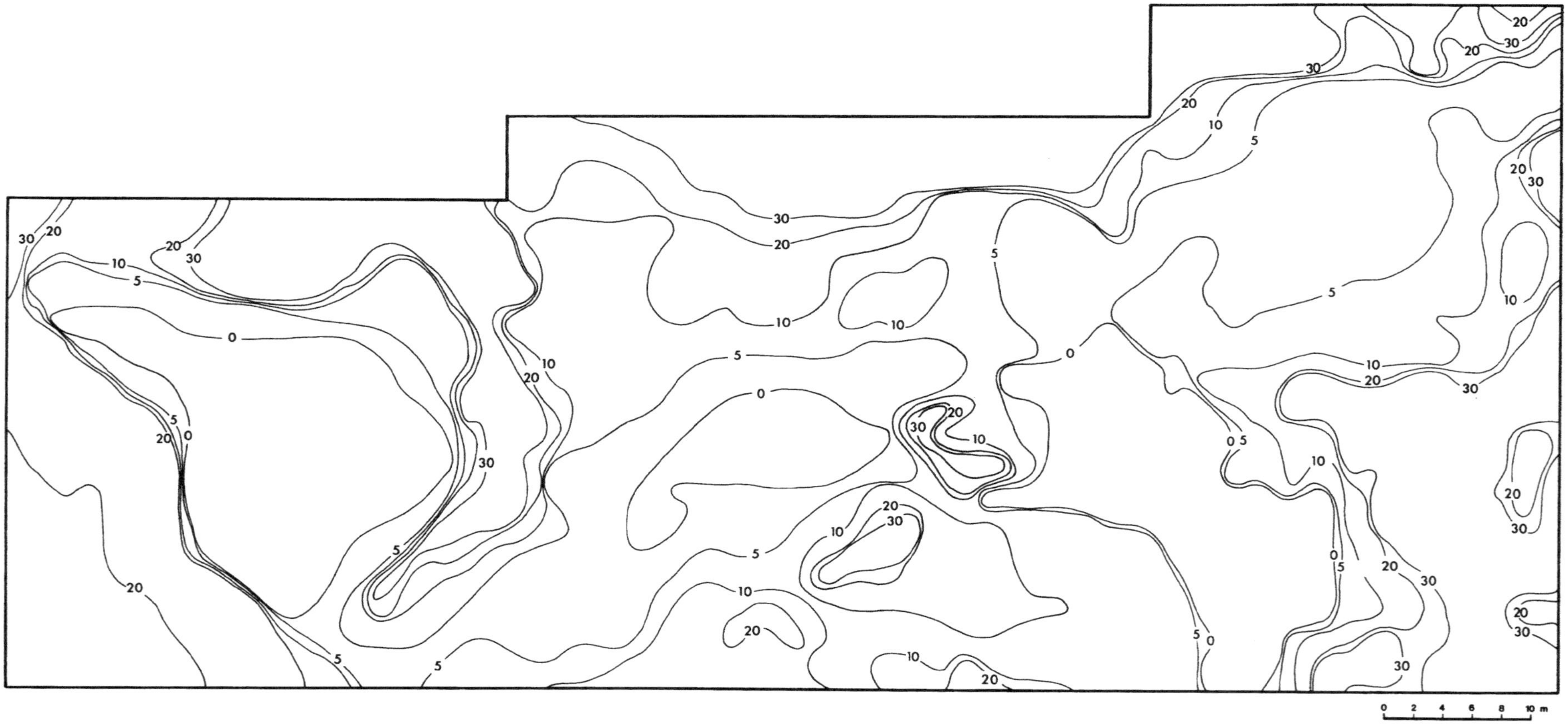


*Fig. 45*. Map showing the groundwater level (cm below the surface) in the Nellie Humps mapping area (cf. Fig. 43).

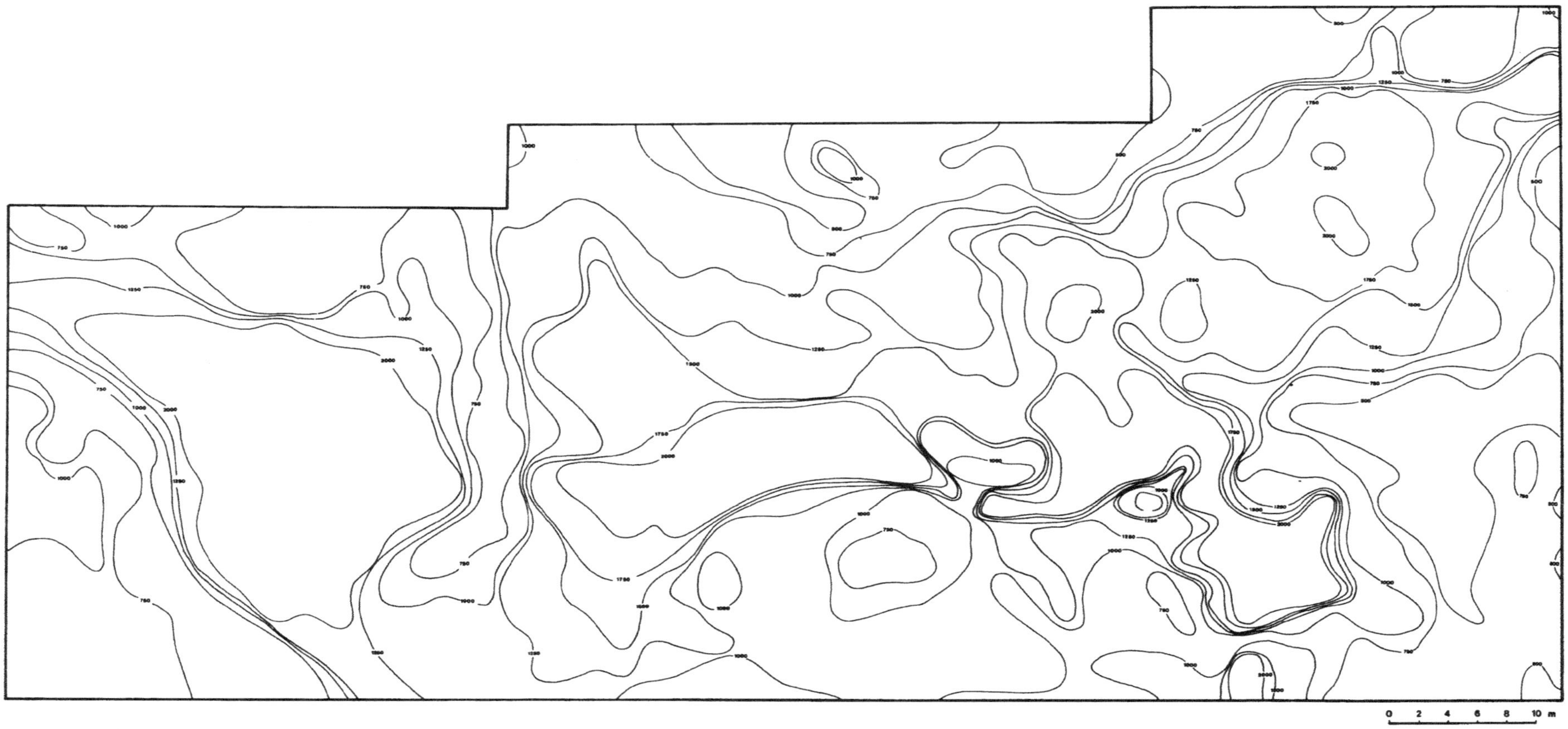

*Fig. 46.* Map showing the soil water content (% of dry weight) in the Nellie Humps mapping area (cf. Fig. 43).

*Fig. 47. Ranunculus biternatus* variant of the Community of *Juncus scheuchzerioides* in a small bog pond. The dominant species are *Juncus scheuchzerioides* and *Ranunculus biternatus.*

Subassociation 4.2.1 juncetosum scheuchzerioidis (type relevé 587)

This community, which is differentiated by *Juncus scheuchzerioides,* is found in bog pools in 25–70 cm deep water. The bryophyte layer is submerged, not attached to the soil, but floating near or at the surface of the water. The herb layer of the vegetation emerges above the water surface.

Subassociation 4.2.2 montietosum fontanae (type relevé 429)

Stands of this subassociation, which does not possess any differential species (Table 26), are found on extremely wet peat soils, e.g. along the edge of bog ponds where the water level is nearly at the soil surface.

Subassociation 4.2.3 clasmatocoleetosum humilis (type relevé 488)

On less extremely wet peat soils, with the groundwater level somewhat further below the surface than at the sites where the subassociation *montietosum* occurs, the subassociation *clasmatocoleetosum humilis* is found. This community contains appreciably more species than the other two subassociations (about 11 species/relevé, compared to c. 6 species/relevé in the other subassociations). Differential species are *Distichophyllum fasciculatum, Blepharidophyllum densifolium, Clasmatocolea humilis, Lepidozia laevifolia, Leptoscyphus expansus* and *Peltigera polydactyla* (Table 26). Towards the drier end of the range of habitats occupied by this community *Uncinia compacta* locally reaches dominance, sometimes accompanied by *Blechnum penna-marina* (e.g. relevé 217 and 90). The three subassociations of the Junco scheuchzerioidis-Drepanocladetum uncinati can be found forming successive zones in the order in which they have been described here along a gradient from c. 50 cm deep water to the shore of bog ponds. In the same order this can be interpreted as a successional series in the terrestrialisation of such pools.

Association 4.3 Distichophylletum fasciculati
(type relevé 20)

On slightly drier peat soils than those on which the Junco-Drepanocladetum occurs, the moss *Distichophyllum fasciculatum* reaches dominance, forming the Distichophylletum fasciculati, of which it is the character-species. In a series from wet to dry this community follows towards the drier side of the Junco scheuchzerioidis-Drepanocladetum uncinati clasmatocoleetosum humilis, but only in situations where the wet-dry gradient is very gradual well-developed stands were found. When this gradient is steepter stands of Blepharidophyllo densifolii-Clasmatocoleetum humilis follow immediately towards the dry side of the Junco scheuchzerioidis-Drepanocladetum uncinati zone.

Association 4.4 Blepharidophyllo densifolii-Clasmatocoleetum humilis
(type relevé 24)

Following the Distichophylletum fasciculati in the series from wet to dry is the Blepharidophyllo densifolii-Clasmatocoleetum humilis, in which the bryophyte layer is dominated by the hepatics *Blepharidophyllum densifolium* and *Clasmatocolea humilis,* both character-species of this association. *Campylopus arboricola, Lepidozia laevifolia* and *Peltigera polydactyla* differentiate this community from the ones described sofar in this complex. Three subassociations are distinguished:

Subassociation 4.4.1 ranunculetosum biternati
(type relevé 24)

*Blepharidophyllum densifolium,* which species achieves dominance throughout this community, differentiates it and the subassociation blechnetosum penna-marinae against the subassociation clasmatocoleetosum humilis. Stands of this subassociation are found on very wet peat soils, with a groundwater level usually between 5 and 10 cm below the surface.

Subassociation 4.4.2 clasmatocoleetosum humilis
(type relevé 411)

This subassociation is differentiated against the other two by *Clasmatocolea humilis,* the dominant bryophyte. Otherwise the floristic composition of the vegetation is very similar to that of the subassociation ranunculetosum biternati. This community occurs on somewhat less wet soils than the subassociation ranunculetosum. Along a gradient from wet to dry a belt of this community is often found between zones of the subassocitions ranunculetosum and blechnetosum, but in other, apparently similar situations this subassociation is not present. No explanation for this phenomenon was found.

Subassociation 4.4.3 blechnetosum penna-marinae
(type relevé 404)

At the drier end of the range of habitats occupied by the Blepharidophyllo-Clasmatocoleetum, on peat soils with a groundwater level at about 10–20 cm below the surface, *Blechnum penna-marina* invades the vegetation, often reaching dominance (Figs. 48, 49). In addition to *Blechnum, Fossombronia australis, Jamesoniella colorata, Jensenia pisicolor, Lophozia cylindriformis, Plagiochila heterodonta, Racomitrium lanuginosum* and *Uncinia compacta* differentiate this subassociation and indicate somewhate drier soil conditions.

In a series from wet to dry this subassociation follows towards the dry side of the subassociations ranunculetosum and clasmatocoleetosum. In a successional series the subassociation blechnetosum can be interpreted as the stage following after the subassociation ranunculetosum. The position of the subassociation clasmatocoleetosum humilis in the succession of the mire vegetation is not clear.

*Table 26a.* The *Juncus scheuchzerioides* – *Blepharidophyllum densifolium* complex (first part).

| | Community of Juncus scheuchzerioides | | | | | | | | | | | | | | | | |
|---|---|---|---|---|---|---|---|---|---|---|---|---|---|---|---|---|---|
| | Campylopus clavatus variant | | | | | | | | Ranunculus biternatus variant | | | | | | | | |
| RELEVE NUMBER | 114 | 113 | 466 | 115 | 434 | 528 | 339 | 467 | 119 | 435 | 586 | 588 | 446 | 564 | 410 | 236 | 565 |
| LOCATION | | | | | | | | | | | | | | | | | |
| SOUTHERN LATITUDE (46°) MIN. | 53 | 53 | 52 | 53 | 53 | 52 | 38 | 52 | 53 | 52 | 52 | 52 | 53 | 52 | 52 | 52 | 52 |
| SEC. | 00 | 00 | 40 | 00 | 42 | 21 | 40 | 30 | 00 | 40 | 28 | 40 | 00 | 27 | 40 | 00 | 15 |
| EASTERN LONGITUDE (37°) MIN. | 50 | 50 | 51 | 50 | 52 | 49 | 00 | 51 | 50 | 51 | 51 | 51 | 51 | 51 | 51 | 50 | 51 |
| SEC. | 50 | 50 | 30 | 50 | 12 | 55 | 20 | 30 | 49 | 30 | 27 | 28 | 30 | 25 | 30 | 40 | 25 |
| ELEVATION (M A.S.L.) | 55 | 55 | 35 | 55 | 20 | 130 | 30 | 35 | 60 | 16 | 20 | 30 | 50 | 20 | 16 | 115 | 20 |
| INCLINATION (°) | 3 | 5 | | 3 | | | 5 | | | | | | | | | | |
| ASPECT | SE | SE | | S | | | N | | | | | | | | | | |
| SOIL DEPTH (CM) | | | | | | 25 | | | | | | | | | | | |
| DEPTH OF GROUNDWATER LEVEL (CM) | -20 | -29 | + 3 | -20 | -10 | - 1 | 0. | - 1 | +35 | +80 | +60 | +80 | - 2 | +60 | +80 | +30 | +20 |
| SOIL WATER CONTENT (X10 %) | 152 | 130 | 247 | 154 | 107 | 194 | 87 | 239 | | | | | 329 | | | | |
| SOIL PH | 4,5 | 4,7 | 4,7 | 4,9 | 4,8 | 4,0 | | 5,0 | 5,3 | 5,3 | 5,0 | 5,2 | 4,9 | 5,2 | | 5,9 | 5,2 |
| SOIL LOSS ON IGNITION (%) | 73 | 75 | 81 | 77 | 95 | 85 | 95 | 85 | | | | | | | | | |
| INFLUENCE OF SALTSPRAY | | | | | | | | | | | | | | | | | |
| INFLUENCE OF TRAMPLING | | | 1 | | 2 | | | | | | | | | | | | |
| INFLUENCE OF MANURING | | | | | 1 | | | | | | | | | | | | |
| HEIGHT OF HERB LAYER (CM) | 5 | 4 | 10 | 5 | 5 | 5 | 20 | 8 | 5 | 5 | 5 | 10 | 10 | 20 | 25 | 10 | 20 |
| COVER OF HERB LAYER (%) | 70 | 40 | 30 | 40 | 10 | 5 | 5 | 60 | 100 | 10 | 20 | 7 | 30 | 25 | 70 | 60 | 99 |
| HEIGHT OF BRYOPHYTE LAYER (CM) | | 1 | | 1 | 1 | 1 | | 1 | | | | | 1 | | | 2 | |
| COVER OF BRYOPHYTE LAYER (%) | | <1 | | 10 | 1 | <1 | | <1 | | | | | <1 | | | <1 | |
| COVER OF ALGAE (%) | | | 100 | | | | | | 100 | | 80 | | 100 | 100 | | 100 | |
| COVER OF BARE SOIL (%) | 30 | 60 | | 50 | 90 | 95 | 95 | 40 | | | | | | | | | |
| COVER OF ROCKS (%) | | | | | | | | | | | | | | | | | |
| QUADRAT SIZE (M2) | 1 | 1 | 25 | 1 | 6 | 16 | 9 | 25 | 1 | 10 | 10 | 10 | 6 | 15 | 3 | 4 | 15 |
| TOTAL NUMBER OF SPECIES | 1 | 3 | 3 | 4 | 4 | 4 | 5 | 9 | 2 | 3 | 3 | 3 | 3 | 4 | 2 | 3 | 3 |
| DIAGNOSTIC TAXA OF ASSOCIATIONS AND SUBASSOCIATIONS: | | | | | | | | | | | | | | | | | |
| JUNCUS SCHEUCHZERIOIDES | 4 | 3 | 3 | 3 | 2A | 2M | 1 | 4 | 5 | 2A | 1 | 1 | 3 | 2A | | 1 | 1 |
| RANUNCULUS BITERNATUS | | | | | + | + | | | 2A | 1 | 2A | + | | 1 | 1 | | 5 |
| DREPANOCLADUS UNCINATUS | | | | | 1 | | | | | | | | + | | | | |
| DISTICHOPHYLLUM FASCICULATUM | | | | | | | | R | | | | | | | | | |
| BLEPHARIDOPHYLLUM DENSIFOLIUM | | | | | | | | | | | | | | | | | |
| CLASMATOCOLEA HUMILIS | | | | | | | | | | | | | | | | | |
| LEITOSCYPHUS EXPANSUS | | | | | | | | | | | | | | | | | |
| LEPIDOZIA LAEVIFOLIA | | | | | | | | | | | | | | | | | |
| CAMPYLOPUS ARBORICOLA | | | | | | | | | | | | | | | | | |
| PELTIGERA POLYDACTYLA | | | | | | | | | | | | | | | | | |
| ANDREWSIANTUS LANCISTIPUS | | | | | | | | | | | | | | | | | |
| UNCINIA COMPACTA | | | R | R | | | | + | | | | | | | | | |
| FOSSOMBRONIA AUSTRALIS | | | | | | | | | | | | | | | | | |
| JENSENIA PISICOLOR | | | | | | | | | | | | | | | | | |
| LOPHOZIA CYLINDRIFORMIS | | | | | | | | | | | | | | | | | |
| PLAGIOCHILA HETERODONTA | | | | | | | | | | | | | | | | | |
| PTYCHOMNION RINGIANUM | | | | | | | | R | | | | | | | | | |
| JAMESONIELLA GRANDIFLORA | | | | | | | | | | | | | | | | | |
| JAMESONIELLA COLORATA | | | | | | | | R | | | | | | | | | |
| RACOMITRIUM LANUGINOSUM | | | | | | | | + | | | | | | | | | |
| LYCOPODIUM MAGELLANICUM | | | | | | | | | | | | | | | | | |
| DISTICHOPHYLLUM IMBRICATUM | | | | | | | | | | | | | | | | | |
| RICCARDIA CF. PINGUIS | | | | | | | | | | | | | | | | | |
| BREUTELIA INTEGRIFOLIA | | | | | | + | | | | | | | | | | | |
| BRYUM LAEVIGATUM | | | | | | | | | | | | | | | | | |
| RICCARDIA CF. MULTIFIDA | | | | | | | | | | | | | | | | + | |
| BLECHNUM PENNA-MARINA | | | | | | | | | | | | | | | | | |
| ACAENA MAGELLANICA | | | | | | | R | | | | | | | | | | |
| SCHISTOCHILA CARNOSA | | | | | | | | | | | | | | | | | |
| PLAGIOTHECIUM PLATYPHYLLUM | | | | | | | | | | | | | | | | | |
| KETZGERIA DECIPIENS | | | | | | | | | | | | | | | | | |
| CATAGONIUM POLITUM | | | | | | | | | | | | | | | | | |
| HYGROAMBLYSTEGIUM FILUM | | | | | | | | | | | | | | | | | |
| LOPHOCOLEA RANDII | | | | | | | | | | | | | | | | | |
| AGROSTIS MAGELLANICA | | 1 | + | 1 | + | + | 2A | 1 | | | 1 | 2A | 2A | + | 4 | 4 | 3 |
| COMPANIONS: | | | | | | | | | | | | | | | | | |
| CAMPYLOPUS SUBNITENS | | | | | | | | | | | | | | | | | |
| DICRANOLOMA BILLARDIERI | | | | | | | | | | | | | | | | | |
| CAMPYLOPUS CLAVATUS | | 1 | | 2A | | | | 1 | | | | | | | | | |
| MONTIA FONTANA | | | | | | | | | | | | | | | | | |
| AZORELLA SELAGO | | | | | | | + | | | | | | | | | | |
| MARCHANTIA BERTEROANA | | | | | | | | | | | | | | | | | |
| ACROBOLBUS OCHROPHYLLUS | | | | | | | | | | | | | | | | | |
| CLASMATOCOLEA VERMICULARIS | | | | | | | | | | | | | | | | | |
| SYMPHYOGYNA MARIONENSIS | | | | | | | | | | | | | | | | | |
| CLADONIA COLL. | | | | | | | | | | | | | | | | | |
| TEMNOMA QUADRIPARTITA | | | | | | | | | | | | | | | | | |
| POA COOKII | | | | | | | | | | | | | | | | | |

| | JUNCO SCHEUCHZERIOIDIS - DREPANOCLADETUM UNCINATI | | | | | | | | | | | | | |
|---|---|---|---|---|---|---|---|---|---|---|---|---|---|---|
| | Juncetosum scheuchzerioidis | | | | | | montietosum fontanae | | | | | | | |
| RELEVE NUMBER | 120 | 589 | 587 | 590 | 105 | 109 | 74 | 81 | 80 | 72 | 429 | 11 | 230 | 471 |
| LOCATION | | | | | | | | | | | | | | |
| SOUTHERN LATITUDE (46°) MIN. | 53 | 52 | 52 | 52 | 52 | 52 | 52 | 52 | 52 | 52 | 53 | 52 | 52 | 52 |
| SEC. | 00 | 45 | 25 | 45 | 35 | 35 | 45 | 45 | 45 | 45 | 43 | 37 | 00 | 15 |
| EASTERN LONGITUDE (37°) MIN. | 50 | 51 | 51 | 51 | 51 | 51 | 52 | 52 | 52 | 52 | 52 | 51 | 50 | 51 |
| SEC. | 49 | 05 | 15 | 05 | 20 | 20 | 45 | 45 | 45 | 45 | 30 | 30 | 40 | 03 |
| ELEVATION (M A.S.L.) | 60 | 30 | 20 | 30 | 25 | 25 | 9 | 15 | 15 | 10 | 50 | 25 | 110 | 50 |
| INCLINATION (°) | | | | | | | | 5 | | 3 | | | | |
| ASPECT | | | | | | | | S | | NK | | | | |
| SOIL DEPTH (CM) | | | | | | | 35 | 15 | 20 | 35 | | | | |
| DEPTH OF GROUNDWATER LEVEL (CM) | +30 | +45 | +25 | +25 | +45 | +70 | -10 | -12 | -12 | -25 | - 5 | -10 | - 2 | - 4 |
| SOIL WATER CONTENT (X10 %) | | | | | | | 131 | 71 | 112 | 204 | 220 | | 318 | 215 |
| SOIL PH | | 5,1 | 5,1 | 4,8 | | | 5,7 | | | 5,1 | 4,9 | | | 4,7 |
| SOIL LOSS ON IGNITION (%) | | | | | | | 95 | 70 | 87 | 89 | 97 | 90 | 85 | 84 |
| INFLUENCE OF SALTSPRAY | | | | | | | 1 | | | 1 | | | | |
| INFLUENCE OF TRAMPLING | | | | | | | | | | | | | 1 | |
| INFLUENCE OF MANURING | | | | | | | | | | | | | | |
| HEIGHT OF HERB LAYER (CM) | 15 | 15 | 5 | 30 | 15 | 15 | 10 | 5 | 10 | 7 | 10 | 15 | 10 | 30 |
| COVER OF HERB LAYER (%) | 25 | 70 | 50 | 10 | 90 | 40 | 100 | 7 | 95 | 70 | 90 | 3 | 11 | 7 |
| HEIGHT OF BRYOPHYTE LAYER (CM) | 3 | 5 | 5 | 10 | 10 | 9 | 3 | 4 | 5 | 2 | 2 | 7 | 4 | 5 |
| COVER OF BRYOPHYTE LAYER (%) | 100 | 70 | 99 | 100 | 100 | 100 | 100 | 100 | 100 | 100 | 35 | 100 | 00 | 100 |
| COVER OF ALGAE (%) | | | | | | | | | | | | | | |
| COVER OF BARE SOIL (%) | | | | | | | | | | | | | | |
| COVER OF ROCKS (%) | | | | | | | | | | | | | | |
| QUADRAT SIZE (M2) | 1 | 25 | 16 | 18 | 1 | 2 | 1 | 1 | 1 | 1 | 9 | 1 | 4 | 16 |
| TOTAL NUMBER OF SPECIES | 3 | 4 | 5 | 5 | 5 | 6 | 3 | 4 | 5 | 5 | 6 | 6 | 7 | 8 |
| DIAGNOSTIC TAXA OF ASSOCIATIONS AND SUBASSOCIATIONS: | | | | | | | | | | | | | | |
| JUNCUS SCHEUCHZERIOIDES | 2A | 4 | 2M | 2M | 3 | 2M | | | | + | 1 | R | + | |
| RANUNCULUS BITERNATUS | | 1 | 3 | + | 2A | 1 | + | 1 | 2B | 1 | R | 2A | 1 | + |
| DREPANOCLADUS UNCINATUS | 5 | 5 | 5 | 5 | 5 | 5 | 5 | 5 | 5 | 5 | 3 | 5 | 5 | 5 |
| DISTICHOPHYLLUM FASCICULATUM | | | | | | | | | | | | | | |
| BLEPHARIDOPHYLLUM DENSIFOLIUM | | | | | | | | | | | | | | |
| CLASMATOCOLEA HUMILIS | | | | | | | | | | | | | | |
| LEITOSCYPHUS EXPANSUS | | | | 1 | | | | | + | | | 1 | | |
| LEPIDOZIA LAEVIFOLIA | | | | | | | | | | | | | | |
| CAMPYLOPUS ARBORICOLA | | | | | | | | | | | | | | |
| PELTIGERA POLYDACTYLA | | | | | | | | | | | | | | R |
| ANDREWSIANTUS LANCISTIPUS | | | | | | | | | | | | | | |
| UNCINIA COMPACTA | | | | | | | | | | | | | | |
| FOSSOMBRONIA AUSTRALIS | | | | | | | | | | | | | | + |
| JENSENIA PISICOLOR | | | | | | | | | | | | | | |
| LOPHOZIA CYLINDRIFORMIS | | | | | | | | | | | | | | |
| PLAGIOCHILA HETERODONTA | | | | | | | | | | | | | | |
| PTYCHOMNION RINGIANUM | | | | | | | | | | | | | | |
| JAMESONIELLA GRANDIFLORA | | | | | | | | | | | | | | |
| JAMESONIELLA COLORATA | | | | | | | | | | | | | | |
| RACOMITRIUM LANUGINOSUM | | | | | | | | | | | | | | |
| LYCOPODIUM MAGELLANICUM | | | | | | | | | | | | | | |
| DISTICHOPHYLLUM IMBRICATUM | | | | | | | | | | | | | | |
| RICCARDIA CF. PINGUIS | | | | | | | | | | | | | | |
| BREUTELIA INTEGRIFOLIA | | | | | | | | | | | + | | + | |
| BRYUM LAEVIGATUM | | | | | | | | | | | 1 | | | |
| RICCARDIA CF. MULTIFIDA | | | | | | | | | | | | | | |
| BLECHNUM PENNA-MARINA | | | | | 3 | | | | | | | | | |
| ACAENA MAGELLANICA | | | | | + | + | | | | | | | | |
| SCHISTOCHILA CARNOSA | | | | | | | | | | | | | | |
| PLAGIOTHECIUM PLATYPHYLLUM | | | | | | | | | | | | | | |
| KETZGERIA DECIPIENS | | | | | | | | | | | | | | |
| CATAGONIUM POLITUM | | | | | | | | | | | | | | |
| HYGROAMBLYSTEGIUM FILUM | | | | | | | | | | | | | | |
| LOPHOCOLEA RANDII | | | | | | | | | | | + | 1 | 1 | |
| AGROSTIS MAGELLANICA | 2B | + | 1 | 2A | | 3 | 5 | 2A | 5 | 5 | 5 | 2A | + | 2A |
| COMPANIONS: | | | | | | | | | | | | | | |
| CAMPYLOPUS SUBNITENS | | | | | | | | | | | | | | |
| DICRANOLOMA BILLARDIERI | | | | | | | | | | | | | | |
| CAMPYLOPUS CLAVATUS | | | | | | | | | | | | | | |
| MONTIA FONTANA | | | 1 | | | R | | + | 1 | 2A | | | 1 | 1 |
| AZORELLA SELAGO | | | | | | | | | | | | | | |
| MARCHANTIA BERTEROANA | | | | | | | | | | | | | | + |
| ACROBOLBUS OCHROPHYLLUS | | | | | | | | | | | | | | |
| CLASMATOCOLEA VERMICULARIS | | | | | | | | | | | | | | |
| SYMPHYOGYNA MARIONENSIS | | | | | | | | | | | | | | |
| CLADONIA COLL. | | | | | | | | | | | | | | |
| TEMNOMA QUADRIPARTITA | | | | | | | | | | | | | | |
| POA COOKII | | | | | | | | | | | | | | |

| | JUNCO SCHEUCHZERIOIDIS - DREPANOCLADETUM UNCINATI | | | | | | | | | DISTICHOPHYLLETUM | | | | | |
|---|---|---|---|---|---|---|---|---|---|---|---|---|---|---|---|
| | clasmatocoleetosum humilis | | | | | | | | | | | | | | |
| RELEVE NUMBER | 407 | 390 | 13 | 12 | 489 | 606 | 472 | 217 | 90 | 607 | 8 | 9 | 408 | 7 | 20 |
| LOCATION | | | | | | | | | | | | | | | |
| SOUTHERN LATITUDE (46°) MIN. | 52 | 52 | 52 | 52 | 49 | 52 | 52 | 52 | 52 | 52 | 52 | 52 | 52 | 52 | 52 |
| SEC. | 40 | 40 | 37 | 37 | 35 | 30 | 15 | 10 | 35 | 40 | 28 | 40 | 40 | 40 | 40 |
| EASTERN LONGITUDE (37°) MIN. | 51 | 51 | 51 | 51 | 42 | 51 | 51 | 51 | 51 | 51 | 51 | 51 | 51 | 51 | 51 |
| SEC. | 30 | 30 | 30 | 30 | 05 | 30 | 03 | 15 | 28 | 31 | 30 | 31 | 30 | 31 | 31 |
| ELEVATION (M A.S.L.) | 16 | 20 | 25 | 25 | 40 | 20 | 50 | 35 | 27 | 20 | 25 | 25 | 16 | 25 | 22 |
| INCLINATION (°) | | | | | | | | | | | | | | | |
| ASPECT | | | | | | | | | | | | | | | |
| SOIL DEPTH (CM) | | | | | | | | | | | | | | | |
| DEPTH OF GROUNDWATER LEVEL (CM) | - 6 | - 7 | -11 | -12 | + 1 | -10 | - 1 | -13 | - 7 | - 5 | - 6 | - 6 | - 4 | - 7 | - 5 |
| SOIL WATER CONTENT (X10 %) | 200 | 250 | | | | | 198 | 129 | 164 | | | | 221 | | |
| SOIL PH | 4,4 | 4,5 | | | | | 4,9 | 4,4 | 4,3 | | | | 4,4 | | |
| SOIL LOSS ON IGNITION (%) | 91 | 97 | | | | | 94 | 85 | 82 | 96 | | 94 | 94 | | 95 |
| INFLUENCE OF SALTSPRAY | | | | | | | | | | | | | | | |
| INFLUENCE OF TRAMPLING | | | | | | | | | | | | | | | |
| INFLUENCE OF MANURING | | | | | | | | | | | | | | | |
| HEIGHT OF HERB LAYER (CM) | 15 | 25 | 12 | 15 | 15 | 25 | 25 | 20 | 20 | 10 | 10 | 15 | 15 | 10 | 15 |
| COVER OF HERB LAYER (%) | 3 | 5 | 3 | 3 | 10 | 40 | 10 | 90 | 95 | 5 | 3 | 3 | 5 | 7 | 10 |
| HEIGHT OF BRYOPHYTE LAYER (CM) | 5 | 5 | 7 | 6 | 5 | 10 | 4 | 2 | 20 | 3 | 4 | 4 | 4 | 4 | 5 |
| COVER OF BRYOPHYTE LAYER (%) | 100 | 100 | 100 | 100 | 100 | 100 | 100 | 100 | 100 | 100 | 100 | 100 | 100 | 100 | 100 |
| COVER OF ALGAE (%) | | | | | 100 | | | | | | | | | | |
| COVER OF BARE SOIL (%) | | | | | | | | | | | | | | | |
| COVER OF ROCKS (%) | | | | | | | | | | | | | | | |
| QUADRAT SIZE (M2) | 4 | 3 | 1 | 1 | 9 | 1 | 16 | 4 | 1 | 1 | 1 | 1 | 4 | 1 | 1 |
| TOTAL NUMBER OF SPECIES | 7 | 9 | 9 | 10 | 10 | 11 | 12 | 11 | 12 | 4 | 6 | 6 | 7 | 7 | 8 |
| DIAGNOSTIC TAXA OF ASSOCIATIONS AND SUBASSOCIATIONS: | | | | | | | | | | | | | | | |
| JUNCUS SCHEUCHZERIOIDES | 1 | + | | R | 2A | + | | | | 1 | + | + | 1 | 1 | + |
| RANUNCULUS BITERNATUS | 1 | 1 | 2A | 1 | 2A | 1 | 1 | 1 | | | | | | | 2B |
| DREPANOCLADUS UNCINATUS | 5 | 5 | 5 | 5 | 5 | 5 | 5 | 5 | 5 | | | | + | | + |
| DISTICHOPHYLLUM FASCICULATUM | | 1 | | + | | | R | | 1 | 5 | 5 | 5 | 5 | 5 | 5 |
| BLEPHARIDOPHYLLUM DENSIFOLIUM | | 1 | + | | + | | + | + | 1 | | + | + | R | + | 2B |
| CLASMATOCOLEA HUMILIS | | + | 2A | 1 | 1 | + | + | + | + | | + | + | | + | 1 |
| LEITOSCYPHUS EXPANSUS | 1 | + | 1 | 2A | | 1 | + | 1 | + | + | 1 | 1 | 1 | 1 | 1 |
| LEPIDOZIA LAEVIFOLIA | 1 | 1 | | | + | | | | 2M | | | | | | |
| CAMPYLOPUS ARBORICOLA | | | | | | | | | | | | | | | |
| PELTIGERA POLYDACTYLA | | | | | | + | + | 1 | | | | | | | |
| ANDREWSIANTUS LANCISTIPUS | | | | | | | + | | 1 | | | | | | |
| UNCINIA COMPACTA | | | | | | | | 4 | 4 | | | | | | |
| FOSSOMBRONIA AUSTRALIS | | | | | + | | | | | | | | | | |
| JENSENIA PISICOLOR | | | | | | | | | | | | | | | |
| LOPHOZIA CYLINDRIFORMIS | | | | | | | | | | | | | | | |
| PLAGIOCHILA HETERODONTA | | | | | | | | | | | | | | | |
| PTYCHOMNION RINGIANUM | | | | | + | | | | | | | | | | |
| JAMESONIELLA GRANDIFLORA | | | | | | | | | | | | | | | |
| JAMESONIELLA COLORATA | | | | | | | | | | | | | | | |
| RACOMITRIUM LANUGINOSUM | | | | | | | | | | | | | | | |
| LYCOPODIUM MAGELLANICUM | | | | | | | | | | | | | | | |
| DISTICHOPHYLLUM IMBRICATUM | | | | | | | | | | | | | | | |
| RICCARDIA CF. PINGUIS | | | | | | | | | | | | | | | |
| BREUTELIA INTEGRIFOLIA | | | | | | | | | | | | | | | |
| BRYUM LAEVIGATUM | | | | | | | | | | | | | | | |
| RICCARDIA CF. MULTIFIDA | | | | | | | | | 1 | | | | | | |
| BLECHNUM PENNA-MARINA | | | | R | | | | 4 | R | | | | | | |
| ACAENA MAGELLANICA | | | | | | | | | | | | | | | |
| SCHISTOCHILA CARNOSA | | | | | | | | | | | | | | | |
| PLAGIOTHECIUM PLATYPHYLLUM | | | | | | | | + | | | | | | | |
| KETZGERIA DECIPIENS | | | | | | | | | | | | | | | |
| CATAGONIUM POLITUM | | | | | | | | | | | | | | | |
| HYGROAMBLYSTEGIUM FILUM | | | | | | | | | | | | | | | |
| LOPHOCOLEA RANDII | 1 | | + | + | + | + | 1 | | | | | | 1 | + | |
| AGROSTIS MAGELLANICA | 1 | 2A | 1 | 2M | 1 | 3 | 2A | 2A | 3 | 1 | + | 1 | 2A | 2A | 2A |
| COMPANIONS: | | | | | | | | | | | | | | | |
| CAMPYLOPUS SUBNITENS | | | | | | | | | | | | | | | |
| DICRANOLOMA BILLARDIERI | | | | | | | | | | | | | | | |
| CAMPYLOPUS CLAVATUS | | | | | | | | | | | | | | | |
| MONTIA FONTANA | | | 1 | + | | + | + | | + | | | | | | |
| AZORELLA SELAGO | | | | | | | | | | | | | | | |
| MARCHANTIA BERTEROANA | | | | | | R | | | | | | | | | |
| ACROBOLBUS OCHROPHYLLUS | | | | | | | | | | | | | | | |
| CLASMATOCOLEA VERMICULARIS | | | | | | | | | | | | | | | |
| SYMPHYOGYNA MARIONENSIS | | | | | | | | | | | | | | | |
| CLADONIA COLL. | | | | | | | | | | | | | | | |
| TEMNOMA QUADRIPARTITA | | | | | | | | | | | | | | | |
| POA COOKII | | | | | | | | | | | | | | | |

*Table 26a* (continued).

| | FASCICULATI | | | | BLEPHARIDOPHYLLO DENSIFOLII - CLASMATOCOLEETUM HUMILIS: ranunculetosum biternati | | | | | | | | | | | | |
|---|---|---|---|---|---|---|---|---|---|---|---|---|---|---|---|---|---|
| RELEVE NUMBER | 19 | 10 | 392 | 413 | 401 | 17 | 393 | 396 | 403 | 170 | 169 | 297 | 400 | 484 | 14 | 15 | 16 |
| LOCATION | | | | | | | | | | | | | | | | | |
| SOUTHERN LATITUDE (46°) MIN. | 52 | 52 | 52 | 52 | 52 | 52 | 52 | 52 | 52 | 50 | 50 | 51 | 52 | 49 | 52 | 52 | 52 |
| SEC. | 40 | 40 | 40 | 40 | 40 | 40 | 40 | 40 | 40 | 50 | 50 | 40 | 40 | 45 | 40 | 40 | 40 |
| EASTERN LONGITUDE (37°) MIN. | 51 | 51 | 51 | 51 | 51 | 51 | 51 | 51 | 51 | 48 | 48 | 50 | 51 | 51 | 51 | 51 | 51 |
| SEC. | 31 | 31 | 30 | 30 | 30 | 31 | 30 | 30 | 31 | 00 | 00 | 50 | 30 | 30 | 31 | 31 | 31 |
| ELEVATION (M A.S.L.) | 22 | 25 | 20 | 22 | 23 | 25 | 20 | 20 | 19 | 80 | 80 | 60 | 20 | 50 | 25 | 25 | 25 |
| INCLINATION (°) | | | | | | | | | 3 | | | | | | | | |
| ASPECT | | | | | | | | | E | | | | | | | | |
| SOIL DEPTH (CM) | | | | | | | | | | | | | | | | | |
| DEPTH OF GROUNDWATER LEVEL (CM) | - 7 | - 7 | - 4 | - 6 | -11 | -10 | -12 | - 8 | - 7 | -10 | -10 | - 5 | -10 | - 6 | -10 | -15 | - 8 |
| SOIL WATER CONTENT (X10 %) | | | 302 | 177 | 196 | | 152 | 144 | 164 | 237 | 201 | 183 | 137 | | | | |
| SOIL PH | | | 4,5 | 4,5 | 4,1 | | 4,2 | 4,4 | 4,2 | | | 4,1 | 4,4 | | | | |
| SOIL LOSS ON IGNITION (%) | | | 97 | 95 | 95 | | 95 | 95 | 95 | 79 | 97 | 95 | 94 | | | 94 | |
| INFLUENCE OF SALTSPRAY | | | | | | | | | | | | | | | | | |
| INFLUENCE OF TRAMPLING | | | | | | | | | | | | | | 1 | | | |
| INFLUENCE OF MANURING | | | | | | | | | | | | | | 1 | | | |
| HEIGHT OF HERB LAYER (CM) | 15 | 10 | 15 | 20 | 25 | 15 | 15 | 20 | 20 | 15 | 8 | 15 | 20 | 15 | 15 | 15 | 15 |
| COVER OF HERB LAYER (%) | 10 | 3 | 3 | 5 | 20 | 20 | 3 | 5 | 7 | <1 | <1 | 1 | 7 | 7 | 10 | 20 | 10 |
| HEIGHT OF BRYOPHYTE LAYER (CM) | 5 | 5 | 3 | 7 | 5 | 5 | 4 | 4 | 3 | 5 | 5 | 3 | 3 | 1 | 5 | 5 | 5 |
| COVER OF BRYOPHYTE LAYER (%) | 100 | 100 | 100 | 100 | 100 | 100 | 100 | 100 | 100 | 100 | 100 | 100 | 100 | 100 | 100 | 100 | 100 |
| COVER OF ALGAE (%) | | | | | | | | | | | | | | | | | |
| COVER OF BARE SOIL (%) | | | | | | | | | | | | | | | | | |
| COVER OF ROCKS (%) | | | | | | | | | | | | | | | | | |
| QUADRAT SIZE (M2) | 1 | 1 | 4 | 4 | 4 | 1 | 4 | 4 | 4 | 1 | 1 | 4 | 4 | 9 | 1 | 1 | 1 |
| TOTAL NUMBER OF SPECIES | 9 | 9 | 9 | 9 | 5 | 5 | 6 | 6 | 6 | 6 | 7 | 7 | 7 | 7 | 7 | 7 | 7 |
| DIAGNOSTIC TAXA OF ASSOCIATIONS AND SUBASSOCIATIONS: | | | | | | | | | | | | | | | | | |
| JUNCUS SCHEUCHZERIOIDES | + | + | + | 1 | | | | | | | | + | | | + | | |
| RANUNCULUS BITERNATUS | 1 | + | 1 | + | | 1 | + | + | R | | + | 1 | + | 2A | + | 2A | 3 |
| DREPANOCLADUS UNCINATUS | + | + | + | 1 | | R | R | | | | | + | | | | | |
| DISTICHOPHYLLUM FASCICULATUM | 5 | 5 | 4 | 5 | | | | | | 2A | + | | | | | 1 | + |
| BLEPHARIDOPHYLLUM DENSIFOLIUM | 1 | 2B | 3 | | 5 | 5 | 5 | 5 | 5 | 5 | 5 | 5 | 5 | 5 | 5 | 5 | 5 |
| CLASMATOCOLEA HUMILIS | 2M | | 1 | + | + | 2A | + | + | + | 1 | 1 | 2M | 1 | 2B | 2A | 2M | 3 |
| LEPTOSCYPHUS EXPANSUS | 1 | 1 | + | 1 | | | | | | | | | | | 1 | 2A | 1 |
| LEPIDOZIA LAEVIFOLIA | | | | | | | | | | + | | + | + | + | | | |
| CAMPYLOPUS ARBORICOLA | | | | | | | | | R | 2A | 2A | | | | | | |
| PELTIGERA POLYDACTYLA | | | | | | | | R | | | | | R | + | | | |
| ANDREWSIANTHUS LANCISTIPUS | | | | | | | | | | | | | | | | | |
| UNCINIA COMPACTA | | | | | | | | | | | | | | | | | |
| FOSSOMBRONIA AUSTRALIS | | | R | | | | | | | | | | | | | | |
| JENSENIA PISICOLOR | | | | | | | | | | | | | | | | | |
| LOPHOZIA CYLINDRIFORMIS | | | | | R | | | | | | | | | 1 | | | |
| PLAGIOCHILA HETERODONTA | | | | | | | | | | | | | | | | | |
| PTYCHOMNION RINGIANUM | | | | | | | | | | | | | | | | | |
| JAMESONIELLA GRANDIFLORA | | | | | | | | | | | | | | | | | |
| JAMESONIELLA COLORATA | | | | | | | | | | | | | | | | | |
| RACOMITRIUM LANUGINOSUM | | | | | | | | | | | | | | | | | |
| LYCOPODIUM MAGELLANICUM | | | | | | | | | | | | | | | | | |
| DISTICHOPHYLLUM IMBRICATUM | | | | | | | | | | | | | | | | | |
| RICCARDIA CF. PINGUIS | | | | | | | | | | | | | | | | | |
| BREUTELIA INTEGRIFOLIA | | | | | | | | | | | | | | | | | |
| BRYUM LAEVIGATUM | | | | | | | | | | | | | | | | | |
| RICCARDIA CF. MULTIFIDA | | | | | | | | | | | | | | | | | |
| BLECHNUM PENNA-MARINA | | | | | + | | | | 1 | | | | | | | | R |
| ACAENA MAGELLANICA | | | | | | | | | | | | | | | | | |
| SCHISTOCHILA CARNOSA | | | | | | | | | | | | | | | | | |
| PLAGIOTHECIUM PLATYPHYLLUM | | | | | | | | | | | | | | | | | |
| METZGERIA DECIPIENS | | | | | | | | | | | | | | | | | |
| CATAGONIUM POLITUM | | | | | | | | | | | | | | | | | |
| HYGROAMBLYSTEGIUM FILUM | | | | | | | | | | | | | | | | | |
| LOPHOCOLEA RANDII | | + | | 1 | | | | | | | | | | | | | |
| AGROSTIS MAGELLANICA | 2A | 1 | 1 | 2A | 2B | 2A | 1 | 2A | 2A | + | + | + | 2A | + | 2A | 2B | 2A |
| COMPANIONS: | | | | | | | | | | | | | | | | | |
| CAMPYLOPUS SUBNITENS | | | | | | | | | | | | | | | | | |
| DICRANOLOMA BILLARDIERI | | | | | | | | | | | | | | | | | |
| CAMPYLOPUS CLAVATUS | | | | | | | | | | | | | | | | | |
| MONTIA FONTANA | | | | | | | + | R | | | + | | R | | | | |
| AZORELLA SELAGO | | | | | | | | | | | | | | | | | |
| MARCHANTIA BERTEROANA | | | | | | | | | | | | | | | | | 1 |
| ACROBOLBUS OCHROPHYLLUS | | | | | | | | | | | | | | | 1 | | |
| CLASMATOCOLEA VERMICULARIS | | | | | | | | | | | | | | | | | |
| SYMPHYOGYNA MARIONENSIS | | | | | | | | | | | | | | | | | |
| CLADONIA COLL. | | | | | | | | | | | | | | | | | |
| TEMNOMA QUADRIPARTITA | | | | | | | | | | | | | | | | | |
| POA COOKII | | | | | | | | | | | | | | | | | |

| | BLEPHARIDOPHYLLO DENSIFOLII - CLASMATOCOLEETUM HUMILIS: ranunculetosum biternati (continued) | | | | | | | | | | | | | | |
|---|---|---|---|---|---|---|---|---|---|---|---|---|---|---|---|
| RELEVE NUMBER | 23 | 24 | 477 | 21 | 397 | 37 | 391 | 300 | 36 | 395 | 172 | 305 | 18 | 22 | 89 |
| LOCATION | | | | | | | | | | | | | | | |
| SOUTHERN LATITUDE (46°) MIN. | 52 | 52 | 52 | 52 | 52 | 52 | 52 | 51 | 52 | 51 | 50 | 53 | 52 | 52 | 52 |
| SEC. | 40 | 40 | 47 | 40 | 40 | 40 | 40 | 40 | 40 | 40 | 50 | 00 | 40 | 40 | 35 |
| EASTERN LONGITUDE (37°) MIN. | 51 | 51 | 51 | 51 | 51 | 51 | 51 | 50 | 51 | 51 | 48 | 51 | 51 | 51 | 51 |
| SEC. | 31 | 31 | 20 | 31 | 30 | 30 | 30 | 50 | 31 | 30 | 00 | 35 | 31 | 31 | 20 |
| ELEVATION (M A.S.L.) | 22 | 22 | 50 | 22 | 20 | 15 | 20 | 60 | 25 | 21 | 80 | 45 | 25 | 22 | 27 |
| INCLINATION (°) | | | | | | 3 | | | | 5 | | | | | |
| ASPECT | | | | | | SE | | | | N | | | | | |
| SOIL DEPTH (CM) | | | | | | | | | | | | | | | |
| DEPTH OF GROUNDWATER LEVEL (CM) | -13 | -16 | - 4 | -13 | - 8 | -18 | - 6 | - 8 | -18 | -30 | - 4 | - 8 | -13 | - 9 | - 9 |
| SOIL WATER CONTENT (X10 %) | | | 181 | | 184 | | 216 | 174 | | 146 | 300 | 176 | | | 140 |
| SOIL PH | | | 4,4 | | 4,4 | | 4,5 | 4,4 | | 4,1 | | | | | 4,3 |
| SOIL LOSS ON IGNITION (%) | 95 | | 95 | | 94 | | 77 | 94 | | 77 | 91 | 90 | | 95 | |
| INFLUENCE OF SALTSPRAY | | | | | | | | | | | | | | | |
| INFLUENCE OF TRAMPLING | | | | | | | | | | | | | | | |
| INFLUENCE OF MANURING | | | | | | | | | | | | | | | |
| HEIGHT OF HERB LAYER (CM) | 15 | 15 | 25 | 15 | 15 | 10 | 15 | 25 | 10 | 20 | 15 | 10 | 15 | 15 | 10 |
| COVER OF HERB LAYER (%) | 7 | 40 | 7 | 20 | 5 | 80 | 5 | 30 | 70 | 7 | 60 | 15 | 15 | 3 | 60 |
| HEIGHT OF BRYOPHYTE LAYER (CM) | 2 | 3 | 3 | 3 | 6 | 4 | 5 | 3 | 3 | 6 | 4 | 3 | 5 | 3 | 3 |
| COVER OF BRYOPHYTE LAYER (%) | 100 | 100 | 100 | 100 | 100 | 100 | 100 | 100 | 100 | 100 | 100 | 100 | 100 | 100 | 100 |
| COVER OF ALGAE (%) | | | | | | | | | | | | | | | |
| COVER OF BARE SOIL (%) | | | | | | | | | | | | | | | |
| COVER OF ROCKS (%) | | | | | | | | | | | | | | | |
| QUADRAT SIZE (M2) | 1 | 1 | 16 | 1 | 4 | 1 | 4 | 4 | 1 | 4 | 1 | 4 | 1 | 1 | 1 |
| TOTAL NUMBER OF SPECIES | 9 | 8 | 8 | 9 | 8 | 8 | 9 | 9 | 8 | 9 | 9 | 10 | 9 | 10 | 12 |
| DIAGNOSTIC TAXA OF ASSOCIATIONS AND SUBASSOCIATIONS: | | | | | | | | | | | | | | | |
| JUNCUS SCHEUCHZERIOIDES | | | | | | | | | | | | | | + | |
| RANUNCULUS BITERNATUS | 3 | 2B | + | 2A | 1 | + | 2A | | | | 1 | | + | 2B | R |
| DREPANOCLADUS UNCINATUS | | | | | R | 1 | + | | | | | | | + | |
| DISTICHOPHYLLUM FASCICULATUM | | | | | | | | | | | + | 1 | | + | |
| BLEPHARIDOPHYLLUM DENSIFOLIUM | 5 | 5 | 5 | 5 | 5 | 5 | 5 | 5 | 5 | 5 | 5 | 5 | 5 | 5 | 5 |
| CLASMATOCOLEA HUMILIS | 2A | 2A | 1 | 1 | + | + | 1 | 1 | 2A | + | + | 2M | 2M | 2A | 2A |
| LEPTOSCYPHUS EXPANSUS | 1 | 1 | + | 2M | R | | + | | | | | | + | + | |
| LEPIDOZIA LAEVIFOLIA | | | | | R | | | 1 | 2A | + | + | + | | 1 | 1 |
| CAMPYLOPUS ARBORICOLA | | | R | | | | | 2A | 1 | + | 1 | + | | | |
| PELTIGERA POLYDACTYLA | | R | | R | | | | | R | R | | | + | | |
| ANDREWSIANTHUS LANCISTIPUS | | | + | | | | | | | | | + | | | 1 |
| UNCINIA COMPACTA | | | | | | | | 1 | + | + | | | | | + |
| FOSSOMBRONIA AUSTRALIS | | | | | | | | | | | | | | | |
| JENSENIA PISICOLOR | | | | | | | | | | | | | + | | + |
| LOPHOZIA CYLINDRIFORMIS | | | | | | | | + | | | | | | | + |
| PLAGIOCHILA HETERODONTA | | | | | | | | | | | | | | | |
| PTYCHOMNION RINGIANUM | | | | | | | | | | | | | | | |
| JAMESONIELLA GRANDIFLORA | | | | | | | | | | | | | | | 1 |
| JAMESONIELLA COLORATA | | | | | | | | | | | | | | | |
| RACOMITRIUM LANUGINOSUM | | | | | | | | | | | | | | | |
| LYCOPODIUM MAGELLANICUM | | | | | | | | | | | | | | | |
| DISTICHOPHYLLUM IMBRICATUM | | | | | | | | | | | | | | | |
| RICCARDIA CF. PINGUIS | | | | | | | | | | | | | | | |
| BREUTELIA INTEGRIFOLIA | | | | | | | | | | | | | | | |
| BRYUM LAEVIGATUM | | | | | | | | | | | | | | | |
| RICCARDIA CF. MULTIFIDA | | | | | | | | | | | | | | | |
| BLECHNUM PENNA-MARINA | R | + | | R | | | | | + | + | | | R | | R |
| ACAENA MAGELLANICA | | | | | | | | | | | | | | | |
| SCHISTOCHILA CARNOSA | | | | | | + | | | | | | | | | |
| PLAGIOTHECIUM PLATYPHYLLUM | | | | | | | | | | | | | | | |
| METZGERIA DECIPIENS | | | | | | | | | | | | | | | |
| CATAGONIUM POLITUM | | | | | | | | | | | | | | | |
| HYGROAMBLYSTEGIUM FILUM | | | | | | | | | | | | | | | |
| LOPHOCOLEA RANDII | | | | | | + | + | | | | | R | | | |
| AGROSTIS MAGELLANICA | 2A | 3 | 2A | 2B | 1 | 5 | 1 | 3 | 4 | 2A | 4 | 2B | 2B | 2M | 4 |
| COMPANIONS: | | | | | | | | | | | | | | | |
| CAMPYLOPUS SUBNITENS | | | | | | | | | | | | | | | |
| DICRANOLOMA BILLARDIERI | | | | | | | | | | | | | | | |
| CAMPYLOPUS CLAVATUS | | | | | | | | | | | | | | | |
| MONTIA FONTANA | + | R | + | R | + | 1 | + | | | | + | R | R | | |
| AZORELLA SELAGO | | | | | | | | + | | | | | | | |
| MARCHANTIA BERTEROANA | R | | | | | | | | | | | | | | |
| ACROBOLBUS OCHROPHYLLUS | | | | | | | | | | | | | | | + |
| CLASMATOCOLEA VERMICULARIS | | | | | | | | | | | | | | + | |
| SYMPHYOGYNA MARIONENSIS | | | | | | | | | | | | | | | |
| CLADONIA COLL. | | | | | | | | | | | | | | | |
| TEMNOMA QUADRIPARTITA | | | | | | | | | | | | | | | |
| POA COOKII | | | | | | | | | | | | + | | | |

| | clasmatocoleetosum humilis | | | | | | | | | | | | | blechnetosum penna-marinae | | | | | | | |
|---|---|---|---|---|---|---|---|---|---|---|---|---|---|---|---|---|---|---|---|---|---|
| RELEVE NUMBER | 394 | 418 | 412 | 579 | 39 | 345 | 432 | 485 | 6 | 411 | 87 | 344 | 440 | 398 | 29 | 171 | 208 | 402 | 409 | 39 | 404 |
| LOCATION | | | | | | | | | | | | | | | | | | | | | |
| SOUTHERN LATITUDE (46°) MIN. | 52 | 52 | 52 | 53 | 52 | 37 | 53 | 49 | 52 | 52 | 52 | 37 | 53 | 52 | 52 | 50 | 51 | 52 | 52 | 52 | 52 |
| SEC. | 40 | 40 | 40 | 05 | 41 | 30 | 47 | 40 | 40 | 40 | 35 | 25 | 20 | 40 | 40 | 50 | 40 | 40 | 40 | 40 | 40 |
| EASTERN LONGITUDE (37°) MIN. | 51 | 51 | 51 | 51 | 51 | 59 | 52 | 42 | 51 | 51 | 51 | 59 | 51 | 51 | 51 | 48 | 50 | 51 | 51 | 51 | 51 |
| SEC. | 30 | 30 | 30 | 55 | 30 | 20 | 15 | 50 | 25 | 31 | 30 | 15 | 25 | 30 | 31 | 00 | 50 | 30 | 30 | 31 | 30 |
| ELEVATION (M A.S.L.) | 20 | 16 | 22 | 25 | 15 | 165 | 45 | 45 | 20 | 22 | 20 | 165 | 60 | 20 | 25 | 80 | 60 | 23 | 17 | 25 | 23 |
| INCLINATION (°) | 30 | 5 | | | 7 | 2 | | | | | | | | | | | | 3 | 20 | 5 | 7 |
| ASPECT | N | E | | | ESE | W | | | | | | | | | | | | [illegible] | [illegible] | S | [illegible] |
| SOIL DEPTH (CM) | | | | | | | | | | | | | | | | | | | | | |
| DEPTH OF GROUNDWATER LEVEL (CM) | -43 | -10 | - 7 | -16 | -17 | - 1 | - 9 | -13 | - 7 | -15 | - 2 | - 3 | -10 | -20 | -31 | -18 | - | -12 | -20 | -30 | -12 |
| SOIL WATER CONTENT (X10 %) | 98 | 137 | 113 | | | 217 | 225 | | | 116 | 196 | 192 | 176 | 116 | | 130 | 178 | 65 | 79 | | 122 |
| SOIL PH | 4,3 | 4,2 | 4,5 | | | 4,5 | 4,2 | | | 4,5 | | 4,5 | 4,5 | 4,3 | | | 4,0 | 4,2 | 4,1 | | 4,1 |
| SOIL LOSS ON IGNITION (%) | 96 | | 93 | 96 | | 92 | 95 | | | 93 | | 94 | 96 | 95 | | 62 | 74 | 95 | 93 | | 94 |
| INFLUENCE OF SALTSPRAY | | | | | | | | | | | | | | | | | | | | | |
| INFLUENCE OF TRAMPLING | | | | | | 1 | | | | | | | 1 | | | | | | | | |
| INFLUENCE OF MANURING | | 1 | | | | 1 | | | | | | | | | | | | | | | |
| HEIGHT OF HERB LAYER (CM) | 15 | 20 | 15 | 25 | 10 | 30 | 20 | 20 | 15 | 20 | 12 | 35 | 20 | 15 | 15 | 15 | 25 | 7 | 10 | 15 | 20 |
| COVER OF HERB LAYER (%) | 3 | 7 | 1 | 5 | 65 | 3 | 17 | 7 | 49 | 3 | 100 | 45 | 7 | 40 | 90 | 70 | 90 | 40 | 90 | 90 | 50 |
| HEIGHT OF BRYOPHYTE LAYER (CM) | 2 | 2 | 2 | 3 | 4 | 4 | 4 | 2 | 3 | 2 | 3 | 4 | 4 | 3 | 2 | 4 | 3 | 4 | 3 | 2 | [illegible] |
| COVER OF BRYOPHYTE LAYER (%) | 100 | 100 | 100 | 100 | 100 | 100 | 100 | 100 | 100 | 100 | 100 | 100 | 99 | 95 | 100 | 100 | 80 | 90 | 95 | 100 | [illegible] |
| COVER OF ALGAE (%) | | | | | | | | | | | | | | | | | | | | | |
| COVER OF BARE SOIL (%) | | | | | | | | | | | | | 1 | | | | | | | | |
| COVER OF ROCKS (%) | | | | | | | | | | | | | | | | | | | | | |
| QUADRAT SIZE (M2) | 3 | 4 | 1 | 9 | 1 | 9 | 9 | 9 | 1 | 3 | 4 | 6 | 9 | 4 | 1 | 1 | 4 | 4 | 4 | 1 | 4 |
| TOTAL NUMBER OF SPECIES | 5 | 7 | 7 | 8 | 9 | 10 | 10 | 11 | 12 | 14 | 15 | 7 | 10 | 10 | 15 | 15 | 15 | 16 | 17 | 18 | 19 |
| DIAGNOSTIC TAXA OF ASSOCIATIONS AND SUBASSOCIATIONS: | | | | | | | | | | | | | | | | | | | | | |
| JUNCUS SCHEUCHZERIOIDES | + | | | + | | + | 2M | | | + | | | | | | | | | | | |
| RANUNCULUS BITERNATUS | | + | + | 1 | + | | | + | 1 | + | | | | | + | | R | | | | |
| DREPANOCLADUS UNCINATUS | + | + | P | | + | | + | | + | + | 1 | | | | | | 2A | R | + | + | + |
| DISTICHOPHYLLUM FASCICULATUM | | | | | | 2A | 1 | | | | | + | + | + | | 1 | + | + | + | | + |
| BLEPHARIDOPHYLLUM DENSIFOLIUM | 1 | 1 | | 2M | 1 | + | 1 | 4 | 1 | 1 | 1 | | 2A | 5 | 5 | 5 | 4 | 5 | 5 | 5 | 5 |
| CLASMATOCOLEA HUMILIS | 5 | 5 | 5 | 5 | 5 | 5 | 5 | 5 | 5 | 5 | 5 | + | 1 | + | 1 | + | | 1 | 1 | 2B | + |
| LEPTOSCYPHUS EXPANSUS | | + | + | | + | | + | | 1 | + | [illegible] | | | P | 1 | | | + | + | 1 | + |
| LEPIDOZIA LAEVIFOLIA | | | + | | 1 | 2M | + | | | + | 1 | 1 | + | 1 | 2B | + | 2B | [illegible] | [illegible] | [illegible] | 1 |
| CAMPYLOPUS ARBORICOLA | | | | | | 2A | 2A | | | | | 5 | 5 | | | 1 | | + | | | 1 |
| PELTIGERA POLYDACTYLA | | | | + | | | | + | + | + | | | | R | + | [illegible] | [illegible] | + | + | + | + |
| ANDREWSIANTHUS LANCISTIPUS | | | + | | | | + | | | + | | | + | | | 1 | 1 | + | + | | + |
| UNCINIA COMPACTA | | | | | | | | | | | 1 | | | 1 | | + | 2B | 1 | 1 | | 2A |
| FOSSOMBRONIA AUSTRALIS | | | | | | | | | | + | | | | | 2A | | | + | [illegible] | + | |
| JENSENIA PISICOLOR | | | | | | | | | | | 1 | | | | 1 | + | | | | + | + |
| LOPHOZIA CYLINDRIFORMIS | | | | | | | | | | | 1 | | | | | + | | | + | 1 | + |
| PLAGIOCHILA HETERODONTA | | | | | | | | | | | + | | | | + | | | + | | 1 | |
| PTYCHOMNION RINGIANUM | | | | | | | | | | | 2A | | | | | | | | | | |
| JAMESONIELLA GRANDIFLORA | | | | | | | | | | | 2A | | | | | | | | | | |
| JAMESONIELLA COLORATA | | | | | | + | | | | R | - | | | | + | | 1 | | + | | 1 |
| RACOMITRIUM LANUGINOSUM | | | | | | | | | | | R | | | | | | + | + | | | + |
| LYCOPODIUM MAGELLANICUM | | | | | | | | | | | | | | | | | | | | | |
| DISTICHOPHYLLUM IMBRICATUM | | | | | | | | | | | | | | | | | | | | | |
| RICCARDIA CF. PINGUIS | | | | | | | | | | | + | | | | | | | | | | |
| BREUTELIA INTEGRIFOLIA | | | | | | | | | | | | | | | | | | | | | |
| BRYUM LAEVIGATUM | | | | | | | | | | | | | | | | | | | | | |
| RICCARDIA CF. MULTIFIDA | | | | | | | | | | | + | | | | | | | | | | |
| BLECHNUM PENNA-MARINA | | | | | | | | | R | | | | | 3 | 4 | 3 | 1 | 3 | 5 | 4 | 3 |
| ACAENA MAGELLANICA | | | | | | | | | | | | | | | | | | | | | |
| SCHISTOCHILA CARNOSA | | | | | | + | | + | | | | 1 | 1 | | | | | | | | |
| PLAGIOTHECIUM PLATYPHYLLUM | | | | | | | | | | | | | | | | | | | | | |
| METZGERIA DECIPIENS | | | | | | | | | | | | | | | | | | 1 | 1 | 1 | 1 |
| CATAGONIUM POLITUM | | | | | | | | | | | | | | | | | | | | | |
| HYGROAMBLYSTEGIUM FILUM | | | | | | | | | | | | | | | | | | | | | |
| LOPHOCOLEA RANDII | | | | | + | | | + | + | | | | | | | | | | | | |
| AGROSTIS MAGELLANICA | 1 | 2A | 1 | 1 | 4 | 1 | 2A | 1 | 5 | 1 | 5 | 3 | 2A | 1 | 3 | 3 | 5 | 2A | + | 2B | 1 |
| COMPANIONS: | | | | | | | | | | | | | | | | | | | | | |
| CAMPYLOPUS SUBNITENS | | | | | | | | | | | | | | | | | | | | | |
| DICRANOLOMA BILLARDIERI | | | | | | | | | | | | | | | | | | | | | |
| CAMPYLOPUS CLAVATUS | | | | | | | | | | | | | | | | | | | | | |
| MONTIA FONTANA | | + | | 1 | + | + | | + | + | R | | | + | | + | + | | | | + | |
| AZORELLA SELAGO | | | | | | | | + | | | | | R | | | + | 2A | | R | | |
| MARCHANTIA BERTEROANA | | | | | | | | + | + | + | | | | | + | | | | | + | |
| ACROBOLBUS OCHROPHYLLUS | | | | | | | | | + | | | | | | | | | | | + | |
| CLASMATOCOLEA VERMICULARIS | | | | 1 | | | | | | | | | | | | | | | | | |
| SYMPHYOGYNA MARIONENSIS | | | | | | | | | | | | | | | | | | | + | | + |
| CLADONIA COLL. | | | | | | | | | | | | | | | | | | | | | |
| TEMNOMA QUADRIPARTITA | | | | | | | | | | | | | | | | | | | | | |
| POA COOKII | | | | | | | | 1 | | | | | | | | | | | | | |

*Table 26b*. The *Juncus scheuzerioides – Blepharidophyllum densifolium* complex (second part).

| UNCINIO COMPACTAE - PTYCHOMNIETUM RINGIANI | | | | | | | | | | | | | | | | | |
|---|---|---|---|---|---|---|---|---|---|---|---|---|---|---|---|---|---|
| | | | | | | | | | | | | | UNCINIA variant | | | BLECHNUM variant | |
| RELEVE NUMBER | 491 | 468 | 489 | 490 | 492 | 475 | 486 | 483 | 494 | 511 | 487 | 234 | 101 | 94 | 88 | 493 | 495 |
| LOCATION | | | | | | | | | | | | | | | | | |
| SOUTHERN LATITUDE (46°) MIN. | 49 | 52 | 49 | 49 | 49 | 52 | 49 | 49 | 49 | 53 | 49 | 50 | 52 | 52 | 52 | 49 | 49 |
| SEC. | 50 | 25 | 40 | 48 | 50 | 35 | 37 | 40 | 50 | 15 | 43 | 40 | 35 | 35 | 35 | 50 | 50 |
| EASTERN LONGITUDE (37°) MIN. | 42 | 51 | 42 | 42 | 42 | 50 | 42 | 42 | 42 | 37 | 42 | 52 | 51 | 51 | 51 | 42 | 42 |
| SEC. | 15 | 28 | 15 | 25 | 15 | 49 | 05 | 20 | 15 | 45 | 20 | 00 | 20 | 20 | 20 | 15 | 15 |
| ELEVATION (M A.S.L.) | 75 | 40 | 45 | 70 | 75 | 50 | 55 | 50 | 75 | 60 | 60 | 115 | 25 | 25 | 25 | 75 | 75 |
| INCLINATION (°) | | | | | | | | | | | | | 3 | | 5 | | 35 |
| ASPECT | | | | | | | | | | | | | SW | | E | | S |
| SOIL DEPTH (CM) | | 50 | | | | | | | 35 | | | | | | | | 40 |
| DEPTH OF GROUNDWATER LEVEL (CM) | -2 | -2 | -7 | -7 | -2 | -2 | -6 | -3 | -3 | -1 | -17 | -6 | -10 | -14 | -15 | -19 | -28 |
| SOIL WATER CONTENT (X10 %) | | 182 | | | | 221 | | | | | | 131 | 160 | 156 | 140 | | |
| SOIL PH | | 4,9 | | | | 4,3 | | | | | | 4,4 | 4,1 | 4,5 | 4,3 | | |
| SOIL LOSS ON IGNITION (%) | | 87 | | | | 88 | | | | | | 89 | 86 | 80 | 90 | | |
| INFLUENCE OF SALTSPRAY | | | | | | | | | | | | | | | | | |
| INFLUENCE OF TRAMPLING | | | | | | | | | | | | | | | | | |
| INFLUENCE OF MANURING | | | | | | | | | | | | | | | | | |
| HEIGHT OF HERB LAYER (CM) | 20 | 25 | 15 | 20 | 25 | 20 | 20 | 20 | 25 | 25 | 20 | 15 | 15 | 20 | 20 | 25 | 15 |
| COVER OF HERB LAYER (%) | 20 | 95 | 30 | 30 | 30 | 17 | 30 | 7 | 35 | 15 | 30 | 60 | 95 | 95 | 90 | 60 | 65 |
| HEIGHT OF BRYOPHYTE LAYER (CM) | 5 | 9 | 7 | 7 | 6 | 8 | 5 | 7 | 5 | 5 | 5 | 4 | 4 | 7 | 5 | 6 | 9 |
| COVER OF BRYOPHYTE LAYER (%) | 100 | 99 | 100 | 100 | 100 | 100 | 100 | 100 | 100 | 99 | 100 | 60 | 95 | 100 | 100 | 100 | 95 |
| COVER OF ALGAE (%) | | | | | | | | | | | | | | | | | |
| COVER OF BARE SOIL (%) | | | | | | | | | | | | | | | | | |
| COVER OF ROCKS (%) | | | | | | | | | | | | | | | | | |
| QUADRAT SIZE (M2) | 15 | 25 | 9 | 9 | 9 | 16 | 9 | 9 | 9 | 9 | 9 | 4 | 1 | 1 | 1 | 9 | 18 |
| TOTAL NUMBER OF SPECIES | 8 | 9 | 9 | 8 | 10 | 11 | 12 | 12 | 13 | 13 | 14 | 14 | 14 | 16 | 15 | 9 | 17 |
| DIAGNOSTIC TAXA OF ASSOCIATIONS AND SUBASSOCIATIONS: | | | | | | | | | | | | | | | | | |
| JUNCUS SCHEUCHZERIOIDES | | | | | | | | + | | 1 | | 1 | | | | | |
| RANUNCULUS BITERNATUS | + | + | | | | | + | R | | | R | | | | | | |
| DREPANOCLADUS UNCINATUS | + | | | | | + | | | | | | + | | + | 1 | | |
| DISTICHOPHYLLUM FASCICULATUM | + | | + | | | + | | | | | 1 | | 1 | + | 1 | | |
| BLEPHARIDOPHYLLUM DENSIFOLIUM | | | + | + | | + | + | + | + | | | + | 1 | 1 | 1 | + | + |
| CLASMATOCOLEA HUMILIS | | | | + | + | 1 | + | 1 | + | + | | + | | + | 1 | | |
| LEPTOSCYPHUS EXPANSUS | + | | | | | | 1 | + | + | | | | | + | | | |
| LEPIDOZIA LAEVIFOLIA | | + | + | 1 | 1 | + | 1 | + | 1 | 1 | + | + | 2M | 2M | 2B | 1 | 1 |
| CAMPYLOPUS ARBORICOLA | | | | | | + | | | | | 1 | + | 1 | | 1 | | |
| PELTIGERA POLYDACTYLA | | | | | | | + | + | + | + | R | | | | | | + |
| ANDREWSIANTHUS LANCISTIPUS | | | | | | | | | | | + | | + | | 1 | | |
| UNCINIA COMPACTA | | 1 | | + | + | + | 1 | | + | 1 | 1 | 2A | 5 | 5 | 5 | 1 | |
| FOSSOMBRONIA AUSTRALIS | | | | | | | | + | + | | | | | R | | | |
| JENSENIA PISICOLOR | | | | | | | | | | | | + | | + | | | |
| LOPHOZIA CYLINDRIFORMIS | | | | | | | | | | | | | | | 1 | + | + |
| PLAGIOCHILA HETERODONTA | | | | | | | | | | | | | 1 | 2A | | | + |
| PTYCHOMNION RINGIANUM | 5 | 5 | 5 | 5 | 5 | 5 | 5 | 5 | 5 | 5 | 5 | 5 | 5 | 5 | 4 | 5 | 5 |
| JAMESONIELLA GRANDIFLORA | | | | | + | | | | + | | | | | + | | + | |
| JAMESONIELLA COLORATA | | | | + | | | | | | + | | + | | | | | + |
| RACOMITRIUM LANUGINOSUM | | R | | | | + | | | | | | | 1 | 1 | 2A | | + |
| LYCOPODIUM MAGELLANICUM | | | | | | | | | | | | | | | | | |
| DISTICHOPHYLLUM IMBRICATUM | | | | | | | | | | | | | | | | | |
| RICCARDIA CF. PINGUIS | | | | | | | | | | + | 1 | | | + | | | |
| BREUTELIA INTEGRIFOLIA | | | | | | | | | | | | + | | | | | |
| BRYUM LAEVIGATUM | | | | | | | | | | 1 | | | | | | | |
| RICCARDIA CF. MULTIFIDA | | | | | | | | | | + | | | | | | | |
| BLECHNUM PENNA-MARINA | + | R | + | 1 | 2A | + | | + | | | 2A | | | | 1 | 3 | 4 |
| ACAENA MAGELLANICA | | | | | | | | | | R | | | | | | | + |
| SCHISTOCHILA CARNOSA | | | + | | | | | | | | + | | | | | | |
| PLAGIOTHECIUM PLATYPHYLLUM | | | | | R | | | | | | | | | | | | |
| METZGERIA DECIPIENS | | | | | | | | | | | | | | | | | |
| CATAGONIUM POLITUM | | | | | | | | | | | | | | | | | |
| HYGROAMBLYSTEGIUM FILUM | | | | | | | | | | | | | | | | | |
| LOPHOCOLEA RANDII | | | | | | | | | | | | | | | | | |
| AGROSTIS MAGELLANICA | 2B | 5 | 3 | 2B | 3 | 2B | 3 | 2A | 3 | 2B | 2B | 4 | 5 | 5 | 3 | 3 | + |
| COMPANIONS: | | | | | | | | | | | | | | | | | |
| CAMPYLOPUS SUBNITENS | | | | | | | | | | | | | | | | | |
| DICRANOLOMA BILLARDIERI | | | | | | | | | | | | | | | | | |
| CAMPYLOPUS CLAVATUS | | | | | | | | | | | | | | | | | |
| MONTIA FONTANA | R | | R | . | | | + | R | | | + | | | + | + | | |
| AZORELLA SELAGO | | + | | | R | | + | | + | + | + | 3 | 1 | | | + | + |
| MARCHANTIA BERTEROANA | | | | | | | | | | | | | | | | | |
| ACROBOLBUS OCHROPHYLLUS | | | | | + | | | | + | | | | 1 | | | | |
| CLASMATOCOLEA VERMICULARIS | | | | | | | | | | | | | | | + | | |
| SYMPHYOGYNA MARIONENSIS | | | | | | | | | | | | | + | | | | |
| CLADONIA COLL. | | | | | | | | | | | | | | | | | + |
| TEMNOMA QUADRIPARTITA | | | | | | | | | | | | + | + | | | | |
| POA COOKII | | | | | | | | | | | | | | | | | + |

| LYCOPODIO MAGELLANICI - JAMESONIELLETUM COLORATAE | | | | | | | | | | | | | | | | | | | | | | | | | | | |
|---|---|---|---|---|---|---|---|---|---|---|---|---|---|---|---|---|---|---|---|---|---|---|---|---|---|---|---|
| | ranunculetosum biternati | | | | | | | | | | | | | transition | | | | | catagonietosum politi | | | | | | | | |
| RELEVE NUMBER | 331 | 91 | 314 | 687 | 335 | 433 | 536 | 302 | 470 | 469 | 461 | 303 | 161 | 447 | 86 | 549 | 534 | 532 | 157 | 526 | 304 | 539 | 545 | 98 | 415 | 301 | 449 |
| LOCATION | | | | | | | | | | | | | | | | | | | | | | | | | | | |
| SOUTHERN LATITUDE (46°) MIN. | 38 | 52 | 53 | 52 | 38 | 53 | 52 | 52 | 52 | 52 | 51 | 52 | 57 | 53 | 52 | 52 | 52 | 52 | 57 | 52 | 52 | 52 | 52 | 52 | 52 | 51 | 50 |
| SEC. | 25 | 35 | 10 | 35 | 48 | 27 | 21 | 10 | 15 | 20 | 00 | 10 | 20 | 05 | 35 | 21 | 21 | 21 | 15 | 21 | 15 | 21 | 21 | 35 | 40 | 40 | 30 |
| EASTERN LONGITUDE (37°) MIN. | 00 | 51 | 48 | 51 | 00 | 52 | 49 | 50 | 51 | 51 | 48 | 50 | 51 | 51 | 51 | 49 | 49 | 49 | 51 | 49 | 51 | 49 | 49 | 51 | 51 | 50 | 53 |
| SEC. | 10 | 20 | 45 | 20 | 10 | 03 | 55 | 50 | 00 | 10 | 00 | 55 | 10 | 25 | 20 | 55 | 55 | 55 | 10 | 55 | 00 | 55 | 55 | 20 | 30 | 50 | 00 |
| ELEVATION (M A.S.L.) | 70 | 27 | 270 | 25 | 25 | 40 | 130 | 55 | 50 | 40 | 110 | 50 | 80 | 50 | 20 | 130 | 130 | 130 | 60 | 130 | 40 | 130 | 130 | 27 | 26 | 60 | 100 |
| INCLINATION (°) | 1 | | | | | 2 | | 7 | | 3 | 4 | 10 | 7 | | | | | | 30 | | 7 | | 3 | 5 | 17 | 10 | |
| ASPECT | E | | | | | W | | ESE | | NW | N | NE | E | | | | | | ENE | | SSW | | S | W | ENE | SE | |
| SOIL DEPTH (CM) | | | | | | | | | | | | | | | 20 | | | | 30 | | | | | | | | 30 |
| DEPTH OF GROUNDWATER LEVEL (CM) | -5 | -7 | -10 | -20 | -12 | -20 | -11 | -8 | -35 | -20 | - | - | -25 | -20 | -18 | -5 | -30 | -13 | — | -17 | -25 | -18 | -16 | -35 | -18 | -12 | -25 |
| SOIL WATER CONTENT (X10 %) | 34 | 127 | 94 | 75 | 177 | 98 | 118 | 171 | 88 | 95 | 54 | 109 | 113 | 120 | 62 | 104 | 54 | 78 | 35 | 55 | 90 | 42 | 65 | 139 | 91 | 110 | 66 |
| SOIL PH | 4,9 | | 4,7 | 4,7 | | 4,4 | 4,9 | 4,3 | 4,3 | 4,7 | 5,3 | 4,4 | 4,7 | 4,8 | 4,7 | 5,0 | 5,3 | 5,2 | 4,3 | 5,4 | 4,6 | 5,3 | 5,5 | 4,1 | 3,8 | 4,1 | 4,6 |
| SOIL LOSS ON IGNITION (%) | 80 | 74 | 68 | 84 | 89 | 91 | 72 | 85 | 97 | 75 | 41 | 85 | 68 | 67 | 59 | 65 | 55 | 44 | 69 | 43 | 76 | | 51 | 90 | 73 | 82 | 65 |
| INFLUENCE OF SALTSPRAY | | | | | | | | | | | | | | | | | | | | | | | | | | | |
| INFLUENCE OF TRAMPLING | | | | | | | | | | | | | | | | | | | | | | | | | | | |
| INFLUENCE OF MANURING | | | | | | | | | | | | | | | | | | | | | | | | | | | |
| HEIGHT OF HERB LAYER (CM) | 10 | 12 | 20 | 10 | 15 | 20 | 25 | 30 | 20 | 10 | 20 | 30 | 10 | 15 | 10 | 25 | 20 | 25 | 15 | 25 | 30 | 25 | 15 | 15 | 25 | 25 | 20 |
| COVER OF HERB LAYER (%) | 90 | 70 | 95 | 99 | 90 | 30 | 95 | 85 | 40 | 97 | 65 | 60 | 90 | 95 | 60 | 95 | 80 | 80 | 99 | 50 | 25 | 50 | 95 | 35 | 40 | 5 | 60 |
| HEIGHT OF BRYOPHYTE LAYER (CM) | 3 | 5 | 1 | 2 | 2 | 4 | 3 | 3 | 4 | 4 | 6 | 3 | 4 | 7 | 3 | 4 | 5 | 5 | 15 | 10 | 8 | 7 | 5 | 5 | 8 | 7 | 6 |
| COVER OF BRYOPHYTE LAYER (%) | 100 | 100 | 100 | 100 | 99 | 99 | 95 | 97 | 100 | 65 | 99 | 99 | 85 | 99 | 90 | 95 | 97 | 80 | 99 | 70 | 99 | 80 | 75 | 100 | 95 | 100 | 99 |
| COVER OF ALGAE (%) | | | | | | | | | | | | | | | | | | | | | | | | | | | |
| COVER OF BARE SOIL (%) | | | | | | | | | | | | | 5 | | | | | | | | | | | | | | |
| COVER OF ROCKS (%) | | | | | | | | | | | | | | | | | | | | | | | | | | | |
| QUADRAT SIZE (M2) | 4 | 1 | 4 | 1 | 4 | 4 | 9 | 4 | 16 | 16 | 9 | 4 | 4 | 9 | 4 | 9 | 16 | 16 | 9 | 16 | 4 | 16 | 16 | 1 | 4 | 4 | 9 |
| TOTAL NUMBER OF SPECIES | 10 | 12 | 13 | 16 | 16 | 17 | 20 | 13 | 13 | 13 | 14 | 17 | 19 | 15 | 16 | 16 | 17 | 21 | 11 | 14 | 15 | 15 | 17 | 17 | 18 | 19 | 21 |
| DIAGNOSTIC TAXA OF ASSOCIATIONS AND SUBASSOCIATIONS: | | | | | | | | | | | | | | | | | | | | | | | | | | | |
| JUNCUS SCHEUCHZERIOIDES | | | + | | | | 1 | | | | R | | 1 | | | | + | 1 | | + | | + | | | | | |
| RANUNCULUS BITERNATUS | | | R | R | | | | | + | + | 1 | R | | | | | | | | | R | | | | | | |
| DREPANOCLADUS UNCINATUS | | | | | | + | | + | | | + | | | | | | + | | | | | + | | | | + | |
| DISTICHOPHYLLUM FASCICULATUM | | 1 | | | | | | | | | | + | | + | | | | 1 | | | | | | + | R | | + |
| BLEPHARIDOPHYLLUM DENSIFOLIUM | 1 | 1 | + | 3 | 1 | + | 2A | 1 | 2A | + | | 1 | + | 1 | | + | | 1 | | | | | | + | + | + | |
| CLASMATOCOLEA HUMILIS | | 1 | | 2A | + | | 1 | + | + | | | + | + | | | + | + | 1 | | + | | | | 1 | 1 | + | 1 |
| LEPTOSCYPHUS EXPANSUS | | | | + | + | | + | | | | + | | | | | | | | | | + | | | | + | | |
| LEPIDOZIA LAEVIFOLIA | 1 | 2M | | 1 | + | 1 | 2M | 1 | 1 | 1 | 1 | 2M | 1 | 1 | 1 | 2M | 1 | 1 | 1 | + | 1 | + | | 1 | 2M | 1 | 1 |
| CAMPYLOPUS ARBORICOLA | 1 | | | | 1 | 1 | | | | | | | | | R | | | | | | | | | 1 | + | + | |
| PELTIGERA POLYDACTYLA | | | | 1 | + | | | R | + | R | + | | | | | | | + | R | | | | + | | | | |
| ANDREWSIANTHUS LANCISTIPUS | | | | + | 1 | + | | R | | | | + | | | | + | | | | | | | | | 1 | + | + |
| UNCINIA COMPACTA | 1 | 1 | + | + | 1 | | 1 | + | 1 | 2A | | | 1 | 4 | 3 | 1 | 1 | + | | 1 | 1 | 1 | | | + | 1 | 1 |
| FOSSOMBRONIA AUSTRALIS | | | | R | | R | | | | | | | | | | | | | | | | | | | + | | |
| JENSENIA PISICOLOR | | 1 | 2B | + | 1 | 1 | + | | | + | + | | 1 | + | R | + | | | | + | + | | | | + | | + |
| LOPHOZIA CYLINDRIFORMIS | | | | | | | | | | | | | | | | | | + | | | + | | | 1 | + | | + |
| PLAGIOCHILA HETERODONTA | | 1 | | 1 | + | | 1 | | | + | | R | 1 | 2A | 2A | 1 | 1 | + | 2M | 1 | + | 1 | 2B | | | 1 | 1 |
| PTYCHOMNION RINGIANUM | | | | | | R | | | | | | | | | + | | | | | | | | | + | | | |
| JAMESONIELLA GRANDIFLORA | | | | | | | | | | | | | | | | | | + | | + | | | | | | | |
| JAMESONIELLA COLORATA | 5 | 5 | 5 | 4 | 5 | 5 | 5 | 5 | 5 | 4 | 5 | 5 | 5 | 4 | 5 | 5 | 4 | 4 | + | | 1 | 1 | + | + | 2A | 1 | + |
| RACOMITRIUM LANUGINOSUM | | + | + | | | 2A | 1 | + | + | + | + | + | + | 4 | 2A | 3 | 4 | 3 | 5 | 4 | 5 | 4 | 4 | 5 | 5 | 5 | 5 |
| LYCOPODIUM MAGELLANICUM | | | | | | | + | | | | | | + | + | + | + | + | | | | | + | | | | | + |
| DISTICHOPHYLLUM IMBRICATUM | | | | | | + | | | | | | | | | | | | | | | | | | | | R | + |
| RICCARDIA CF. PINGUIS | + | | + | | + | | 1 | | | | + | | R | | R | 1 | + | + | | + | | | | | | | |
| BREUTELIA INTEGRIFOLIA | | | + | | | | + | | | | | | 1 | | | | | + | | | | | | | | | |
| BRYUM LAEVIGATUM | | | + | | | | | | | | | | 1 | | | | | | | | | | | | | | |
| RICCARDIA CF. MULTIFIDA | | | | | | | | | | | | | R | | + | | | | | | | | | | | | |
| BLECHNUM PENNA-MARINA | + | | | | | 1 | 1 | 4 | 2B | 5 | 4 | 4 | 2B | 3 | 3 | + | 2A | 4 | 5 | 3 | 2A | 2A | 5 | 2A | 3 | 1 | 4 |
| ACAENA MAGELLANICA | | | | | | | | | | | + | | R | | | | | | + | + | 1 | | 2A | | | | |
| SCHISTOCHILA CARNOSA | | | | | + | | | | | | | | | | | | | | | | | | | | | | |
| PLAGIOTHECIUM PLATYPHYLLUM | | | | | | | + | | | | | + | | | | | | | | | + | | + | | + | | |
| METZGERIA DECIPIENS | | | | | | | | | | | | + | | | | | | | | | | + | 1 | | 1 | | 1 |
| CATAGONIUM POLITUM | | | | | | | | | | | | | | + | | + | + | + | 1 | + | | + | | | | | + |
| HYGROAMBLYSTEGIUM FILUM | | | | | | | | | | | | | | | | | | | | | | | | | | | |
| LOPHOCOLEA RANDII | | | | | | | | | | | | + | | | | | | | | | | | | | | | |
| AGROSTIS MAGELLANICA | 5 | 4 | 5 | 5 | 5 | 2A | 5 | 4 | 2B | + | 2A | 2A | 5 | 4 | 2B | 5 | 5 | 4 | + | 2B | 2A | 2B | + | 1 | 1 | 1 | 2A |
| COMPANIONS: | | | | | | | | | | | | | | | | | | | | | | | | | | | |
| CAMPYLOPUS SUBNITENS | | | | | | | | | | | | | | R | | | | | | | | | | | | | + |
| DICRANOLOMA BILLARDIERI | | | | | | | | | | | | | | | | | | | | | | | | | | | |
| CAMPYLOPUS CLAVATUS | | | + | | | | + | | | | | + | | | | | + | | | | | | | | | | |
| MONTIA FONTANA | | + | | + | | + | | R | + | + | | R | | R | | | | | | | | | | + | | + | + |
| AZORELLA SELAGO | + | R | + | | + | 1 | 1 | R | + | 1 | 4 | + | 1 | + | 2A | 2A | 1 | 1 | + | 2B | R | 2B | 2B | | | + | + |
| MARCHANTIA BERTEROANA | | | | | | | | | | | | | | | | | | | | | | | | | | | |
| ACROBOLBUS OCHROPHYLLUS | | | | + | | + | 1 | | + | | | | | | | | | | | | + | | + | + | | + | |
| CLASMATOCOLEA VERMICULARIS | | | | | | | | | | | | | 1 | | | | | | | | | | | + | | | |
| SYMPHYOGYNA MARIONENSIS | + | | | | + | | + | | | | | | | | | + | + | + | | | | | + | | | | |
| CLADONIA COLL. | | | | | | | | | | | | | | | + | | + | + | | | | + | | + | | | + |
| TEMNOMA QUADRIPARTITA | | | | | | | | | | | | | | | R | | | | | | | | | | | | |
| POA COOKII | | | | | | R | | | | | | | | | | | | | | | | | + | 2B | | R | |

*Table 26b* (continued).

**Degradation stages of Lycopodio - Jamesonielletum of other mire communities**

| | | | | | | | | | | | | | |
|---|---|---|---|---|---|---|---|---|---|---|---|---|---|
| RELEVE NUMBER | 259 | 254 | 64 | 188 | 131 | 99 | 73 | 340 | 97 | 85 | 100 | 102 | 117 |
| LOCATION | | | | | | | | | | | | | |
| SOUTHERN LATITUDE (46°) MIN. | 53 | 53 | 52 | 51 | 57 | 52 | 52 | 38 | 52 | 51 | 52 | 52 | 53 |
| SEC. | 05 | 05 | 45 | 00 | 40 | 35 | 45 | 40 | 35 | 20 | 35 | 35 | 00 |
| EASTERN LONGITUDE (37°) MIN. | 51 | 51 | 52 | 47 | 51 | 51 | 52 | 00 | 51 | 52 | 51 | 51 | 50 |
| SEC. | 55 | 55 | 45 | 30 | 15 | 20 | 45 | 25 | 20 | 35 | 20 | 20 | 50 |
| ELEVATION (M A.S.L.) | 20 | 20 | 10 | 45 | 75 | 25 | 9 | 15 | 25 | 20 | 25 | 25 | 55 |
| INCLINATION (°) | 10 | | 5 | | 5 | | | | | | | | |
| ASPECT | SW | | NNE | | SSE | | | | | | | | |
| SOIL DEPTH (CM) | | | 7 | | | | 22 | | | | | | |
| DEPTH OF GROUNDWATER LEVEL (CM) | -35 | -22 | --- | -35 | | + 2 | -20 | -11 | - 1 | + 3 | - 2 | - 8 | -18 |
| SOIL WATER CONTENT (x10 %) | 111 | 101 | 92 | 86 | 101 | 327 | 140 | 140 | 234 | 277 | 160 | 120 | 120 |
| SOIL PH | 4,3 | 4,1 | 4,8 | | | | | | 4,5 | | 4,9 | 4,4 | 4,4 |
| SOIL LOSS ON IGNITION (%) | 99 | 97 | | 90 | 85 | 94 | | 94 | 90 | 87 | 70 | | |
| INFLUENCE OF SALTSPRAY | | | 2 | | | | 1 | | | | | | |
| INFLUENCE OF TRAMPLING | 2 | 2 | | 1 | 2 | | | | | | | | |
| INFLUENCE OF MANURING | 2 | 2 | | 1 | 1 | | | | | | | | |
| HEIGHT OF HERB LAYER (CM) | 10 | 5 | 5 | 25 | 15 | 10 | 9 | 25 | 10 | 10 | 7 | 20 | 20 |
| COVER OF HERB LAYER (%) | 100 | 100 | 100 | 100 | 97 | 90 | 100 | 85 | 55 | 95 | 100 | 100 | 100 |
| HEIGHT OF BRYOPHYTE LAYER (CM) | | | | 1 | 1 | | 1 | 1 | 1 | 1 | 2 | 1 | 3 |
| COVER OF BRYOPHYTE LAYER (%) | | | | <1 | 3 | | 1 | <1 | 1 | 5 | 20 | 1 | <1 |
| COVER OF ALGAE (%) | | | | | | 100 | | 80 | 100 | 80 | 100 | | |
| COVER OF BARE SOIL (%) | | | | | 3 | | | | | | | | |
| COVER OF ROCKS (%) | | | | | | | | | | | | | |
| QUADRAT SIZE (M2) | 4 | 4 | 1 | 4 | 4 | 1 | 1 | 9 | 1 | 1 | 1 | 1 | 1 |
| TOTAL NUMBER OF SPECIES | 4 | 4 | 3 | 5 | 7 | 1 | 2 | 3 | 5 | 7 | 8 | 10 | 10 |
| DIAGNOSTIC TAXA OF ASSOCIATIONS AND SUBASSOCIATIONS: | | | | | | | | | | | | | |
| JUNCUS SCHEUCHZERIOIDES | | | | | 1 | | | | | | | | + |
| RANUNCULUS BITERNATUS | + | | | 1 | | | | | | | | | |
| DREPANOCLADUS UNCINATUS | | | | 1 | | | + | | | 1 | | + | + |
| DISTICHOPHYLLUM FASCICULATUM | | | | | | | | | | | | | |
| BLEPHARIDOPHYLLUM DENSIFOLIUM | | | | | | | | | | + | | | + |
| CLASMATOCOLEA HUMILIS | | | | | | | | | 1 | 1 | 1 | | + |
| LEPTOSCYPHUS EXPANSUS | | | | | | | | | | | | | |
| LEPIDOZIA LAEVIFOLIA | | | | | | | | | | | | | |
| CAMPYLOPUS ARBORICOLA | | | | | | | | | | | | | |
| PELTIGERA POLYDACTYLA | | | | | | | | | | | | | |
| ANDREWSIANTHUS LANCISTIPUS | | | | | | | | | | | | | |
| UNCINIA COMPACTA | | | | | | | | + | + | + | + | + | 2A |
| FOSSOMBRONIA AUSTRALIS | | | | | + | | | | | | | | |
| JENSENIA PISICOLOR | | | | | | | | | | + | + | | R |
| LOPHOZIA CYLINDRIFORMIS | | | | | | | | | | | | | |
| PLAGIOCHILA HETERODONTA | | | | | | | | | | | | + | |
| PTYCHOMNION RINGIANUM | | | | | | | | | | | R | | |
| JAMESONIELLA GRANDIFLORA | | | | | | | | | | | | | + |
| JAMESONIELLA COLORATA | | | | | | | | + | 1 | | + | 1 | |
| RACOMITRIUM LANUGINOSUM | | | | | | | | | | | | + | + |
| LYCOPODIUM MAGELLANICUM | | | | | | | | | | | | | |
| DISTICHOPHYLLUM IMBRICATUM | | | | | | | | | | | | | |
| RICCARDIA CF. PINGUIS | | | | | | | | | | + | | 1 | |
| BREUTELIA INTEGRIFOLIA | | | | | | | | | | | | | |
| BRYUM LAEVIGATUM | | | | | | | | | | | | | |
| RICCARDIA CF. MULTIFIDA | | | | | | | | | + | | R | + | |
| BLECHNUM PENNA-MARINA | | | | | | | | | | | | | |
| ACAENA MAGELLANICA | | | | 2B | | | | | | | | | |
| SCHISTOCHILA CARNOSA | | | | | | | | | | | | | |
| PLAGIOTHECIUM PLATYPHYLLUM | | | | | | | | | | | | | |
| METZGERIA DECIPIENS | | | | | | | | | | | | | |
| CATAGONIUM POLITUM | | | | | | | | | | | | | |
| HYGROAMBLYSTEGIUM FILUM | | | | | | | | | | | | | |
| LOPHOCOLEA RANDII | | | | | | | | | | | | | |
| AGROSTIS MAGELLANICA | 5 | 5 | 5 | 5 | 5 | 5 | 5 | 5 | 4 | 5 | 5 | 5 | 5 |
| COMPANIONS: | | | | | | | | | | | | | |
| CAMPYLOPUS SUBNITENS | | | | | | | | | | | | | |
| DICRANOLOMA BILLARDIERI | | | | | | | | | | | | | |
| CAMPYLOPUS CLAVATUS | | | | | | | | | | | 2B | + | |
| MONTIA FONTANA | 2B | + | 2A | 1 | 1 | | | | | | | | |
| AZORELLA SELAGO | | | | | | | | | | | | | + |
| MARCHANTIA BERTEROANA | | | | | 1 | | | | | | | | |
| ACROBOLBUS OCHROPHYLLUS | | | | | | | | | | | | | |
| CLASMATOCOLEA VERMICULARIS | | | | | | | | | | | | | |
| SYMPHYOGYNA MARIONENSIS | | | | | | | | | | | | | |
| CLADONIA COLL. | | | | | | | | | | | | | |
| TEMNOMA QUADRIPARTITA | | | | | | | | | | | | | |
| POA COOKII | | | | | | | | | | | | | |

**JAMESONIELLETUM GRANDIFLORAE**

| | | | | | | | | | | | |
|---|---|---|---|---|---|---|---|---|---|---|---|
| RELEVE NUMBER | 213 | 218 | 503 | 107 | 84 | 204 | 215 | 212 | 185 | 214 | 108 |
| LOCATION | | | | | | | | | | | |
| SOUTHERN LATITUDE (46°) MIN. | 52 | 52 | 52 | 52 | 52 | 51 | 52 | 52 | 51 | 52 | 52 |
| SEC. | 10 | 10 | 45 | 35 | 35 | 00 | 10 | 10 | 00 | 10 | 35 |
| EASTERN LONGITUDE (37°) MIN. | 51 | 51 | 37 | 51 | 51 | 48 | 51 | 51 | 47 | 51 | 51 |
| SEC. | 15 | 15 | 55 | 20 | 20 | 00 | 15 | 15 | 30 | 15 | 20 |
| ELEVATION (M A.S.L.) | 35 | 35 | 50 | 25 | 20 | 130 | 35 | 35 | 45 | 35 | 25 |
| INCLINATION (°) | | | | | 5 | | | | | | |
| ASPECT | | | | | S | | | | | | |
| SOIL DEPTH (CM) | | | | | | | | | | | 40 |
| DEPTH OF GROUNDWATER LEVEL (CM) | -18 | -15 | -27 | -20 | -40 | -20 | - 8 | - 7 | -35 | -18 | --- |
| SOIL WATER CONTENT (x10 %) | 123 | 100 | | 64 | 140 | 39 | 71 | 94 | 64 | 33 | 51 |
| SOIL PH | 4,3 | 4,6 | | 4,6 | 4,6 | 5,0 | 4,7 | 4,6 | 4,6 | 53 | 4,8 |
| SOIL LOSS ON IGNITION (%) | 77 | 50 | | 62 | 87 | 70 | 63 | 58 | 33 | 34 | 33 |
| INFLUENCE OF SALTSPRAY | | | | | | | | | | | |
| INFLUENCE OF TRAMPLING | | | | | | | | | 1 | | |
| INFLUENCE OF MANURING | | | | | | | | | 1 | | |
| HEIGHT OF HERB LAYER (CM) | 25 | 20 | 25 | 20 | 20 | 15 | 4 | 15 | 5 | 15 | 20 |
| COVER OF HERB LAYER (%) | 70 | 75 | 35 | 60 | 80 | 85 | 90 | 70 | 70 | 85 | 95 |
| HEIGHT OF BRYOPHYTE LAYER (CM) | 4 | 3 | 5 | 4 | 3 | 4 | 3 | 3 | 1 | 3 | 2 |
| COVER OF BRYOPHYTE LAYER (%) | 100 | 85 | 70 | 100 | 100 | 95 | 100 | 100 | 90 | 93 | 100 |
| COVER OF ALGAE (%) | | | | | | | | | | | |
| COVER OF BARE SOIL (%) | | | | | | | | | | | |
| COVER OF ROCKS (%) | | | | | | | | | | | |
| QUADRAT SIZE (M2) | 4 | 4 | 10 | 1 | 1 | 4 | 4 | 4 | 4 | 4 | 1 |
| TOTAL NUMBER OF SPECIES | 11 | 12 | 12 | 13 | 13 | 17 | 6 | 7 | 8 | 9 | 9 |
| DIAGNOSTIC TAXA OF ASSOCIATIONS AND SUBASSOCIATIONS: | | | | | | | | | | | |
| JUNCUS SCHEUCHZERIOIDES | | | | + | + | + | | | | | |
| RANUNCULUS BITERNATUS | 2A | 1 | 1 | R | 2A | + | 5 | 4 | 2B | 2B | |
| DREPANOCLADUS UNCINATUS | + | 1 | | + | + | 1 | | | | | 1 |
| DISTICHOPHYLLUM FASCICULATUM | | | | | | | | | | | |
| BLEPHARIDOPHYLLUM DENSIFOLIUM | + | + | + | + | | | | | | | |
| CLASMATOCOLEA HUMILIS | | | 1 | + | 1 | + | | | | | |
| LEPTOSCYPHUS EXPANSUS | + | 1 | + | + | 1 | + | + | | | | |
| LEPIDOZIA LAEVIFOLIA | + | | | | 2A | + | | | | | |
| CAMPYLOPUS ARBORICOLA | | | | | | | | | | | |
| PELTIGERA POLYDACTYLA | + | R | + | + | | R | | R | R | | |
| ANDREWSIANTHUS LANCISTIPUS | | | | | | | | | | | |
| UNCINIA COMPACTA | | 3 | | | | | | | | | |
| FOSSOMBRONIA AUSTRALIS | | | | | | | | + | | | |
| JENSENIA PISICOLOR | | | | + | 4 | | | | | | + |
| LOPHOZIA CYLINDRIFORMIS | | + | | | 1 | + | | | | | |
| PLAGIOCHILA HETERODONTA | | | | | | | | | + | | |
| PTYCHOMNION RINGIANUM | | | | | | + | | | | | |
| JAMESONIELLA GRANDIFLORA | 5 | 5 | 4 | 5 | 4 | 5 | 5 | 5 | 5 | 5 | 5 |
| JAMESONIELLA COLORATA | | | | | | | | | | | |
| RACOMITRIUM LANUGINOSUM | | + | | | | | | | | | |
| LYCOPODIUM MAGELLANICUM | | | | | | | | | | | |
| DISTICHOPHYLLUM IMBRICATUM | | | | | | | | | | | |
| RICCARDIA CF. PINGUIS | | | | | | | | | | | |
| BREUTELIA INTEGRIFOLIA | | | | | | | | | | | + |
| BRYUM LAEVIGATUM | | | | | | | | | | | |
| RICCARDIA CF. MULTIFIDA | | | | | | | | | | | |
| BLECHNUM PENNA-MARINA | | | + | | | 2B | | | | | + |
| ACAENA MAGELLANICA | | | | 2A | | 1 | | | 4 | 5 | 5 |
| SCHISTOCHILA CARNOSA | | | | | | | | | | | |
| PLAGIOTHECIUM PLATYPHYLLUM | | | | | | | | | | | |
| METZGERIA DECIPIENS | | | | | | | | | | | |
| CATAGONIUM POLITUM | | | | | | | | | | | |
| HYGROAMBLYSTEGIUM FILUM | | | | | | | | | | | |
| LOPHOCOLEA RANDII | 1 | | | | | + | | | | | 1 |
| AGROSTIS MAGELLANICA | 4 | 3 | 2A | 3 | 4 | 5 | + | + | + | | 4 |
| COMPANIONS: | | | | | | | | | | | |
| CAMPYLOPUS SUBNITENS | | | | | | | | | | | |
| DICRANOLOMA BILLARDIERI | | | | | | | | | | | |
| CAMPYLOPUS CLAVATUS | | | | | | | | | | | |
| MONTIA FONTANA | + | R | + | + | + | | 2B | + | 1 | + | + |
| AZORELLA SELAGO | R | 2B | 3 | + | | 2A | | R | + | 2A | |
| MARCHANTIA BERTEROANA | | | + | | | | | | | 1 | |
| ACROBOLBUS OCHROPHYLLUS | | | | | + | | | | | | |
| CLASMATOCOLEA VERMICULARIS | | | | | | 1 | 1 | | | 2B | |
| SYMPHYOGYNA MARIONENSIS | | | | | | | | | | | |
| CLADONIA COLL. | | | | | | | | | | | |
| TEMNOMA QUADRIPARTITA | | | | | | | | | | | |
| POA COOKII | | | R | | | | | | | R | |

**BRYO LAEVIGATI - BREUTELIETUM INTEGRIFOLIAE**

| | uncinietosum compactae | | | | | | | | | | | | | | | RICCARDIA - variant | | | acaenetosum magellanicae | | | | |
|---|---|---|---|---|---|---|---|---|---|---|---|---|---|---|---|---|---|---|---|---|---|---|---|
| RELEVE NUMBER | 521 | 550 | 103 | 527 | 531 | 491 | 541 | 452 | 525 | 243 | 334 | 455 | 162 | 232 | 242 | 523 | 235 | 524 | 201 | 266 | 202 | 203 | 245 |
| LOCATION | | | | | | | | | | | | | | | | | | | | | | | |
| SOUTHERN LATITUDE (46°) MIN. | 52 | 52 | 52 | 52 | 52 | 50 | 52 | 57 | 52 | 52 | 57 | 50 | 57 | 52 | 52 | 52 | 52 | 52 | 51 | 51 | 51 | 51 | 51 |
| SEC. | 19 | 21 | 35 | 21 | 21 | 05 | 21 | 12 | 21 | 00 | 05 | 45 | 20 | 00 | 00 | 21 | 00 | 21 | 00 | 00 | 00 | 00 | 00 |
| EASTERN LONGITUDE (37°) MIN. | 49 | 49 | 51 | 49 | 49 | 48 | 49 | 50 | 49 | 50 | 51 | 48 | 51 | 50 | 50 | 49 | 50 | 49 | 48 | 48 | 48 | 48 | 48 |
| SEC. | 54 | 55 | 20 | 55 | 55 | 05 | 55 | 30 | 55 | 40 | 15 | 10 | 10 | 40 | 40 | 54 | 40 | 55 | 00 | 00 | 00 | 00 | 00 |
| ELEVATION (M A.S.L.) | 130 | 130 | 25 | 130 | 130 | 110 | 130 | 60 | 130 | 50 | 75 | 100 | 70 | 110 | 50 | 130 | 115 | 130 | 130 | 125 | 130 | 130 | 125 |
| INCLINATION (°) | | | | | | 3 | 1 | 4 | | | 2 | 5 | 5 | | | | | | | 10 | | | 10 |
| ASPECT | | | | | | N | S | E | | | SE | N | E | | | | | | | SE | | | NE |
| SOIL DEPTH (CM) | | | | | 18 | 15 | | | | | | | | | | | | 40 | | | | | 40 |
| DEPTH OF GROUNDWATER LEVEL (CM) | - 2 | - 0 | +30 | - 2 | - 0 | - 4 | - 2 | -12 | - 1 | - 4 | -23 | - 4 | - 7 | - 5 | - 3 | - 5 | - 2 | - 0 | - 2 | -11 | - 8 | - 6 | -15 |
| SOIL WATER CONTENT (x10 %) | 208 | 148 | | 214 | 34 | | 70 | 175 | 223 | 159 | 117 | 167 | 115 | 278 | 240 | 29 | 325 | 115 | 66 | 144 | 99 | 94 | 29 |
| SOIL PH | 5,2 | 4,7 | 4,8 | 5,3 | 5,4 | | 5,7 | 5,3 | 5,2 | 4,8 | 5,0 | 64 | | 4,7 | 4,7 | 5,7 | 4,5 | 5,5 | 4,4 | 4,7 | 5,1 | | |
| SOIL LOSS ON IGNITION (%) | 79 | 53 | 71 | 76 | 52 | | | 79 | 76 | 88 | 67 | 67 | 91 | 66 | 87 | 30 | | 57 | 54 | 61 | 46 | | 60 |
| INFLUENCE OF SALTSPRAY | | | | | | | | | | | | | | | | | | | | | | | |
| INFLUENCE OF TRAMPLING | | | | | | | | | | | | | | 1 | | | | | | | | | |
| INFLUENCE OF MANURING | | | | | | | | | | | | | | | | | | | | | | | |
| HEIGHT OF HERB LAYER (CM) | 10 | 10 | 10 | 25 | 20 | 25 | 15 | 15 | 15 | 15 | 15 | 15 | 10 | 15 | 15 | 10 | 15 | 15 | 12 | 15 | 20 | 15 | 25 |
| COVER OF HERB LAYER (%) | 7 | 20 | 60 | 10 | 85 | 40 | 5 | 95 | 30 | 60 | 85 | 70 | 95 | 50 | 60 | 40 | 60 | 85 | 70 | 20 | 70 | 50 | 30 |
| HEIGHT OF BRYOPHYTE LAYER (CM) | 3 | 3 | 4 | 3 | 5 | 4 | 4 | 5 | 3 | 3 | 5 | 5 | 4 | 4 | 3 | 1 | 2 | 3 | 5 | 4 | 4 | 4 | 4 |
| COVER OF BRYOPHYTE LAYER (%) | 99 | 99 | 90 | 100 | 99 | 100 | 100 | 99 | 100 | 100 | 100 | 100 | 100 | 100 | 100 | 99 | 100 | 100 | 100 | 100 | 100 | 100 | 100 |
| COVER OF ALGAE (%) | | | | | | | | | | | | | | | | | | | | | | | |
| COVER OF BARE SOIL (%) | | 1 | | | | | | | | | | | | | | | | | | | | | |
| COVER OF ROCKS (%) | | | | | | | | | | | | | | | | | | | | | | | |
| QUADRAT SIZE (M2) | 4 | 8 | 1 | 9 | 9 | 6 | 6 | 9 | 4 | 4 | 6 | 9 | 4 | 4 | 4 | 4 | 4 | 9 | 2 | 4 | 2 | 2 | 4 |
| TOTAL NUMBER OF SPECIES | 5 | 5 | 8 | 9 | 9 | 9 | 11 | 12 | 12 | 13 | 13 | 15 | 15 | 17 | 18 | 13 | 13 | 15 | 9 | 9 | 10 | 11 | 11 |
| DIAGNOSTIC TAXA OF ASSOCIATIONS AND SUBASSOCIATIONS: | | | | | | | | | | | | | | | | | | | | | | | |
| JUNCUS SCHEUCHZERIOIDES | 2A | 2M | 1 | 2M | 1 | 1 | 1 | | 1 | 1 | | | + | + | 1 | 1 | 1 | + | | | | + | + |
| RANUNCULUS BITERNATUS | + | 1 | 1 | 1 | | + | + | | | | | | | + | | | | | + | + | | + | |
| DREPANOCLADUS UNCINATUS | | | 2A | | | | | + | | 1 | | | | 1 | 1 | | + | + | 1 | | 1 | 1 | + |
| DISTICHOPHYLLUM FASCICULATUM | | | | | | | | + | | | | + | | | | | | + | | | | | |
| BLEPHARIDOPHYLLUM DENSIFOLIUM | | | | | | | | | | | | | | | | + | + | 1 | | | | | |
| CLASMATOCOLEA HUMILIS | | | | | | | | | + | + | 1 | + | + | 1 | 1 | + | 1 | + | | | | | + |
| LEPTOSCYPHUS EXPANSUS | | | | | | | | + | + | | | + | + | | | | + | | | | | | + |
| LEPIDOZIA LAEVIFOLIA | | | | | | | | | | | | | | | 1 | + | | | | | | | |
| CAMPYLOPUS ARBORICOLA | | | | | | | | | | | | | | | | | | | | | | | |
| PELTIGERA POLYDACTYLA | | | | | | | | | | | | | | R | | | | | | | | | |
| ANDREWSIANTHUS LANCISTIPUS | | | | | | | | | | | | | | | R | | | | | | | | |
| UNCINIA COMPACTA | | | | 1 | + | + | + | + | + | + | 2B | 1 | 1 | 1 | 1 | 1 | R | + | | | | 2A | |
| FOSSOMBRONIA AUSTRALIS | | | | | | | | | | | | | | | | | | | | | | | |
| JENSENIA PISICOLOR | | | R | | | | 1 | + | + | | 1 | | 1 | 1 | + | + | | + | | | | | |
| LOPHOZIA CYLINDRIFORMIS | | | R | | | | | | | | | + | | + | | | | | | | | | |
| PLAGIOCHILA HETERODONTA | | | | | | | | | | | | | | | + | | | | | | | | |
| PTYCHOMNION RINGIANUM | | | | | | | | | | | | | | | | | + | | | | | | |
| JAMESONIELLA GRANDIFLORA | | | R | | | | 1 | | | | | | 2A | | | | | + | | | | | |
| JAMESONIELLA COLORATA | | | | | | | | | | | + | R | | | 1 | + | + | + | | | | | |
| RACOMITRIUM LANUGINOSUM | | | | | | | | | | | + | | | + | | | | R | | | | | |
| LYCOPODIUM MAGELLANICUM | | | | | | | | | | | | | | | | | | | | | | | |
| DISTICHOPHYLLUM IMBRICATUM | | | | | | | | | | | | | | | | | | | | | | | |
| RICCARDIA CF. PINGUIS | | | | + | 1 | | + | 1 | + | + | + | 1 | 1 | | + | + | | + | | | | | |
| BREUTELIA INTEGRIFOLIA | 5 | 5 | 5 | 5 | 5 | 5 | 5 | 5 | 5 | 5 | 5 | 5 | 4 | 5 | 5 | 1 | | 1 | 5 | 5 | 5 | 5 | 5 |
| BRYUM LAEVIGATUM | + | 1 | | 1 | + | | 2A | 1 | 2A | 1 | + | 1 | 2A | 2A | 1 | | | + | 1 | | 1 | 1 | |
| RICCARDIA CF. MULTIFIDA | | | | | | + | | 1 | 1 | | + | + | + | + | R | 5 | 5 | 5 | + | + | + | + | + |
| BLECHNUM PENNA-MARINA | | | | | | + | | | | | | + | | | | | | | | + | | | 1 |
| ACAENA MAGELLANICA | | | | | | + | 1 | | | | | R | | | | | | | 4 | 2A | 2A | 3 | 1 |
| SCHISTOCHILA CARNOSA | | | | | | | | | | | | | | 1 | | | | | | | | | |
| PLAGIOTHECIUM PLATYPHYLLUM | | | | | | | | | | | | | | | | | | | | | | | |
| METZGERIA DECIPIENS | | | | | | | | | | | | | | | | | | | | | | | |
| CATAGONIUM POLITUM | | | | | | | | | | | | | | | | | | | | | | | |
| HYGROAMBLYSTEGIUM FILUM | | | | | | | | | | | | | | | | | | | | 1 | + | + | + |
| LOPHOCOLEA RANDII | | | | | + | | | | | + | | | | | + | | | | | + | | | |
| AGROSTIS MAGELLANICA | + | 2A | 4 | 2A | 5 | 3 | 1 | 5 | 3 | 4 | 4 | 4 | 5 | 3 | 4 | 3 | 4 | 5 | 2A | 2A | 4 | 2A | 2B |
| COMPANIONS: | | | | | | | | | | | | | | | | | | | | | | | |
| CAMPYLOPUS SUBNITENS | | | | | | | | | + | | | | | | | | | | | | | | |
| DICRANOLOMA BILLARDIERI | | | | | | | | | | | | | | | | | | | | | | | |
| CAMPYLOPUS CLAVATUS | | | | | | | | | | | | | | | | + | 1 | | | | | | |
| MONTIA FONTANA | | | | + | + | | + | | | + | | | | 2A | 1 | | | | 1 | | + | + | |
| AZORELLA SELAGO | | | | R | + | + | | + | R | + | + | + | + | R | 1 | + | R | | | R | R | | |
| MARCHANTIA BERTEROANA | | | | | | | | | | | | | | | | | | | | | | | |
| ACROBOLBUS OCHROPHYLLUS | | | | | | | | | | | | | | | | | | | | | | | |
| CLASMATOCOLEA VERMICULARIS | | | | | | | | | | | | | | | | | | | | | | | |
| SYMPHYOGYNA MARIONENSIS | | | | | | | | | | + | + | R | + | | + | | + | | | | | | |
| CLADONIA COLL. | | | | | | | | | | | | | | | | | | | | | | | |
| TEMNOMA QUADRIPARTITA | | | | | | | | + | | | + | | | | | | | | | | | | |
| POA COOKII | | | | | | | | | | | | | | | | | | | | | | | |

*Fig. 48.* Blepharidophyllo densifolii – Clasmatocoleetum humilis. At the centre of the photograph the subassociation ranunculetosum occurs, changing towards the subassociation blechnetosum towards the left and bottom of the picture.

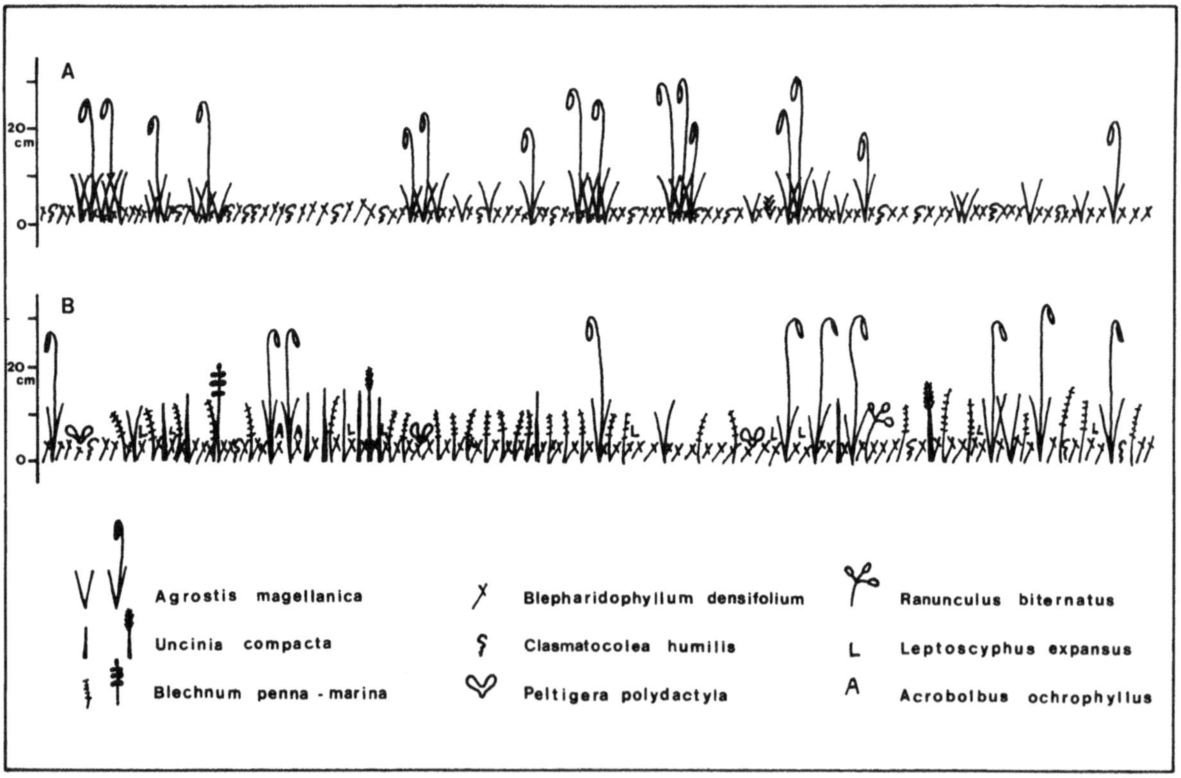


*Fig. 49.* Structure-diagram of the Blepharidophyllo densifolii – Clasmatocoleetum humilis. a = subassociation ranunculetosum biternati; b = subassociation blechnetosum penna-marinae.

## Association 4.5 Uncinio compactae-Ptychomnietum ringiani
(type relevé 475)

Character-species of this association is *Ptychomnion ringianum*, the dominant in the bryophyte layer (Fig. 50). *Jamesoniella colorata, Jamesoniella grandiflora* and *Uncinia compacta* are the differential species of this association against all other mire communities described sofar. Stands of Uncinio compactae-Ptychomnietum ringiani occur on wet peat soils with a groundwater level at about 10 cm below the surface. The soils contain a higher proportion of mineral matter than those under most communities of this complex. On sites where the groundwater level is somewhat deeper below the surface a variant dominated by *Uncinia compacta* occurs (relevés 101, 94 and 88; Table 26), while on still drier soils *Blechnum penna-marina* reaches dominance in this vegetation (e.g. relevés 493 and 495).

## Association 4.6 Lycopodio magellanici-Jamesonielletum coloratae
(type relevé 545)

On drier soils than those on which he communities of the mire complex described so far occur, the hepatic *Jamesoniella colorata* and the moss *Racomitrium lanuginosum* reach dominance. These two bryophytes are transgredient character-species of this association: each is also the character-species of one of the subassociations of this community. A further character-species is *Lycopodium magellanicum*. Differential species of this association against all other communities of this complex are *Catagonium politum, Metzgeria decipiens, Plagiothecium platyphyllum* and *Plagiochila heterodonta. Symphyogyna marionensis* and *Riccardia* cf. *pinguis* differentiate the Lycopodio magellanici-Jamesonielletum coloratae and the Bryo laevigati-Breutelietum integrifoliae together against the other mire communities.

*Fig. 50.* Uncinio compactae – Ptychomnietum ringiani.

*Fig. 51.* Part of the north-eastern coastal plain of Marion Island with Van Den Boogaard River. Most of the plain is covered by peat deposits. The predominant community here is the Lycopodio magellanici – Jamesonielletum coloratae (most of the light coloured areas).

The soils on which stands of this community (Fig. 51) are found are drier than those under the communities of this complex which have been described sofar, with a groundwater level usually 15–25 cm below the surface. The proportion of organic matter in the soil is very high (loss-on-ignition generally between 70 and 90%), but in exceptional cases a much lower loss-on-ignition was found (less than 45%).

Two subassociations are discerned.

Subassociation 4.6.1 ranunculetosum biternati
(type relevé 561)

The bryophyte stratum in this subassociation is dominated by *Jamesoniella colorata*, the character-species of this community. Differential species are *Agrostis magellanica* and *Ranunculus biternatus*. This subassociation occurs on soils with a higher water content than those under stands of the subassociation catagonietosum politi. On comparatively wet sites *Agrostis magellanica* dominates the herb layer of this community. On sites which are somewhat drier than average *Blechnum penna-marina* reaches dominance.

Subassociation 4.6.2 catagonietosum politi
(type relevé 545)

On even drier peat soils *Racomitrium lanuginosum* attains dominance (Fig. 52). It is the character-species of this subassociation. Differential species are *Catagonium politum, Lophozia cylindriformis, Metzgeria decipiens* and *Acaena magellanica* (Table 26). The great abundance of *Blechnum penna-marina* is indicative for the comparatively dry soil conditions of the sites where this community is found. Transitional stands between the two subassociations occur in intermediate habitats (e.g. relevés 532–447; Table 26).

In the series of mire vegetation from wet to dry, the Lycopodio magellanici-Jamesonielletum coloratae catagonietosum politi is the final community at the dry side of the gradient, preceeded by the subassociation ranunculetosum biternati (cf. Table 33). In a successional sequence this association is interpreted as the stage following on the Blepharidophyllo-Clasmatocoleetum blechnetosum or the Uncinio-Ptychomnietum. The subassociation catagonietosum is seen as a later developmental stage than the subassociation ranunculetosum biternati. When further accumulation of organic matter or changes in the water regime of the site result in still less wet conditions, the vegetation develops towards the Isopterygio pulchelli-Blechnetum pennamarinae uncinietosum compactae (Par. 4.5).

Retrogression of Lycopodio-Jamesonielletum vegetation occurs locally, caused by animal action or by changes in the water regime. The effect of fouraging birds on the vegetation of mire areas has already been described earlier in this chapter. In areas where the bryophyte layer of the vegetation has been destroyed, and where some enrichment of the soil with animal excreta has taken place, *Montia fontana* starts colonising the bare soil (e.g. relevés 259–131) and the vegetation develops towards the Montio fontanae-Callitrichetum antarcticae agrostietosum magellanicae (cf. Par. 4.2). In the absence of such enrichment of the soil *Juncus scheuchzerioides* spreads into the open places, and the vegetation develops into the *Campylopus clavatus* variant of the Community of *Juncus scheuchzerioides*. Retrogression also takes place when, as a result from changes in the drainage system of an area, the water level regularly rises to stand aboven the surface. A gelatinous mat of algae, up to 10 cm thick, usually develops (cf. Croome 1973), smothering the bryophyte layer, of which generally only fragments remain (relevés 99–100; Table 26).

Association 4.7 Jamesonielletum grandiflorae
(type relevé 213)

Character-species of this association, and dominant in the bryophyte layer, is *Jamesoniella grandiflora*. The *Jamesonielletum grandiflorae* forms no part of the series of mire communities usually found along wet-dry gradients in the mires of Marion and Prince Edward Islands. Its characteristic habitat is restricted nearly exclusively to the banks of streams, where it occurs on well-drained soils consisting of a shallow peat layer overlying sandy or gravelly deposits, and with a groundwater level of on the average c. 20 cm below the surface. In rare cases stands of the Jamesonielletum grandiflorae are associated with wind-blown volcanic ash deposits occurring on sites where drainage water from the surrounding areas collects. The mineral content of the top layer of the soils on which this association is

*Fig. 52. Racomitrium lanuginosum, Jamesoniella colorata, Blechnum penna-marina* and *Agrostis magellanica* in a stand of Lycopodio magellanici – Jamesonielletum coloratae.

found is higher than in the soils under the other communities of the mire complex (usually less than 70% loss-on-ignition). *Agrostis magellanica* generally dominates the herb layer. On comparatively wet sites, e.g. close to the edge of a stream, *Ranunculus biternatus* may attain dominance in this community (e.g. relevé 215 and 212), while on more mineral soils, e.g. on elevated embankments along streams, *Acaena magellanica* becomes dominant in the herb layer (relevés 185–108; Table 26).

Association 4.8 Bryo laevigati-Breutelietum integrifoliae
(type relevé 384)

In water tracks in the mires, the vegetation differs strongly from that of the mire expanse. These water tracks are sites with a comparatively strong lateral water transport through the peat soil, but the water flow is not as strong as in springs and flushes (Par. 4.3). An open water channel is usually absent or very small, but after heavy rains the water may rise above the surface. On average the depth of the groundwater level is not more than 10 cm below the surface, but this depth is, as is the rate of water flow, closely correlated with the amount of rainfall. Fluctuations in the water flow are larger here than in springs and flushes overgrown by stands of Bracythecietum subplicati or related communities, as the catchment area feeding these water tracks generally is much smaller.

The plant-nutrient status of the water tracks differs from that of the mire expanse: the somewhat higher pH of the soil may indicate more minerotrophic conditions in the water tracks, but may also be the result from water movement (Sparling 1966), and the water flow through the soil enlarges the availability of nutrients to the plants (cf. Chapman 1965; Ingram 1967). The soils contain more mineral matter than those of the surrounding mire expanse, probably as a result of deposition of mineral particles brought in from surrounding areas by the water.

Characteristic for these water tracks is a vegetation with a bryophyte stratum dominated by *Breutelia integrifolia, Bryum laevigatum* and *Riccardia* cf. *multifida,* the character-species of this association (Fig. 53).

Subassociation 4.8.1 uncinietosum compactae (type relevé 384)

Differential species of this subassociation are *Clasmatocolea humilis, Jensenia pisicolor, Riccardia* cf. *pinguis, Uncinia compacta, Jamesoniella colorata* and *Jamesoniella grandiflora.* This community forms the characteristic vegetation of water tracks in mires. Usually *Breutelia integrifolia* dominates the bryophyte stratum, but at some sites a variant dominated by *Riccardia* cf. *multifida* was found, possibly related to a greater deposition of mineral particles by the water. One small stand dominated by *Bryum laevigatum* was met with in a narrow, deep drainage channel on a black lava slope where the water flow after heavy rains was very fast.

Subassociation 4.8.2 acaenetosum magellanicae (type relevé 202)

This subassociation, which is differentiated by *Acaena magellanica* and *Hygroamblystegium filum,* shows a floristic resemblance to the Brachythecietum subplicati breutelietosum integrifoliae (Par. 4.3), and is found in similar habitats, viz. associated with small springs or drainage lines on grey lava slopes. The Brachythecietum subplicati, however, occurs on sites with a more continuous water supply, where variation in the amount of rainfall results in only small fluctuations of the water level.

On the western coastal plain of Marion Island and on the escarpment going up from this plain to the central highland a type of vegetation occurs which consists of a mosaic of large *Azorella selago* cushions and of mats of bryophytes. These mats are fragments of associations of the *Juncus scheuchzerioides-Blepharidophyllum densifolium* complex, notably the Blepharidophyllo densifolii-Clasmatocoleetum humilis (relevés 502, 504 and 498; Table 27) and the Lycopodio magellanici-Jamesonielletum coloratae (relevé 501). This vegetation occurs on loamy soils with a thin top layer of organic matter, and is found at elevations up to 500 m. No groundwater level occurs within 40 cm from the surface. By preventing a quick run-off of the rainwater, the *Azorella* cushions help to create a comparatively constant water regime at these sites. The abundance of such stands in the western part of Marion may be related to the probably more constantly wet moisture regime there.

Not all communities of the *Juncus scheuchzerioides-Blepharidophyllum densifolium* complex were found on both Marion and Prince Edward Island. No stands of Junco scheuchzerioidis-Drepanocladetum uncinati, Uncinio compactae-Ptychomnietum ringiani, Jamesonielletum grandiflorae and Bryo laevigati-Breutelietum integrifoliae were met with on the latter island. Insufficient data on the mire vegetation of Prince Edward Island are available to allow for a detailed comparison of the species compositions of all mire communities on Marion with those of their counterparts on Prince Edward. Conspicuous in the mires of the latter island is the high presence and abundance of *Campylopus subnitens* and *Schistochila carnosa,* and the absence of a number of species which play an important role in the vegetation of the mires on Marion Island, e.g. *Ptychomnion ringianum, Racomitrium lanuginosum* and *Breutelia integrifolia.*

*Fig. 53.* Mire area near Skua's Ridge, Marion Island. Most of the light-colored areas are covered with Lycopodio magellanici – Jamesonielletum coloratae. From the right-hand side, near the bottom of the picture, towards the left a pattern of communities can be seen along a water track through the mire (Bryo laevigati – Breutelietum integrifoliae).

*Synchorological remarks*

Most Subantarctic islands seem to possess mire and bog communities similar in structure and ecology to those of Marion and Prince Edward Islands. Insufficient data, however, are available for any detailed comparison.

Hébrard (1970b) reports the presence of mire communities on Kerguelen and Crozet. Important species here include *Anisothecium hookeri* (C. Müll.) Broth., *Makednothallus crassifrons* Steph., *Breutelia chrysura* (C. Müll.) Broth., *Breutelia pendula, Bryum laevigatum* and *Drepanocladus uncinatus*. Aubert de la Rue (1962), however, states that the main peat-forming plants in the bogs of Kerguelen are phanerogams, notably *Juncus scheuchzerioides, Acaena magellanica* and *Azorella selago*.

The species composition of the bog communities dominated by bryophytes on Macquarie differs considerably from the one of the bogs of Marion and Prince Edward, but on a generic level the resemblance is conspicuous: the most abundant bryophytes in the bogs on Macquarie are *Breutelia pendula, Bryum laevigatum, Drepanocladus aduncus* and *Ptychomnion aciculare* (Taylor 1955a). In bog ponds on this island, and on extremely wet peat soils, communities corresponding to the *Ranunculus biternatus* variant of the Community of *Juncus scheuchzerioides* occur,

*Table 27.* Mosaic of *Azorella selago* and bryophyte mats.

| Relevé number | 506 | 504 | 498 | 501 | |
|---|---|---|---|---|---|
| Location: | | | | | |
| Southern Latitude (46°) min. | 52 | 52 | 52 | 52 | |
| sec. | 35 | 16 | 00 | 20 | |
| Eastern Longitude (37°) min. | 38 | 38 | 39 | 38 | |
| sec. | 10 | 30 | 30 | 15 | |
| Elevation (m a.s.l.) | 60 | 70 | 450 | 30 | |
| Herb layer: height (cm) | 20 | 20 | 10 | 25 | |
| cover (%) | 70 | 65 | 65 | 65 | |
| Bryophyte layer: height (cm) | 3 | 2 | 10 | 3 | |
| cover (%) | 30 | 40 | 35 | 40 | |
| Quadrat size ($m^2$) | 9 | 9 | 25 | 9 | |
| Number of species | 12 | 12 | 12 | 18 | Soc. |
| **Azorella selago** | 4 | 4 | 4 | 4 | 3–4 |
| Blepharidophyllum densifolium | 3 | 1 | 3 | + | 3–4 |
| Clasmatocolea humilis | 1 | 3 | + | + | 2–4 |
| Peltigera polydactyla | + | + | + | + | 2 |
| Ranunculus biternatus | 1 | + | | + | 1 |
| Agrostis magellanica | 2A | 1 | | 2B | 2–3 |
| Marchantia berteroana | + | + | | + | 2 |
| Schistochila carnosa | + | + | + | | 2 |
| Lepidozia laevifolia | | 1 | + | + | 2 |
| Jamesoniella colorata | | | + | 3 | 2–4 |
| Montia fontana | + | + | | | 1–2 |
| Lophozia cylindriformis | + | | | + | 2 |
| Lophocolea randii | + | | | | 2 |
| Clasmatocolea vermicularis | 1 | | | | 3 |
| Colobanthus kerguelensis | | + | + | | 1–2 |
| Campylopus polytrichoides | | + | | | 2 |
| Distichophyllum fasciculatum | | | + | | 3 |
| Racomitrium lanuginosum | | | + | | 2 |
| Metzgeria decipiens | | | + | + | 2 |
| Ditrichum strictum | | | 1 | | 2 |
| Leptoscyphus expansus | | | | + | 2 |
| Campylopus arboricola | | | | + | 2 |
| Uncinia compacta | | | | + | 1 |
| Blechnum penna-marina | | | | 1 | 2–3 |
| Plagiothecium platyphyllum | | | | + | 2 |
| Ptychomnion ringianum | | | | R | 2 |
| Campylopus introflexus | | | | + | 2 |

e.g. the *Juncus scheuchzerioides* association, the *Juncus scheuchzerioides-Scirpus aucklandicus* association and the *Ranunculus biternatus* association (Taylor 1955a).

On South Georgia the rush *Rostkovia magellanica* is one of the most important mire species. Bryophytes play an important role in this vegetation, and on wet sites extensive pure stands of *Drepanocladus uncinatus* occur (Greene 1964), possibly resembling the Junco scheuchzerioidis-Drepanocladetum uncinati.

No mire communities have been reported from Heard and Macdonald Islands (Schenck 1905; Werth 1906; Law & Burstall 1953). These islands possess a very small flora, in which nearly all species which are important in the mires of Marion and Prince Edward Islands are lacking (cf. Tables 10, 11, 12).

In the Maritime Antarctic extensive mats of bryophytes occur, forming peat deposits of up to several metres in thickness (see e.g. Wace 1965; Gimingham & Smith 1970; Lindsay 1971; Smith 1972; Allison & Smith 1973), but these are not comparable in species composition or in ecology to the communities of the *Juncus scheuchzerioides-Blepharidophyllum densifolium* complex. On the South Shetland Islands, however, spongy carpets of *Drepanocladus uncinatus* occur in areas where melt water accumulates during summer (Lindsay 1971); these seem to resemble somewhat the Junco scheuchzerioidis-Drepanocladetum uncinati. Similar carpets on Signy Island, South Orkney Islands, seem to occur in appreciably drier habitats (cf. Smith 1972).

In the southern cool temperate regions the most important peatforming plants in ombrogenous mires are cushion plants (e.g. *Donatia, Gaimardia, Oreobolus* and *Phyllachne*) and *Sphagnum* spp. (Wace 1965, Godley 1978), and no communities resembling any of the associations of the mire complex of Marion and Prince Edward Islands have been reported from these regions. Communities with *Juncus scheuchzerioides* do occur on several of the southern temperate islands, e.g. on the Falkland Islands, where a '*Juncus scheuchzerioides*-association' grows on damp ground at the margins of streams and in dune slacks (Skottsberg 1909, 1913; Moore 1968), but these are floristically as well as ecologically different from the *Juncus scheuchzerioides* community described in this study.

There is some floristic affinity between the communities of this complex and a number of páramo mire communities described recently by Cleef (1981) under the name Calamagrostion ligulatae; this affinity is especially noteworthy in the Bryo-Breutelietum but cannot yet be evaluated properly with the available data.

### 4.5 Communities of well-drained lowland slopes: the Blechnum penna-marina complex

The aspect of the vegetation of well-drained, relatively dry lowland slopes is determined by the dominance of *Blechnum penna-marina* (Fig. 54). This fern forms dense swards in which bryophytes usually do not reach a high cover (Fig. 55). The complex consists of only one association, which is divided into three subassociations (Table 28).

Association 5.1 Isopterygio pulchelli-Blechnetum penna-marinae
(type relevé 539)

Stands of this community cover the largest part of the slopes in the lowlands of Marion and Prince Edward Islands. On these well-drained, relatively dry slopes no peat formation takes place. The soil consists of a 5–30 cm deep, dark brown layer of humus and plant litter, overlying deposits of loamy clay, which usually are more than 40 cm thick (Fig. 15; Table 7). No groundwater level is usually found within 40 cm from the surface, and on many sites it is not present at all.

Character-taxa of the Isopterygio pulchelli-Blechnetum penna-marinae are *Isopterygium pulchellum* var. *antarcticum*, cf. *Amblystegium* (cf. Zanten 1971), *Peltigera canina, Campylopus polytrichoides, Metzgeria decipiens, Plagiothecium*

*Fig. 54.* Lowland slopes with Isopterygio pulchelli – Blechnetum penna-marinae (dark slopes). The more level areas are covered by stands of Lycopodio magellanicae – Jamesonielletum coloratae (near Junior's Kop, Marion).

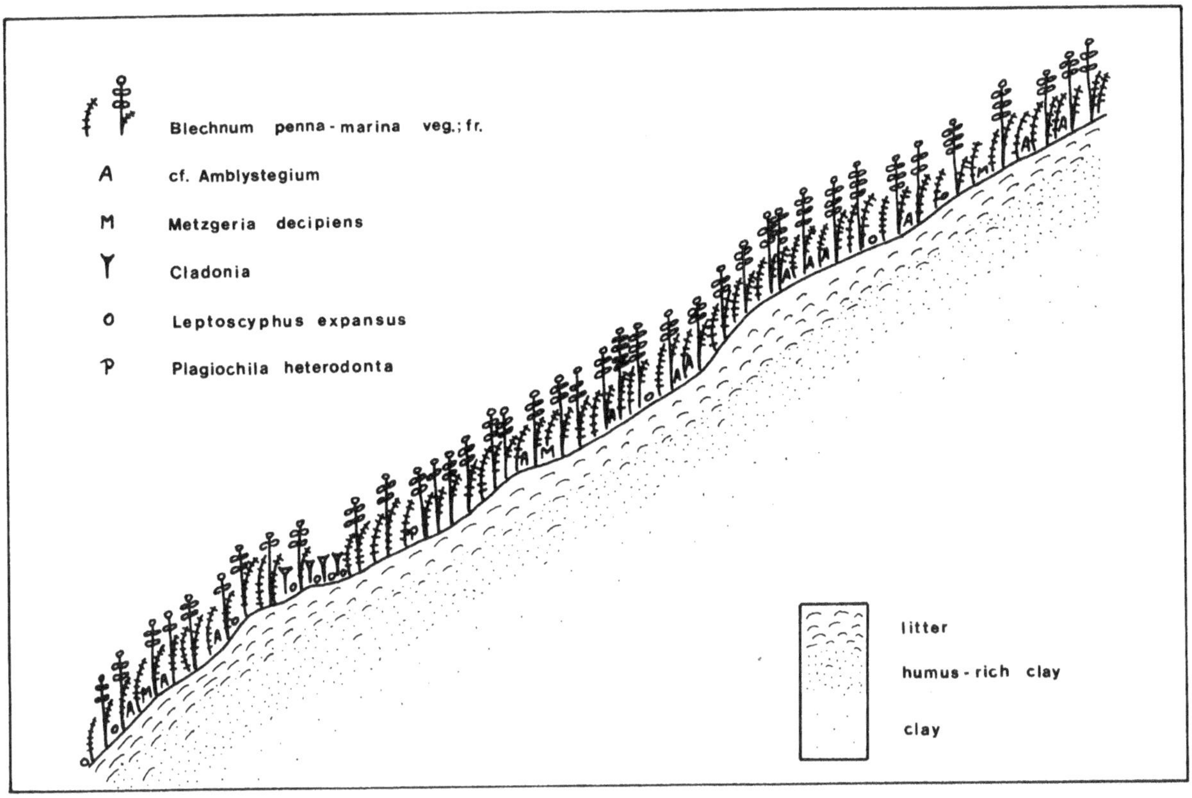


*Fig. 55.* Structure-diagram of Isopterygio pulchelli – Blechnetum penna-marinae typicum.

*platyphyllum* and *Blechnum penna-marina*. The latter species is commonly met with in many of the plant communities on the islands, but reaches very high presence and cover values only in this association (cf. Table 32). Three subassociations are distinguished:

Subassociation 5.1.1 uncinietosum compactae (type relevé 299)

Many of the species which occur in this subassociation have their optimum in communities of the mire complex, e.g. *Uncinia compacta, Symphyogyna marionensis, Jensenia pisicolor* and *Fossombronia australis*. Together with *Distichophyllum imbricatum, Jungermannia coniflora* and *Cladonia* spp. these species differentiate the subassociation uncinietosum against the other two subassociations. The moss *Distichophyllum imbricatum* is rare on Marion Island, where it was found in two communities only: this one and the Lycopodio magellanici-Jamesonielletum coloratae. It has not been found on Prince Edward Island. *Jungermannia coniflora* has its optimum in this subassociation and in the Jungermannio coniflorae-Racomitrietum crispuli, the fjaeldmark vegetation of lowland areas (Par. 4.6). Differential species of the subassociations uncinietosum compactae and brachythecietosum rutabuli together against the subassociation typicum are *Lepidozia laevifolia, Clasmatocolea humilis, Acrobolbus ochrophyllus, Peltigera polydactyla* and *Blepharidophyllum densifolium*. These species have their optimum in communities of the mire complex, and their presence here indicates comparatively moist conditions.

Stands of this subassociation are usually found on soils with a thick layer of plant litter and humus, and an appreciably higher water content than those on which stands of the other two subassociations occur. Generally this community is found on slopes, but in some cases it occurs on more or less level sites

*Table 28*. The *Blechnum penna-marina* complex.

ISOPTERYGIO PULCHELLI - BLECHNETUM PENNA-MARINAE

| uncinietosum compactae | | | | | | | | | |
|---|---|---|---|---|---|---|---|---|---|
| RELEVE NUMBER | 441 | 137 | 551 | 552 | 417 | 299 | 445 | 295 | 290 |
| LOCATION: | | | | | | | | | |
| SOUTHERN LATITUDE (46°) MIN. | 52 | 57 | 52 | 52 | 52 | 51 | 53 | 53 | 53 |
| SEC. | 12 | 45 | 21 | 21 | 40 | 40 | 15 | 20 | 10 |
| EASTERN LONGITUDE (37°) MIN. | 50 | 51 | 49 | 49 | 51 | 50 | 51 | 51 | 51 |
| SEC. | 45 | 00 | 55 | 55 | 30 | 50 | 33 | 15 | 15 |
| ELEVATION (M A.S.L.) | 70 | 80 | 140 | 135 | 17 | 60 | 50 | 60 | 60 |
| INCLINATION (°) | 7 | 10 | 30 | 35 | 20 | 2 | 15 | | 25 |
| ASPECT | NE | SSW | E | N | E | N | N | | N |
| SOIL DEPTH (CM) | 20 | | | | | | | | |
| DEPTH OF GROUNDWATER LEVEL (CM) | --- | - | - | - | - | --- | - | - | - |
| SOIL WATER CONTENT (X10 %) | 53 | 102 | 69 | 64 | 100 | 45 | 52 | 91 | 65 |
| SOIL PH | 4,8 | | 4,1 | 4,3 | 4,2 | 4,5 | 4,5 | 4,3 | 3,9 |
| SOIL LOSS ON IGNITION (%) | 67 | | 90 | 91 | | 76 | 83 | 79 | 73 |
| INFLUENCE OF SALTSPRAY | | | | | | | | | |
| INFLUENCE OF TRAMPLING | | | | | | | | | |
| INFLUENCE OF MANURING | | | | | | | | | |
| HEIGHT OF HERB LAYER (CM) | 10 | 15 | 15 | 15 | 8 | 10 | 10 | 20 | 15 |
| COVER OF HERB LAYER (%) | 100 | 99 | 100 | 100 | 99 | 99 | 99 | 90 | 100 |
| HEIGHT OF BRYOPHYTE LAYER (CM) | 1 | 1 | 1 | 1 | 2 | 1 | 1 | 1 | 1 |
| COVER OF BRYOPHYTE LAYER (%) | <1 | 5 | 3 | 1 | 5 | <1 | <1 | 5 | <1 |
| COVER OF ALGAE (%) | | | | | | | | | |
| COVER OF BARE SOIL (%) | | 1 | | | 1 | 1 | | 5 | |
| COVER OF ROCKS (%) | | | | | | | | | |
| QUADRAT SIZE (M2) | 25 | 9 | 16 | 15 | 4 | 9 | 16 | 16 | 16 |
| TOTAL NUMBER OF SPECIES | 25 | 23 | 19 | 18 | 17 | 17 | 16 | 16 | 15 |
| CHARACTER-TAXA OF THE ASSOCIATION: | | | | | | | | | |
| BLECHNUM PENNA-MARINA | 5 | 4 | 5 | 5 | 5 | 4 | 5 | 5 | 5 |
| CF. AMBLYSTEGIUM | | | 1 | 1 | | | | | |
| PLAGIOTHECIUM PLATYPHYLLUM | | + | + | + | | + | | | |
| METZGERIA DECIPIENS | 1 | | + | | 1 | + | + | | + |
| ISOPTERYGIUM PULCHELLUM | + | | | 1 | | | | | |
| CAMPYLOPUS POLYTRICHIIDES | + | + | | | | | + | + | |
| PELTIGERA CANINA | | | | R | 1 | | | | |
| DIFFERENTIAL TAXA OF THE SUBASSOCIATIONS: | | | | | | | | | |
| UNCINIA COMPACTA | + | 2A | R | R | R | + | R | 1 | + |
| SYMPHYOGYNA MARIONENSIS | + | | R | | | + | + | | + |
| FOSSOMBRONIA AUSTRALIS | | + | | | R | | + | + | + |
| JENSENIA PISICOLOR | R | | + | | + | | | + | |
| CLADONIA COLL. | + | | + | + | | + | + | | |
| DISTICHOPHYLLUM IMBRICATUM | R | + | | R | | | + | | |
| JUNGERMANNIA CONIFLORA | + | + | | | | | | | |
| LEPIDOZIA LAEVIFOLIA | + | 1 | + | + | 1 | + | + | + | |
| CLASMATOCOLEA HUMILIS | + | 1 | + | + | 1 | + | + | | |
| ACROBOLBUS OCHROPHYLLUS | | + | | | R | + | | | |
| PELTIGERA POLYDACTYLA | | + | + | | | | | | + |
| BLEPHARIDOPHYLLUM DENSIFOLIUM | | + | | | + | | | | |
| DREPANOCLADUS UNCINATUS | | + | + | | + | + | | | |
| ACAENA MAGELLANICA | | | | | | | | | |
| BRACHYTHECIUM RUTABULUM | | | | | | | | | |
| LOPHOCOLEA RANDII | | | + | | | | | | |
| COMPANION SPECIES | | | | | | | | | |
| AZORELLA SELAGO | 2A | 4 | 1 | 1 | | 3 | 1 | 2A | 2B |
| AGROSTIS MAGELLANICA | R | 2A | + | + | + | 1 | | 2A | |
| POA COOKII | R | + | + | | | 1 | + | R | R |
| LEPTOSCYPHUS EXPANSUS | | + | 1 | + | + | R | | | |
| LOPHOZIA CYLINDRIFORMIS | 1 | + | 1 | + | 1 | | + | | 1 |
| PLAGIOCHILA HETERODONTA | + | + | + | | + | | + | | + |
| MONTIA FONTANA | R | + | | | | | | R | + |
| JAMESONIELLA COLORATA | R | 1 | | + | 1 | + | | | |
| RANUNCULUS BITERNATUS | R | R | | | | | | | |
| RACOMITRIUM LANUGINOSUM | R | | | + | | | | + | |
| LEPTODONTIUM PROLIFERUM | | | | | R | | R | + | |
| DISTICHOPHYLLUM FASCICULATUM | | | | R | | | | + | + |
| HYPNUM CUPRESSIFORME | | | | | | + | | | |

| brachythecietosum rutabuli | | | | | | | | | | | | |
|---|---|---|---|---|---|---|---|---|---|---|---|---|
| RELEVE NUMBER | 537 | 269 | 464 | 268 | 544 | 285 | 555 | 480 | 156 | 135 | 121 | 124 |
| LOCATION: | | | | | | | | | | | | |
| SOUTHERN LATITUDE (46°) MIN. | 52 | 51 | 51 | 51 | 52 | 52 | 52 | 51 | 57 | 57 | 53 | 53 |
| SEC. | 21 | 00 | 00 | 00 | 21 | 55 | 21 | 00 | 30 | 45 | 00 | 00 |
| EASTERN LONGITUDE (37°) MIN. | 49 | 48 | 48 | 48 | 49 | 49 | 49 | 48 | 51 | 51 | 50 | 50 |
| SEC. | 55 | 00 | 00 | 00 | 55 | 45 | 55 | 00 | 00 | 00 | 45 | 45 |
| ELEVATION (M A.S.L.) | 130 | 120 | 110 | 160 | 140 | 200 | 130 | 110 | 100 | 50 | 60 | 60 |
| INCLINATION (°) | 25 | 45 | 4 | 45 | 30 | 10 | 40 | | 15 | 25 | 25 | |
| ASPECT | SE | SSE | N | N | S | N | SSE | | E | S | E | |
| SOIL DEPTH (CM) | | | | 25 | | | | 12 | | 10 | | |
| DEPTH OF GROUNDWATER LEVEL (CM) | - | - | - | --- | - | - | - | --- | - | --- | --- | --- |
| SOIL WATER CONTENT (X10 %) | 97 | 48 | 43 | 91 | 42 | 23 | 79 | | 39 | 83 | 54 | 48 |
| SOIL PH | 4,5 | 4,8 | | 4,4 | 5,4 | 5,5 | 4,2 | | 4,6 | 4,6 | 4,0 | 4,8 |
| SOIL LOSS ON IGNITION (%) | 82 | 50 | 48 | 74 | 50 | 25 | 80 | | 74 | 79 | 92 | 66 |
| INFLUENCE OF SALTSPRAY | | | | | | | | | | | | |
| INFLUENCE OF TRAMPLING | | | | | | | | | | | | |
| INFLUENCE OF MANURING | | | | | | | | | | | | |
| HEIGHT OF HERB LAYER (CM) | 15 | 20 | 12 | 20 | 20 | 15 | 15 | 15 | 20 | 12 | 25 | 20 |
| COVER OF HERB LAYER (%) | 99 | 100 | 99 | 97 | 100 | 99 | 100 | 100 | 100 | 100 | 100 | 100 |
| HEIGHT OF BRYOPHYTE LAYER (CM) | 3 | 3 | 2 | 2 | 4 | 4 | 3 | 2 | 5 | 2 | 1 | |
| COVER OF BRYOPHYTE LAYER (%) | 25 | 70 | 85 | 25 | 40 | 40 | 20 | 20 | 35 | 1 | <1 | |
| COVER OF ALGAE (%) | | | | | | | | | | | | |
| COVER OF BARE SOIL (%) | | | | | | 1 | | | | | | |
| COVER OF ROCKS (%) | | | | | | | | | | | | |
| QUADRAT SIZE (M2) | 16 | 4 | 16 | 4 | 49 | 25 | 10 | 16 | 20 | 4 | 2 | 4 |
| TOTAL NUMBER OF SPECIES | 19 | 19 | 14 | 14 | 15 | 10 | 8 | 8 | 6 | 5 | 5 | 4 |
| CHARACTER-TAXA OF THE ASSOCIATION: | | | | | | | | | | | | |
| BLECHNUM PENNA-MARINA | 5 | 5 | 5 | 5 | 3 | 4 | 5 | 5 | 5 | 5 | 5 | 5 |
| CF. AMBLYSTEGIUM | | 1 | 1 | 1 | | 1 | | 1 | + | | R | |
| PLAGIOTHECIUM PLATYPHYLLUM | 1 | | + | + | + | | | | | | | |
| METZGERIA DECIPIENS | | + | | | 1 | | | | | | | |
| ISOPTERYGIUM PULCHELLUM | | | | + | | 2A | | + | | | | |
| CAMPYLOPUS POLYTRICHIIDES | | | | | | | | | | | | |
| PELTIGERA CANINA | | | R | | | | | | | | | |
| DIFFERENTIAL TAXA OF THE SUBASSOCIATIONS: | | | | | | | | | | | | |
| UNCINIA COMPACTA | | | | R | | | | | | | | |
| SYMPHYOGYNA MARIONENSIS | | | | | | | | | | | | |
| FOSSOMBRONIA AUSTRALIS | + | | | | | | | | | | | |
| JENSENIA PISICOLOR | | | | | | | | | | | | |
| CLADONIA COLL. | | | | | | | | | | | | |
| DISTICHOPHYLLUM IMBRICATUM | | | | | | | | | | | | |
| JUNGERMANNIA CONIFLORA | | | | | | | | | | | | |
| LEPIDOZIA LAEVIFOLIA | 1 | 2M | 1 | | + | | | | | | | |
| CLASMATOCOLEA HUMILIS | + | 1 | 2B | | | | | | | | | |
| ACROBOLBUS OCHROPHYLLUS | + | + | | | | | | | | | | |
| PELTIGERA POLYDACTYLA | + | | | R | | | | | | | | |
| BLEPHARIDOPHYLLUM DENSIFOLIUM | 2A | + | | | | | | | | | | |
| DREPANOCLADUS UNCINATUS | 2A | 3 | 4 | 2B | 1 | + | 2B | 2B | 2A | 1 | | |
| ACAENA MAGELLANICA | 1 | + | 2A | 2A | 3 | 2A | 1 | 2B | 2A | 1 | 2A | 3 |
| BRACHYTHECIUM RUTABULUM | + | | 1 | + | 2B | | 1 | 1 | 3 | + | | |
| LOPHOCOLEA RANDII | R | + | 1 | + | | | + | + | | | | |
| COMPANION SPECIES | | | | | | | | | | | | |
| AZORELLA SELAGO | 2B | 2B | + | + | + | 1 | | | | + | + | + |
| AGROSTIS MAGELLANICA | 2A | + | | 2A | 4 | 3 | | | | | | 2A |
| POA COOKII | | | | | | + | 1 | R | R | | | |
| LEPTOSCYPHUS EXPANSUS | 1 | 1 | 1 | | + | | | | | | | |
| LOPHOZIA CYLINDRIFORMIS | + | | | | + | | | | | | R | |
| PLAGIOCHILA HETERODONTA | | | | | 1 | | | | | | | |
| MONTIA FONTANA | | | | | | | + | | | | | |
| JAMESONIELLA COLORATA | + | | | | R | | | | | | | |
| RANUNCULUS BITERNATUS | + | | | | + | | | | | | | |
| RACOMITRIUM LANUGINOSUM | | + | | | | 3 | | | | | | |
| LEPTODONTIUM PROLIFERUM | | | | | | | + | | | | | |
| DISTICHOPHYLLUM FASCICULATUM | 1 | | | | | | | | | | | |
| HYPNUM CUPRESSIFORME | | | | | | | | | | | | |

| typicum | | | | | | | | | | | | |
|---|---|---|---|---|---|---|---|---|---|---|---|---|
| RELEVE NUMBER | 533 | 539 | 158 | 5 | 379 | 32 | 406 | 414 | 399 | 294 | 289 | 31 |
| LOCATION: | | | | | | | | | | | | |
| SOUTHERN LATITUDE (46°) MIN. | 52 | 52 | 57 | 52 | 58 | 52 | 52 | 52 | 52 | 35 | 53 | 52 |
| SEC. | 21 | 22 | 15 | 30 | 10 | 40 | 40 | 40 | 40 | 15 | 10 | 40 |
| EASTERN LONGITUDE (37°) MIN. | 49 | 49 | 51 | 51 | 50 | 51 | 51 | 51 | 51 | 57 | 51 | 51 |
| SEC. | 55 | 54 | 10 | 35 | 15 | 33 | 30 | 30 | 30 | 15 | 15 | 32 |
| ELEVATION (M A.S.L.) | 130 | 130 | 60 | 25 | 90 | 30 | 24 | 25 | 20 | 60 | 55 | 25 |
| INCLINATION (°) | 3 | 3 | 40 | 35 | 45 | 35 | 25 | 25 | 35 | 10 | 5 | 40 |
| ASPECT | SE | E | E | ESE | NW | S | N | NE | SSE | W | E | N |
| SOIL DEPTH (CM) | | | 40 | | 40 | | | | 45 | | | |
| DEPTH OF GROUNDWATER LEVEL (CM) | - | - | --- | | --- | - | - | -50 | --- | - | - | - |
| SOIL WATER CONTENT (X10 %) | 41 | 37 | 35 | | 49 | | 60 | 65 | 58 | 51 | 53 | |
| SOIL PH | 4,2 | 5,0 | 4,5 | | 4,6 | | 3,9 | 4,0 | 4,0 | 4,5 | 3,7 | |
| SOIL LOSS ON IGNITION (%) | | 59 | 91 | | | | 91 | 95 | 93 | 66 | 90 | |
| INFLUENCE OF SALTSPRAY | | | | | | | | | | | | |
| INFLUENCE OF TRAMPLING | | | | | | | | | | | | |
| INFLUENCE OF MANURING | | | | | | | | | | | | |
| HEIGHT OF HERB LAYER (CM) | 15 | 15 | 10 | 10 | 10 | 10 | 15 | 15 | 10 | 15 | 20 | 10 |
| COVER OF HERB LAYER (%) | 100 | 100 | 100 | 100 | 100 | 100 | 100 | 100 | 100 | 100 | 100 | 100 |
| HEIGHT OF BRYOPHYTE LAYER (CM) | 2 | 3 | 3 | 1 | 2 | 1 | 1 | 5 | 1 | 1 | 1 | 1 |
| COVER OF BRYOPHYTE LAYER (%) | <5 | 7 | 7 | <1 | 1 | 1 | <1 | 20 | <1 | <1 | < | 2 |
| COVER OF ALGAE (%) | | | | | | | | | | | | |
| COVER OF BARE SOIL (%) | | | | | | | | | | | | |
| COVER OF ROCKS (%) | | | | | | | | | | | | |
| QUADRAT SIZE (M2) | 25 | 25 | 4 | 1 | 9 | 1 | 6 | 4 | 4 | 9 | 9 | 1 |
| TOTAL NUMBER OF SPECIES | 18 | 16 | 12 | 12 | 11 | 11 | 8 | 8 | 8 | 7 | 6 | 6 |
| CHARACTER-TAXA OF THE ASSOCIATION: | | | | | | | | | | | | |
| BLECHNUM PENNA-MARINA | 5 | 5 | 5 | 5 | 5 | 5 | 5 | 5 | 5 | 5 | 5 | 5 |
| CF. AMBLYSTEGIUM | + | + | 2M | + | + | 1 | 1 | 2A | | | + | 1 |
| PLAGIOTHECIUM PLATYPHYLLUM | + | + | | | | | | | + | + | | |
| METZGERIA DECIPIENS | + | + | 1 | + | | + | 1 | | | 1 | | |
| ISOPTERYGIUM PULCHELLUM | 1 | + | + | | + | + | | | | | | + |
| CAMPYLOPUS POLYTRICHIIDES | + | + | | | | | | | | + | | |
| PELTIGERA CANINA | | | | + | | + | | R | | | | |
| DIFFERENTIAL TAXA OF THE SUBASSOCIATIONS: | | | | | | | | | | | | |
| UNCINIA COMPACTA | | | | | | | | | | | | |
| SYMPHYOGYNA MARIONENSIS | | | | | | | | | | | | |
| FOSSOMBRONIA AUSTRALIS | | | | | | | | | | | | |
| JENSENIA PISICOLOR | | | | | | | | | | | | |
| CLADONIA COLL. | | | | | | | | | | | | |
| DISTICHOPHYLLUM IMBRICATUM | | | | | + | | | | | | | |
| JUNGERMANNIA CONIFLORA | | | | | | | | | | | | |
| LEPIDOZIA LAEVIFOLIA | + | | | | | 1 | | | | | | |
| CLASMATOCOLEA HUMILIS | | | | | | + | | | | | | |
| ACROBOLBUS OCHROPHYLLUS | | | | | | | | | | | | |
| PELTIGERA POLYDACTYLA | | + | | | | | | | | | | |
| BLEPHARIDOPHYLLUM DENSIFOLIUM | | | | | | | | | | | | |
| DREPANOCLADUS UNCINATUS | + | + | R | R | | | | + | + | | | |
| ACAENA MAGELLANICA | 1 | | R | | | | | | | | + | |
| BRACHYTHECIUM RUTABULUM | | | | | | | | | | | | |
| LOPHOCOLEA RANDII | | | + | | | | | | | | | |
| COMPANION SPECIES | | | | | | | | | | | | |
| AZORELLA SELAGO | + | + | R | R | R | R | R | | R | | | + |
| AGROSTIS MAGELLANICA | 1 | 2B | | + | | | R | | | 2A | R | |
| POA COOKII | + | + | | | | | R | R | | + | + | |
| LEPTOSCYPHUS EXPANSUS | 1 | | + | + | + | 1 | R | 1 | 1 | | | 1 |
| LOPHOZIA CYLINDRIFORMIS | 1 | 1 | 1 | + | | | R | 1 | 1 | + | | |
| PLAGIOCHILA HETERODONTA | | 2A | | + | 1 | + | | | | | | |
| MONTIA FONTANA | | | | + | | R | | | | | R | |
| JAMESONIELLA COLORATA | + | | | | | | | R | | | | |
| RANUNCULUS BITERNATUS | | | | R | | | | | R | | | + |
| RACOMITRIUM LANUGINOSUM | | | + | | + | | | | | | | |
| LEPTODONTIUM PROLIFERUM | + | | | | | | | | + | | | |
| DISTICHOPHYLLUM FASCICULATUM | | + | | | | | | | | | | |
| HYPNUM CUPRESSIFORME | | | | | | | | | | | | |

| fragments | | | | | | |
|---|---|---|---|---|---|---|
| RELEVE NUMBER | 132 | 334 | 133 | 330 | 316 | 122 |
| LOCATION: | | | | | | |
| SOUTHERN LATITUDE (46°) MIN. | 57 | 38 | 57 | 38 | 52 | 53 |
| SEC. | 40 | 35 | 40 | 22 | 40 | 00 |
| EASTERN LONGITUDE (37°) MIN. | 51 | 00 | 51 | 00 | 51 | 50 |
| SEC. | 15 | 18 | 15 | 25 | 30 | 45 |
| ELEVATION (M A.S.L.) | 80 | 25 | 75 | 60 | 30 | 60 |
| INCLINATION (°) | 5 | 35 | 5 | 10 | 30 | 70 |
| ASPECT | SSE | N | SSE | NNE | NE | ESE |
| SOIL DEPTH (CM) | | | | | | 35 |
| DEPTH OF GROUNDWATER LEVEL (CM) | | --- | | - | --- | --- |
| SOIL WATER CONTENT (X10 %) | 77 | 70 | 78 | 68 | 114 | 45 |
| SOIL PH | 3,7 | | | 4,1 | 4,4 | 3,8 |
| SOIL LOSS ON IGNITION (%) | 86 | 93 | | 85 | | 92 |
| INFLUENCE OF SALTSPRAY | | | | | | |
| INFLUENCE OF TRAMPLING | 2 | | 1 | | | |
| INFLUENCE OF MANURING | 2 | | 1 | | | |
| HEIGHT OF HERB LAYER (CM) | 30 | 7 | 30 | 10 | 10 | 25 |
| COVER OF HERB LAYER (%) | 70 | 100 | 100 | 100 | 100 | 100 |
| HEIGHT OF BRYOPHYTE LAYER (CM) | 2 | 3 | | 2 | 1 | 1 |
| COVER OF BRYOPHYTE LAYER (%) | <1 | 1 | | <1 | <1 | <1 |
| COVER OF ALGAE (%) | | | | | | |
| COVER OF BARE SOIL (%) | 30 | | | | | |
| COVER OF ROCKS (%) | | | | | | |
| QUADRAT SIZE (M2) | 16 | 9 | 9 | 9 | 16 | 4 |
| TOTAL NUMBER OF SPECIES | 6 | 6 | 5 | 4 | 3 | 3 |
| CHARACTER-TAXA OF THE ASSOCIATION: | | | | | | |
| BLECHNUM PENNA-MARINA | 5 | 5 | 5 | 5 | 5 | 5 |
| CF. AMBLYSTEGIUM | | | | | | + |
| PLAGIOTHECIUM PLATYPHYLLUM | | | | | | |
| METZGERIA DECIPIENS | | | | | | |
| ISOPTERYGIUM PULCHELLUM | | | | | | |
| CAMPYLOPUS POLYTRICHIIDES | | | | | | |
| PELTIGERA CANINA | | | | | | |
| DIFFERENTIAL TAXA OF THE SUBASSOCIATIONS: | | | | | | |
| UNCINIA COMPACTA | | | | | | |
| SYMPHYOGYNA MARIONENSIS | | | | | | |
| FOSSOMBRONIA AUSTRALIS | | | | | | |
| JENSENIA PISICOLOR | | | | | | |
| CLADONIA COLL. | | | | | | |
| DISTICHOPHYLLUM IMBRICATUM | | | | | + | |
| JUNGERMANNIA CONIFLORA | | | | | | |
| LEPIDOZIA LAEVIFOLIA | | | | | | |
| CLASMATOCOLEA HUMILIS | | | | | | |
| ACROBOLBUS OCHROPHYLLUS | | | | | | |
| PELTIGERA POLYDACTYLA | | | | | | |
| BLEPHARIDOPHYLLUM DENSIFOLIUM | | | | | | |
| DREPANOCLADUS UNCINATUS | | | | | | |
| ACAENA MAGELLANICA | | | + | | | 1 |
| BRACHYTHECIUM RUTABULUM | | | | | | |
| LOPHOCOLEA RANDII | | + | | | | |
| COMPANION SPECIES | | | | | | |
| AZORELLA SELAGO | + | | + | | R | |
| AGROSTIS MAGELLANICA | | | | | | |
| POA COOKII | 2A | R | 2B | R | | |
| LEPTOSCYPHUS EXPANSUS | | | | | | |
| LOPHOZIA CYLINDRIFORMIS | | + | | | | |
| PLAGIOCHILA HETERODONTA | | + | | 1 | | |
| MONTIA FONTANA | + | | 1 | | | |
| JAMESONIELLA COLORATA | | | | | | |
| RANUNCULUS BITERNATUS | | | | R | | |
| RACOMITRIUM LANUGINOSUM | | | | | | |
| LEPTODONTIUM PROLIFERUM | | | | | | |
| DISTICHOPHYLLUM FASCICULATUM | | | | | | |
| HYPNUM CUPRESSIFORME | | | | | | |

*Fig. 56.* Strongly exposed fjaeldmark area at about 400 m above sea level, covered by Andreaeo acutifoliae - Racomitrietum crispuli (Feldmark Plateau, Marion).

on well-drained peat soils. Here it can be considered as the stage in the succession of mire vegetation which follows after the Lycopodio magellanici-Jamesonielletum coloratae catagonietosum (cf. Par. 4.4).

Subassociation 5.1.2 brachythecietosum rutabuli (type relevé 537)

Conspicuous in this subassociation is the abundance of *Acaena magellanica* and the comparatively well-developed bryophyte stratum, in which *Drepanocladus uncinatus* plays an important role (Table 28). In addition to these two species *Brachythecium rutabulum* and *Lophocolea randii* are differential species of this subassociation against both other ones. The differential species of the subassociations brachythecietosum rutabuli and uncinietosum compactae together against the subassociation typicum have already been listed above.

The differential species of this subassociation indicate some resemblance between this community and the Acaeno magellanicae-Brachythecietum rutabuli, a community found in comparatively dry drainage lines (Par. 4.3). On sites where stands of Isopterygio pulchelli-Blechnetum penna-marinae brachythecietosum rutabuli are found subsurface drainage of water through the soil seems to occur, although on a much smaller scale than in the drainage lines.

Stands of this community are more common on grey lava than on black lava, where the more porous nature of the substratum permits the drainage water to flow through holes and cracks in the lava (cf. Verwoerd 1971) at depths where it does not influence the vegetation. Sites where this community occur are usually somewhat sheltered from the wind.

The large variation in the number of species/relevé in this subassociation (Table 28) could not be correlated with differences in habitat. In some stands the bryophyte stratum is very fragmentarily developed (e.g. relevé 135, 121 and 124). *Azorella selago* forms large cushions in some stands of this community, locally even achieving dominance (e.g. relevé 544).

Subassociation 5.1.3 typicum
(type relevé 539)

This is the most common of the subassociations of the Isopterygio pulchelli-Blechnetum penna-marinae, and it covers an appreciable part of the lowlands of Marion and Prince Edward Islands. It does not possess any character- or differential taxa. The number of species/relevé is smaller on the average than in both other subassociations, and the total cover of the bryophytes is very low (cf. Tables 28, 32). The subassociation typicum occurs on drier soils than both other subassociations, and it is apparently more tolerant to exposure to strong winds than the subassociation brachythecietosum rutabuli.

Fragmentarily developed stands, containing only very few species, were commonly met with (e.g. relevés 132–122). Except for *Blechnum penna-marina* no diagnostic species are present here. In some cases the influence of animals seems to account for this (e.g. relevé 132, 133 and 330); in other cases the cause for this strongly inpoverished development is not clear.

The Isopterygio pulchelli-Blechnetum penna-marinae is a very common vegetation type on Marion and on Prince Edward Island, but on the latter island only the subassociation typicum was met with.

*Synchorological remarks*

Although *Blechnum penna-marina* as well as many of the other species which make up the Isopterygio pulchelli-Blechnetum penna-marinae are widely spread in Subantarctic and southern temperate regions (cf. Tables 10, 11, 12), communities similar to this association do not seem to play an important role in the vegetation of these areas. Moseley (1877) and Schenck (1905) report the occurrence of pure stands of *Blechnum penna-marina* on Kerguelen, without giving any further particulars about this type of vegetation, but in more recent descriptions of the vegetation of this island group (e.g. Aubert de la Rue 1932, 1964, Chastain 1958, Cour 1958, 1959) no communities dominated by *Blechnum penna-marina,* or any other vegetation type similar to the Isopterygio pulchelli-Blechnetum penna-marinae is mentioned. Werth (1906) observed a community of *Blechnum penna-marina, Lycopodium magellanicum* and mosses on Crozet, but as no further description of this vegetation is given, the resemblance with any of the communities found on Marion and Prince Edward Islands can not be judged.

No communities resembling the Isopterygio pulchelli-Blechnetum penna-marinae seem to occur on Macquarie (cf. Taylor 1955a), Heard and Macdonald Islands (cf. Schenck 1905; Law & Burstall 1953), or on South Georgia (cf. Skottsberg 1912; Greene 1964; Smith & Walton 1975). A type of vegetation in which *Blechnum penna-marina* is abundant and which seems similar in ecology to the Isopterygio-Blechnetum has been reported from the Falkland Islands, viz. the '*Empetrum rubrum-Blechnum penna-marina* association', one of the dwarf-shrub heath communities occurring on these islands (Moore 1968; cf. Skottsberg 1909, 1913). On Tristan da Cunha stands of almost pure *Blechnum penna-marina* cover cliff slopes of between 45 and 60°, growing on dry, stable peat (Wace & Holdgate 1958); these stands seem to resemble the *Blechnum* vegetation of Marion and Prince Edward Islands, but the data are insufficient to judge the similarity of these communities. No similar vegetation has been reported from Gough Island (cf. Wace 1961).

### 4.6 Fjaeldmark vegetation: the Andreaea - Racomitrium crispulum complex

Fjaeldmark vegetation, i.e. an open vegetation of isolated, stunted flowering plants, accompanied by mosses and lichens, which occurs where the climate is not sufficiently favourable to permit the development of a closed vegetation of any kind (Holttum 1922; cf. Warming 1887; Huntley 1971), covers a large part of Marion and Prince Edward Islands. Closed vegetation here is largely restricted to the lowlands, but even there, on sites fully exposed to the force of the winds, e.g. on grey lava ridges, an open fjaeldmark vegetation is found. At elevations above c. 250 m only sheltered sites usually bear a closed plant-cover.

At higher altitudes climatic conditions are more severe than at sea level. With each 1000 m increase in altitude the temperature decreases about 4.5 °C (Schulze 1971). Consequently the frequency of nights with frost increases strongly: Zinderen Bakker (1978) estimates that at 250 m above sea level the temperature just above the soil surface falls below 0 °C on about 50% of the nights, compared to c. 30% at sea level. To the severity of the climate strong winds are added; these are alleged to be more forceful at high elevations than at sea level (Taylor 1955a; Schulze 1971), causing drought stress and wind abrasion to the plants. The importance of this factor in the ecology of the fjaeldmark vegetation is stressed by the use of the word 'wind-desert' for the fjaeldmark areas of the Subantarctic islands by e.g. Schenck (1905) and Werth (1906).

The severe climatological conditions are reflected in the vegetation by the dominance of species with a cushion-forming habit (cf. Fig. 57, 58), which possess a relatively high resistance to water loss to a desiccating atmosphere (see e.g. Gimingham 1967; Gimingham & Smith 1971), have a higher heat retention capacity (cf. Zinderen Bakker 1978) and an increased resistance to wind abrasion. The most conspicuous plants in the fjaeldmark vegetation of the islands Marion and Prince Edward are *Azorella selago* and several species of the genera *Andreaea* and *Ditrichum* (Fig. 58).

Fjaeldmark vegetation usually occurs on loamy soils containing an often large amount of rock debris (cf. Fig. 16). Rocks usually cover a large part of the surface (40–100%; cf. Table 29, Fig. 59), as

*Fig. 57.* Lowland fjaeldmark vegetation (Jungermannio coniflorae – Racomitrietum crispuli).

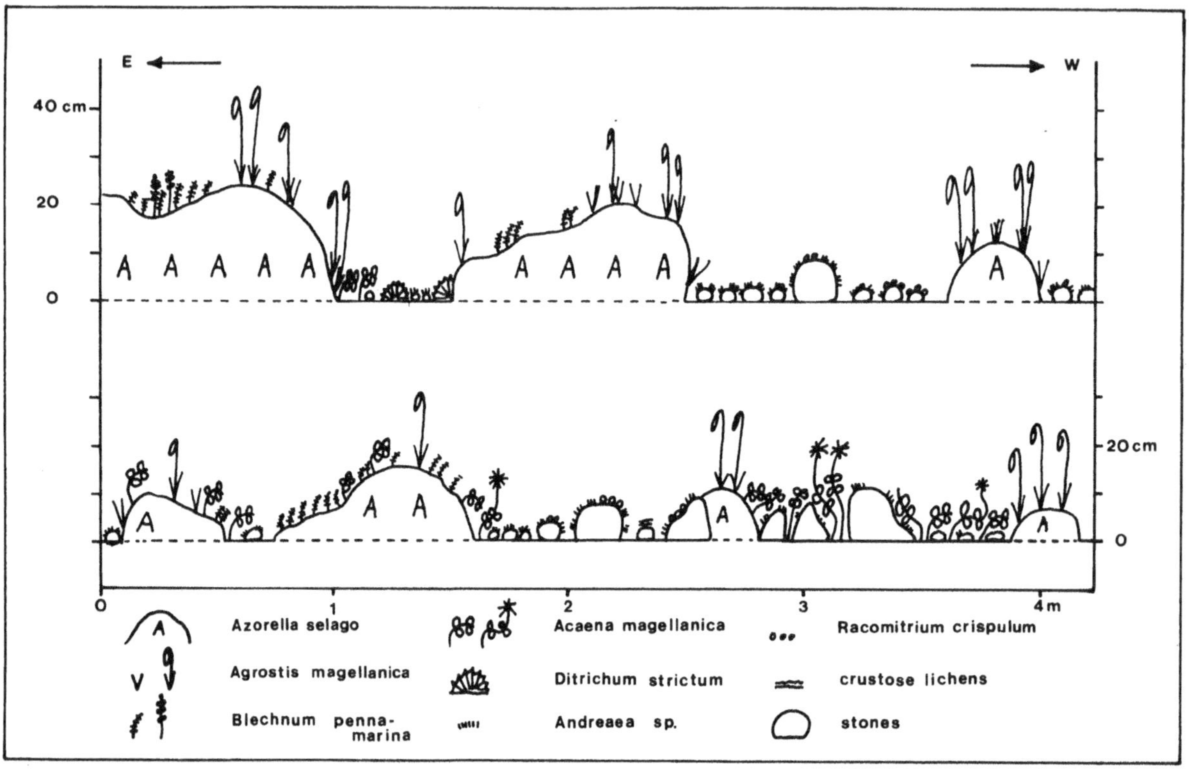


*Fig. 58.* Structure-diagram of Jungermannio coniflorae – Racomitrietum crispuli, the lowland fjaeldmark community.

the finer particles at the surface have been blown away by the strong winds, or washed away by the rain. At high elevations cryoturbatic processes locally result in a sorting of the larger stones from the fine soil particles. Stone lines and at some sites soil polygons are formed by such processes (cf. Zinderen Bakker 1978).

Fjaeldmark soils contain little organic matter and in comparison with other soils on the islands the water content is low, ranging from 20–385% (Table 29). The concentration of plant nutrients in the soil is very low (Smith 1976a,c; Table 9). Smith (loc. cit.) explains this by saying that the rapid percolation of the incoming rainwater through these soils, which have a very low water retention capacity, causes a continuous leaching of minerals from the soil.

Despite the severe environmental conditions the number of species in stands of this type of vegetation is high: the number of species/relevé is on average 16, amongst the highest of any of the plant communities on the islands (cf. Table 32). To some extent this is a result of the complex structure of the substratum, usually consisting of a mosaic of lava rocks and fine-particulate material, forming a mosaic of microhabitats differing in substrate and microclimate. This complexity is further enhanced by the cushion-habit of many of the species in these communities; as a consequence the cushions offer sheltered microhabitats on their leeward sides and in the cushions. Many of the species found in the fjaeldmark communities are more or less restricted to one type of microhabitat, and different microcoenoses may be distinguished; these have not been studied separately.

Association 6.1 Jungermannio coniflorae-Racomitrietum crispuli
(type relevé 442)

This community forms the vegetation of most lowland fjaeldmark areas, where climatological conditions are less severe than at higher elevations.

*Fig. 59.* Andreaeo acutifoliae – Racomitrietum crispuli hymenophylletosum peltati on a grey lava plateau at about 250 m above sea level (Talfelberg, Marion).

The soils contain more organic matter and more water than those under stands of the other association of this complex. The total cover of the vegetation here is appreciably higher than in the fjaeldmark communities of high elevations (on average about 45%, compared with 5–10% at high elevations). A further indication of the less severe environmental conditions of the sites where this association occurs is the high proportion of plants with other growth forms than cushions, e.g. *Blechnum penna-marina* and *Acaena magellanica*, in this vegetation (cf. Table 29). These species, as well as *Uncinia compacta* and *Jungermannia coniflora*, are the differential species of the Jungermannio coniflorae-Racomitrietum crispuli against the Andreaeo acutifoliae-Racomitrietum crispuli, the other association of this complex.

Association 6.2 Andreaeo acutifoliae-Racomitrietum crispuli
(type relevé 553)

Character-taxa of this association are *Andreaea acutifolia*, *Ditrichum strictum*, *Grimmia kerguelensis*, *Philonotis scabrifolia*, *Psilopilum* cf. *tristaniense*, *Hymenophyllum peltatum*, *Bartramia patens* var. *papillata*, *Verrucidens microcarpus* and *Holodontium pumilum*. This community covers a large part of Marion and Prince Edward Islands between c. 100 and 500 m above sea level. It occurs on mineral soils, which show the smallest loss-on-ignition (usually less than 20%) and water content (generally less than 70% of the dry weight) values of any of the islands soils.

The total plant cover in this community rarely exceeds 10% and species which do not form cushions form only a small proportion of the biomass. They are mostly restricted to sheltered microhabitats, such as on the leeward side of stones

*Table 29.* The *Andreaea–Racomitrium crispulum* complex.

| | JUNGERMANNIO CONIFLORAE – RACOMITRIETUM CRISPULI | | | | | | | ANDREAEO ACUTIFOLIAE - RACOMITRIETUM CRISPULI | | | | | | | | | | | | | | |
|---|---|---|---|---|---|---|---|---|---|---|---|---|---|---|---|---|---|---|---|---|---|---|
| | | | | | | | | hymenophylletosum peltati | | | | | | | | hypnetosum cupressiformis | | | | fragments | | |
| RELEVE NUMBER | 306 | 288 | 351 | 443 | 296 | 163 | 442 | 529 | 166 | 165 | 553 | 281 | 315 | 307 | 451 | 282 | 283 | 313 | 372 | 286 | 373 | 284 |
| LOCATION | | | | | | | | | | | | | | | | | | | | | | |
| SOUTHERN LATITUDE (46°) MIN. | 52 | 53 | 39 | 53 | 53 | 57 | 53 | 52 | 57 | 57 | 52 | 53 | 53 | 52 | 52 | 53 | 53 | 53 | 58 | 52 | 57 | 53 |
| SEC. | 50 | 10 | 00 | 20 | 07 | 20 | 12 | 12 | 15 | 15 | 21 | 00 | 20 | 55 | 42 | 00 | 00 | 10 | 05 | 55 | 30 | 00 |
| EASTERN LONGITUDE (37°) MIN. | 51 | 51 | 59 | 51 | 51 | 51 | 51 | 50 | 51 | 51 | 49 | 49 | 48 | 48 | 50 | 49 | 49 | 48 | 50 | 49 | 50 | 49 |
| SEC. | 35 | 15 | 43 | 05 | 30 | 10 | 05 | 55 | 10 | 10 | 56 | 45 | 30 | 45 | 27 | 45 | 45 | 40 | 46 | 45 | 20 | 45 |
| ELEVATION (M A.S.L.) | 40 | 55 | 40 | 70 | 50 | 70 | 70 | 130 | 100 | 90 | 140 | 280 | 300 | 280 | 100 | 280 | 280 | 280 | 160 | 200 | 160 | 280 |
| INCLINATION (°) | | | | | 13 | 7 | 4 | | 7 | 5 | | 25 | | | | 35 | 35 | 35 | 45 | 10 | 4 | 30 |
| ASPECT | | | | | NW | ESE | NW | | E | SSE | | SSE | | | | S | SW | W | SW | N | WNW | N |
| SOIL DEPTH (CM) | | | | 20 | | | 13 | | | | 7 | | 5 | 3 | | 3 | 2 | | | | | 3 |
| DEPTH OF GROUNDWATER LEVEL (CM) | --- | - | --- | --- | - | --- | --- | - | --- | --- | --- | --- | --- | --- | --- | --- | --- | --- | --- | --- | --- | --- |
| SOIL WATER CONTENT (X10 %) | 37 | 38 | 37 | 29 | 22 | 19 | 38 | 28 | 1 | 3 | 4 | 5 | | | 11 | 18 | | | | 6 | 6 | 3 |
| SOIL PH | 5,3 | 5,2 | 5,1 | 5,8 | 4,8 | 5,3 | 5,8 | 5,9 | 5,5 | 5,5 | 6,1 | 5,6 | | | 6,2 | 5,2 | | | | 5,6 | 4,9 | 5,3 |
| SOIL LOSS ON IGNITION (%) | 50 | | 54 | 49 | 9 | 43 | 46 | | 10 | 10 | 13 | 8 | | | 18 | 28 | | | | 19 | 18 | 22 |
| INFLUENCE OF SALTSPRAY | | | | | | | | | | | | | | | | | | | | | | |
| INFLUENCE OF TRAMPLING | | | | | | | | | | | | | | | | | | | | | | |
| INFLUENCE OF MANURING | | | | | | | | | | | | | | | | | | | | | | |
| HEIGHT OF HERB LAYER (CM) | 35 | 10 | 15 | 15 | 20 | 15 | 15 | 25 | 25 | 20 | 20 | 30 | 20 | 15 | 15 | 20 | 20 | 25 | 15 | 25 | 25 | 20 |
| COVER OF HERB LAYER (%) | 37 | 35 | 70 | 45 | 60 | 50 | 15 | 7 | 5 | 5 | 5 | 15 | 5 | 3 | 7 | 30 | 7 | 5 | 3 | 15 | 3 | 5 |
| HEIGHT OF BRYOPHYTE LAYER (CM) | 1 | 1 | 3 | 1 | 1 | 3 | 1 | 3 | 2 | 5 | 3 | 3 | 3 | 1 | 1 | 2 | 1 | 1 | 2 | 5 | 1 | 1 |
| COVER OF BRYOPHYTE LAYER (%) | 3 | <1 | 10 | 3 | <1 | 7 | 10 | 1 | <1 | 3 | 5 | <1 | 20 | 7 | 7 | 1 | <1 | 1 | 3 | <1 | <1 | <1 |
| COVER OF ALGAE (%) | | | | | | | | | | | | | | | | | | | | | | |
| COVER OF BARE SOIL (%) | 10 | | 5 | 17 | 5 | 10 | | 50 | 1 | 2 | | 5 | 55 | | 60 | | | | | 70 | | |
| COVER OF ROCKS (%) | 55 | 90 | 10 | 60 | 40 | 40 | 85 | 45 | 95 | 97 | 99 | 80 | 40 | 99 | 35 | 99 | 97 | 99 | 97 | 15 | 97 | 97 |
| QUADRAT SIZE (M2) | 16 | 49 | 25 | 25 | 9 | 9 | 25 | 25 | 25 | 100 | 25 | 49 | 64 | 64 | 25 | 49 | 49 | 100 | 25 | 49 | 25 | 49 |
| TOTAL NUMBER OF SPECIES | 13 | 14 | 15 | 17 | 14 | 14 | 14 | 14 | 13 | 15 | 19 | 15 | 17 | 16 | 20 | 18 | 20 | 22 | 14 | 8 | 8 | 8 |
| CHARACTER-SPECIES OF THE COMPLEX: | | | | | | | | | | | | | | | | | | | | | | |
| RACOMITRIUM CRISPULUM | 1 | + | + | + | 1 | + | 1 | + | + | + | 1 | + | + | R | + | + | 1 | + | 1 | + | + | + |
| ANDREAEA ACUMINATA | + | + | | 2M | 1 | | 2A | | | | 2A | + | | 1 | | | 1 | + | | | 1 | |
| HERZOGOBRYUM VERMICULARE | | + | | R | | | | | | | | | | | | | 2M | + | | | | + |
| ANDREAEA REGULARIS | 2M | | | | | + | | | | | | | | + | | | | + | | | | |
| GRAMMITIS KERGUELENSIS | + | + | | | | | | | | | | | | + | | + | | | | | | |
| LYCOPODIUM SAURURUS | R | | | R | | | R | | | | | | | | | | | | | R | | |
| DIAGNOSTIC TAXA OF ASSOCIATIONS AND SUBASSOCIATIONS | | | | | | | | | | | | | | | | | | | | | | |
| BLECHNUM PENNA-MARINA | 2B | 2B | 1 | 2B | 3 | 4 | + | | 1 | 1 | | | | | | | | | R | | | |
| ACAENA MAGELLANICA | 1 | 2A | 3 | 3 | + | + | + | + | R | | | | | | | | | + | | + | | |
| JUNGERMANNIA CONIFLORA | | + | + | + | | | + | | | | + | | | | | | | | | | | |
| UNCINIA COMPACTA | | | | + | 1 | | + | | | | | | | | | | | | | | | |
| ANDREAEA ACUTIFOLIA | | | | | | 1 | | | 2M | 2M | 1 | 2A | 2B | 2A | 2A | + | | + | + | + | + | 1 |
| DITRICHUM STRICTUM | | | | | + | | | 1 | + | 2M | 2M | + | 1 | + | 1 | + | 1 | + | | 2M | | + |
| GRIMMIA KERGUELENSIS | | | | | | | | | + | 1 | 1 | | | | | | | | 1 | | + | |
| PHILONOTIS SCABRIFOLIA | | | | | | | | | + | + | | + | + | + | + | + | + | | | | | |
| PSILOPILUM TRISTANIENSE | | | | | | | | + | | | | + | + | + | + | + | + | | | | | |
| HYMENOPHYLLUM PELTATUM | | + | | | | | | | + | + | + | | + | + | + | | + | | | | | |
| BARTRAMIA PATENS | | | | | | | | + | | | | | + | + | + | | | + | | | | |
| VERRUCIDENS MICROCARPUS | | | | | | | | | | | + | | | | | | | + | | | | |
| HOLODONTIUM PUMILUM | | | | | | | | | | | + | | | | | | | | 1 | | | |
| CAMPYLOPUS CLAVATUS | + | | | | + | | + | + | | + | + | 1 | + | | | | | | | 1 | | |
| METZGERIA DECIPIENS | | | + | + | + | + | | | | | | | | | | 1 | 1 | | | | | |
| HYPNUM CUPRESSIFORME | | | | | | | | | + | | | | | | | | + | + | + | | | |
| COLOBANTHUS KERGUELENSIS | | | | | | | | | | | | | | | | + | | + | 1 | | + | |
| ACROBOLBUS OCHROPHYLLUS | | | | | | | | | | | R | | | | | | 1 | | R | | | |
| DITRICHUM CONICUM | | | | | | | | | | | | | | | | | + | | + | | | |
| PRINGLEA ANTISCORBUTICA | | | | | | | | | | | | | | | | | | 1 | + | | | |
| COMPANIONS: | | | | | | | | | | | | | | | | | | | | | | |
| CRUSTOSE LICHENS | + | + | 1 | + | | 1 | + | | 1 | 1 | 1 | + | 1 | 3 | | + | + | + | | | + | |
| AZORELLA SELAGO | 2B | 2A | 3 | 2A | 2B | 2A | 2B | 2A | 2A | 2A | 2A | 2B | 2A | 1 | 2A | 3 | 2A | 1 | 1 | 2B | 1 | 2A |
| AGROSTIS MAGELLANICA | 1 | R | 1 | 1 | 1 | 1 | 2A | + | 1 | 1 | 1 | + | + | + | 1 | | R | + | + | 2A | 1 | R |
| PLAGIOCHILA HETERODONTA | | | 2A | + | 1 | 1 | | + | | + | + | + | R | + | + | 1 | 1 | + | | | | |
| RACOMITRIUM LANUGINOSUM | | 1 | | 1 | + | 2A | | R | 1 | | + | + | R | | + | R | | | + | + | | |
| CLADONIA COLL. | + | | | R | | | + | | | + | + | | + | + | + | | R | | + | | | |
| CATAGONIUM POLITUM | | | | | | + | | + | | | + | | + | | | + | 2M | + | | | | |
| MONTIA FONTANA | | + | R | | | | | | | | | + | | | R | + | + | + | | | | R |
| POA COOKII | | + | + | | | | | | | | | | | | | + | + | + | | | | |
| RANUNCULUS BITERNATUS | | + | + | | | | | | | | | | | | + | | | + | | | | |
| JAMESONIELLA GRANDIFLORA | + | | | | | | + | | | | | | | + | | | | + | | | | |
| LEPIDOZIA LAEVIFOLIA | | | | | | | | | + | + | | | | | | + | + | | | | | |
| CAMPYLOPUS CAVIFOLIUS | | | | | | | | | | | | + | | | | + | | | | | | + |
| JENSENIA PISICOLOR | | | | + | + | | + | | | | + | | | | + | | | | | | | |
| SYMPHYOGYNA MARIONENSIS | | | | | | 1 | | | | + | | | | | R | | | | | | | |
| JAMESONIELLA COLORATA | | | | | | 1 | | | | | | | + | | + | | | | | | | |
| SCHISTOCHILA CARNOSA | | | | | | | | + | | | + | | | | | | | | | | | |

or *Azorella* cushions, or grow in or on top of *Azorella* cushions (cf. Fig. 58).

Two subassociations are discerned, the subassociation hypnetosum cupressiformis, which occurs on deposits of large scoriae, and the subassociation hymenophylletosum peltati, which is found on other types of substratum.

Subassociation 6.2.1 hymenophylletosum peltati (type relevé 553)

This is the most common of the two subassociations, and is usually found on sites with a loamy soil with much rock-debris. Rocks and gravel sometimes completely cover the finer soil material, but between the large stones usually some loamy soil is available for plant growth (cf. Fig. 59). Differential species of this subassociation is *Campylopus clavatus*.

Subassociation 6.2.2 hypnetosum cupressiformis (type relevé 283)

Stands of this community were only found on deposits of comparatively large scoriae, ranging in size from a few cm to over 30 cm in diameter. Only in some of the pockets between the scoriae have some fine soil particles been collected. The nature of the substratum is responsible for a very distinct type of vegetation: only a part of the plants in this community grow above the surface level of the scoriae deposits, i.e. many species of mosses and hepatics growing in the hollows between the scoriae where sufficient light reaches and where the temperature and humidity regime is more favourable than at the surface. *Metzgeria decipiens, Hypnum cupressiformis, Herzogobryum vermiculare, Acrobolbus ochrophyllus,* all differential species of this subassociation, as well as *Catagonium politum* and *Lepidozia laevifolia*, are abundant in the locally very lush bryophyte mats which are found in the spaces between the scoriae, to a depth of about 30 cm below the surface (Fig. 60). Further differential species of this community are *Pringlea antiscorbutica, Ditrichum conicum* and *Colobanthus kerguelensis*.

On sites which offer less sheltered microhabitats, due to the scarcity of rocks on the surface of the soil or to the small sizes of the rock fragments, poorly developed stands of the Andreaeo acutifoliae-Racomitrietum crispuli occur (e.g. relevé 286). Fragmentary stands also occur on sites where cryoturbatic processes prevent the establishment of many species as a result of the movements of the soil particles and rock fragments. The formation of needle ice may also cause the elimination of small plants in these areas (cf. Billings & Mooney 1968). Except for *Azorella selago* and *Agrostis magellanica*, the latter mostly growing on the *Azorella* cushions, the vegetation nearly exclusively consists of epipetric species (e.g. relevé 373). Also on deposits of volcanic ash and of small scoriae, which are not more than a few cm in diameter, fragmentarily developed stands of this community occur (relevé 284).

An interesting phenomenon in the fjaeldmark vegetation of Marion and Prince Edward Islands is the occurrence of completely spherical colonies of mosses (Fig. 61). These moss balls, nearly always consisting of only one species, only in rare cases with a few stems of another species between them, are locally abundant in the fjaeldmark vegetation. Balls of *Andreaea regularis* and *Ditrichum strictum* were commonly met with. Also balls of *Holodontium pumilum* occur (Zanten 1971).

Moss balls have been reported from Arctic and Subarctic regions (Lid 1938; Eythorsson 1951; Benninghoff 1955; Shacklette 1966), Antarctic and Subantarctic areas (Schenck 1905; Aubert de la Rue 1964, 1968; Seppelt & Ashton 1978; Zanten 1971) and the East African mountains (Hedberg 1964). Different explanations for the formation of these globular moss-colonies have been offered. According to Shacklette (1966) and Lid (1938) the wind plays an important role in the formation of these balls, either by undercutting attached moss-colonies, or by overturning moss cushions growing on small stones. It is also an important force in moving these balls around (cf. also James 1876, cited by Schenck 1905; Seppelt & Ashton 1978). Hedberg (1964) and Benninghoff (1955) mention the instability of the substratum as a result of cryoturbatic processes as a force in the formation of moss balls. Hedberg (loc. cit.) suggests that on Mt. Kenia and other East African mountains moss balls originated from moss carpets which were broken up by needle ice formation. Needle ice formation as well as cryoturbatic movements of the substrate particles may cause regular changes in the position of the moss balls, resulting in almost perfectly

*Fig. 60.* Andreaeo acutifoliae – Racomitrietum crispuli hypnetosum cupressiformis on a deposit of scoriae. Only part of the vegetation can be seen at the surface, many bryophytes grow in the hollows between the scoria-fragments.

globular moss colonies in some cases (Hedberg 1964).

On Marion and Prince Edward Islands spherical colonies of *Andreaea* spp., notably of *Andreaea regularis,* occur in large numbers at some localities. These moss balls are found at elevations of more than 150 m above sea level. Here frost occurs on a large proportion of the nights (Zinderen Bakker 1978) and ice formation plays an important role in the process of turning these moss colonies around. At the sites where these moss balls were found periglacial phenomena did not seem to be important. Next to the strong winds, water is a force which causes movement amongst the moss balls on Marion and Prince Edward Islands. During heavy rains much water runs off along the surface of the soil in many fjaeldmark areas, locally forming strong currents. It was observed that colonies of *Andreaea* spec., growing on small rock fragments, were moved around by the water. The formation of balls of *Andreaea* spp. seems to start with such overturning of moss cushions growing on small stones. In all balls formed by this species the centre is formed by a small rock fragment. In globular colonies of *Ditrichum,* however, only organic matter, and rarely some mineral soil material was found in the centre. The formation of balls of this species seems to start with the dislodging of attached cushions.

Seppelt & Ashton (1978) describe the development of sedentary colonies of *Bryum algens* and *Grimmia lawiana* from unattached moss balls on Mawson Rock on the Antarctic continent. Thus moss balls can assist the propagation of species to new sites (cf. also Shacklette 1966). Such a process has not been osberved on Marion and Prince Edward Islands.

Moss balls have also been reported from Europe and New Zealand (e.g. Dixon 1912; Martin 1952). Here trampling and scratching by birds and other

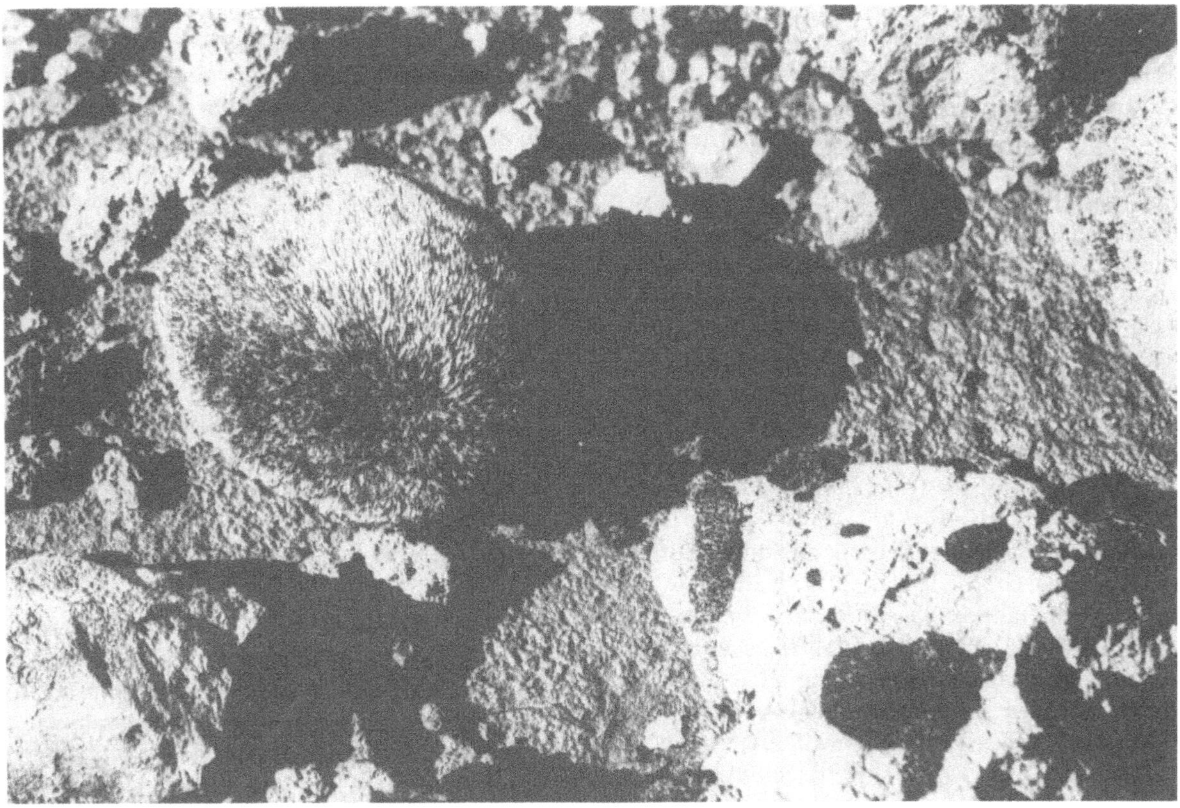

*Fig. 61.* Ball of *Ditrichum strictum* in fjaeldmark vegetation.

animals are important factors in the formation of these unattached colonies.

*Synchorological remarks*

Fjaeldmark vegetation dominated by cushion plants is widely distributed in southern cold and temperate regions (e.g. Schenck 1905; Werth 1906; Taylor 1955a,b; Wace 1960, 1965; Huntley 1971, 1972), as well as in the mountain regions of South America (cf. Rauh 1978; Cleef 1978, 1981), New Zealand (cf. Cockayne 1928; Troll 1948) and Australia (cf. Costin 1957).

All islands of the Kerguelen Province possess plant communities comparable to the Jungermannio coniflorae-Racomitrietum crispuli and Andreaeo acutifoliae-Racomitrietum crispuli. On the islands of the Kerguelen archipelago communities dominated by *Azorella selago* occur on exposed sites (Schenck 1905; Werth 1906; Chastain 1958; Cour 1958, 1959; Aubert de la Rue 1964). They are most common between 600 and 800 m above sea level (Chastain 1958). No detailed data are available, but this '*Azorella*-formation' (Schenck 1905) seems similar in structure and ecology to the communities of the *Andreaea-Racomitrium crispulum* complex. The communities dominated by *Lyallia kerguelensis* and by *Poa kerguelensis* (Chastain 1958, Aubert de la Rue 1964) also seem similar in structure and in habitat to the fjaeldmark vegetation of Marion and Prince Edward Islands.

Large parts of the Iles Crozet are covered by communities dominated by *Azorella selago* cushions, on top of which some grasses often occur (Dreux 1964). There this type of vegetation occurs up to about 300 m above sea level. At high altitudes *Racomitrium* spp. and *Andreaea* spp. dominate the vegetation on basaltic rocks (Hébrard 1970b).

On Heard Island *Azorella selago* grows up to 120 m above sea level (Schenck 1905), forming a vegetation similar to that on Kerguelen (Werth 1906). On Macdonald Island *Azorella selago* covers much of the higher parts (Budd 1972). No furter information on this vegetation is available.

Taylor (1955a) and Ashton & Gill (1965) describe a 'Feldmark formation' on Macquarie. The communities of this formation, which covers most of the island above 200 m above sea level, up to the summits at c. 450 m, resemble those of the fjaeldmark of Marion and Prince Edward Islands. On Macquarie the pattern of distribution of species within a stand as well as of communities within a fjaeldmark area is apparently largely determined by soil movements associated with cryoturbatic processes (Taylor 1955b). In this respect the fjaeldmark of Macquarie differs from that of Marion and Prince Edward Islands.

*Azorella selago* does not occur on South Georgia (Greene 1964, Greene & Walton 1975) and no other cushion plant takes its place there. The cryptogamic communities of rock and scree which occur on this island, and in which *Andreaea* spp. and *Racomitrium crispulum* are prominent (Skottsberg 1912; Greene 1964), seem to correspond in site-ecology to the fjaeldmark of Marion and Prince Edward.

Some cryptogamic communities from the Maritime Antarctic somewhat resemble the communities of the *Andreaea-Racomitrium crispulum* complex. Especially the *Andreaea-Usnea* association (Gimingham & Smith 1970), which occurs on e.g. the South Orkney Islands (Smith 1972) and South Shetland islands (Allison & Smith 1973; Lindsay 1971), resembles the Andreaeo acutifoliae-Racomitrietum crispuli in species composition.

On the Falkland Islands communities in which cushion-forming plants dominate and cryptogams are a conspicuous element in the vegetation occur above 600 m (Skottsberg 1909, 1913, Moore 1968). Insufficient data are available for a comparison with the fjaeldmark communities described in this study, but it seems likely that a resemblance exsists.

Some of the 'montane rock communities' described by Wace (1961) from Gough Island are comparable to communities of the *Andreaea-Racomitrium crispulum* complex, but the 'feldmark' described by this author is very different in structure and in species composition from the Marion Island fjaeldmark. *Azorella selago* does not occur on this island. On Tristan da Cunha, where *Azorella selago* is also lacking, alpine desert vegetation occurs (Wace & Holdgate 1958) of which *Andreaea* spp. and *Racomitrium crispulum* are important constituents. These communities, as well as some forms of the '*Empetrum rubrum-Racomitrium lanuginosum* heath', are comparable to the fjaeldmark associations from Marion and Prince Edward Islands. These communities occur on Tristan from 1500 m above sea level to the summit at 2060 m.

### 4.7 Other communities

In this last group some communities are described, which do not fit well into any of the six community-complexes described sofar.

Community 7.1 Sociation of Limosella australis (type relevé 592)

This community is dominated and characterized by *Limosella australis*. It occurs on the bottom of shallow lakes, usually growing on a thin layer of organic mud overlying mineral deposits (clay and rock fragments). Only one stand was analysed: relevé no. 592; in 20–30 cm deep water of a lake near Macarony Bay, Marion Island; 40 m a.s.l.; total cover of the herb layer: 5%; height of herb layer: 5 cm; algae: 95%; *Limosella australis* 2m.5. Cleef (1981) recently also described communities with a predominance of *Limosella australis* occurring in similar habitats in the Colombian Andes. He grouped these in his order Tillaeetalia.

Community 7.2 Sociation of Potamogeton spec. (type relevé 621)

Only one stand of this type was found, occurring in one of the Albatross lakes, in about 1 m deep water, growing on a soil consisting largely of rock debris and clay, with a thin layer of organic mud: relevé no. 621: in 1 m deep water of one of the Albatross Lakes, Marion Island; 45 m a.s.l.; total cover of the vegetation: 30%; height of herb layer: 100 cm; *Potamogeton* spec. 3.5.

The status of *Potamogeton* spec. on Marion Island has not been ascertained conclusively. Huntley (1971) suggests that it is a recent, naturally introduced species.

*Table 30.* Other communities.

| | Sociation of Azorella selago | | | | | | | | COMMUNITY OF AZORELLA SELAGO AND ACAENA MAGELLANICA | | | |
|---|---|---|---|---|---|---|---|---|---|---|---|---|
| RELEVE NUMBER | 500 | 499 | 512 | 513 | 355 | 514 | 519 | 344 | 244 | 228 | 224 | 548 |
| LOCATION | | | | | | | | | | | | |
| SOUTHERN LATITUDE (46O) MIN. | 51 | 51 | 54 | 54 | 38 | 54 | 56 | 38 | 51 | 50 | 52 | 52 |
| SEC. | 55 | 55 | 15 | 15 | 27 | 15 | 40 | 20 | 50 | 45 | 10 | 21 |
| EASTERN LONGITUDE (37O) MIN. | 39 | 39 | 36 | 36 | 00 | 36 | 35 | 00 | 51 | 51 | 50 | 49 |
| SEC. | 35 | 35 | 33 | 25 | 15 | 25 | 43 | 20 | 05 | 50 | 50 | 55 |
| ELEVATION (M A.S.L.) | 400 | 400 | 50 | 30 | 50 | 30 | 65 | 40 | 65 | 65 | 60 | 145 |
| INCLINATION (O) | | | | | 5 | 3 | 15 | 50 | | 2 | | |
| ASPECT | | | | | E | SW | W | S | | E | | |
| SOIL DEPTH (CM) | | | | 25 | | 30 | | | | | | 50 |
| DEPTH OF GROUNDWATER LEVEL (CM) | --- | --- | - | --- | --- | --- | --- | --- | - | - | - | --- |
| SOIL WATER CONTENT (X10 %) | | | | | | | | | 30 | 57 | 45 | 35 |
| SOIL PH | | | | | | | | | 5,6 | 4,8 | 5,3 | 5,3 |
| SOIL LOSS ON IGNITION (%) | | | | | | | | | 43 | 67 | 39 | |
| INFLUENCE OF SALTSPRAY | | | | | | | | | | | | |
| INFLUENCE OF TRAMPLING | | | | | | | | | | | | |
| INFLUENCE OF MANURING | | | | | | | | 1 | | | | |
| HEIGHT OF HERB LAYER (CM) | 20 | 25 | 25 | 15 | 30 | 15 | 30 | 30 | 30 | 15 | 25 | 25 |
| COVER OF HERB LAYER (%) | 99 | 99 | 100 | 99 | 85 | 99 | 99 | 100 | 99 | 100 | 100 | 97 |
| HEIGHT OF BRYOPHYTE LAYER (CM) | | 5 | | | | | 1 | 1 | 1 | 2 | 3 | 4 |
| COVER OF BRYOPHYTE LAYER (%) | | <1 | | | | | <1 | <1 | 1 | 4 | 5 | 35 |
| COVER OF ALGAE (%) | | | | | | | | | | | | |
| COVER OF BARE SOIL (%) | 1 | 1 | | 1 | 15 | 1 | | | 1 | | | |
| COVER OF ROCKS (%) | | | | | | | | | | | | |
| QUADRAT SIZE (M2) | 100 | 100 | 25 | 25 | 25 | 25 | 25 | 25 | 4 | 9 | 4 | 12 |
| TOTAL NUMBER OF SPECIES | 1 | 3 | 4 | 4 | 4 | 5 | 6 | 7 | 12 | 13 | 12 | 15 |
| AZORELLA SELAGO | 5 | 5 | 5 | 5 | 5 | 5 | 5 | 5 | 5 | 5 | 5 | 4 |
| POA COOKII | | | 1 | + | 1 | 2A | | | + | | | |
| MONTIA FONTANA | | | + | + | + | + | + | | | + | 1 | + |
| PRINGLEA ANTISCORBUTICA | | | | | + | | 1 | | | | + | 1 |
| COLOBANTHUS KERGUELENSIS | | + | 1 | 1 | | 1 | + | + | | | | |
| CAMPYLOPUS CLAVATUS | | + | | | | | | | | | | |
| ACAENA MAGELLANICA | | | | | | 1 | 2B | 3 | 3 | 4 | 5 | 4 |
| RACOMITRIUM CRISPULUM | | | | | | | + | | | | | |
| PLAGIOCHILA HETERODONTA | | | | | | | | R | | | | + |
| RANUNCULUS BITERNATUS | | | | | | | | 1 | + | 1 | + | + |
| PELTIGERA POLYDACTYLA | | | | | | | | R | | R | R | |
| AGROSTIS MAGELLANICA | | | | | | | | | + | 2B | | |
| LOPHOCOLEA RANDII | | | | | | | | | 1 | 1 | + | 1 |
| LEPIDOZIA LAEVIFOLIA | | | | | | | | + | | + | | |
| CLASMATOCOLEA HUMILIS | | | | | | | | | + | 1 | | |
| JAMESONIELLA COLORATA | | | | | | | | | 1 | 1 | | |
| LEPTOSCYPHUS EXPANSUS | | | | | | | | | | + | | |
| DREPANOCLADUS UNCINATUS | | | | | | | | | + | 1 | 2A | 3 |
| METZGERIA DECIPIENS | | | | | | | | | + | + | | 1 |
| FUNARIA LAXA | | | | | | | | | + | | | |
| PHILONOTIS SCABRIFOLIA | | | | | | | | | R | | | |
| BRACHYTHECIUM RUTABULUM | | | | | | | | | | | 1 | 1 |
| AGROSTIS STOLONIFERA | | | | | | | | | | | 1 | |
| BLECHNUM PENNA-MARINA | | | | | | | | | | | + | 2A |
| JAMESONIELLA GRANDIFLORA | | | | | | | | | | | | + |
| CF. AMBLYSTEGIUM | | | | | | | | | | | | 1 |
| RACOMITRIUM LANUGINOSUM | | | | | | | | | | | | + |
| BRACHYTHECIUM PARADOXUM | | | | | | | | | | | | + |
| CERASTIUM HOLOSTEOIDES | | | | | | | | | | | R | |

Community 7.3 Sociation of Azorella selago
(type relevé 513)

This community consists of a more or less continuous mat of *Azorella selago,* in which only few other species occur. It does not possess any diagnostic species, but its total species-combination (Table 30) and its structure (Fig. 62) distinguish it from all other communities on Marion and Prince Edward Islands. Stands of this community occur on well-drained soils, often on deposits of volcanic ash and scoriae, usually on somewhat sheltered sites. On the western part of Marion, on the coastal plain as well as at high elevations, extensive *Azorella* carpets occur. The abundance of this community in this part of the island may be related to the continuous supply of very humid air from the ocean and the probably higher precipitation on this side of the island. Small stands of this community occur between 200 and 500 m above sea level on somewhat sheltered sites, often with a water supply by drainage from surrounding areas (cf. Fig. 62). This community is commonly met with on both Marion and Prince Edward Islands.

Community 7.4 Community of Azorella selago and Acaena magellanica
(type relevé 228)

The dominant species in this community is *Azorella selago,* which forms usually large mats of plants which are appreciably less compact than those in fjaeldmark vegetation and in the *Azorella*-sociation described above. Separate plants are often recognizable by their cushion-form, as opposed to the mats of the *Azorella*-sociation, in which from the outside usually no separate plants are recognizable (cf. Figs. 62, 63). *Acaena magellanica* partly covers the *Azorella,* and a large number of bryophytes

*Fig. 62.* Sociation of *Azorella selago* (foreground) in a depression of the lava flow in a fjaeldmark area at about 400 m above sea level (near Theo, Marion).

*Fig. 63.* Community of *Azorella selago* and *Acaena magellanica* on a scoriae deposit near Gordon's Hill, Marion.

grows on and in between the *Azorella* plants. Many of these bryophytes show their optimum in mire communities, and their presence here indicates comparatively favourable moisture conditions. Stands of this community occur on well-drained soils with a higher water content than fjaeldmark soils, but lower than that of slopes covered by Isopterygio pulchelli-Blechnetum penna-marinae. The proportion of organic matter in the soil is low. This community occurs on lowland slopes as well as on sheltered sites with a favourable water regime at high elevations.

### 4.8 Alien vascular plants in the vegetation of Marion and Prince Edward Islands

The alien vascular plants recorded from Marion and Prince Edward Islands can be grouped into four categories: transient species, persistent species, naturalized species with a restricted distribution and widespread naturalized species (Gremmen 1975; Gremmen & Smith 1980; cf. Walton & Smith 1973; Meurk 1977). These species arrived on the islands by diaspores inadvertently introduced by man (xenophytes in the terminology of Holub & Jirásek, 1967; cf. Schroeder 1969), and are, with exception of the transient species, not restricted to sites influenced by man, but occur in, and sometimes replace, natural communities.

Two of the islands' plant communities are characterized and usually dominated by alien species, viz. the Callitricho antarcticae-Poetum annuae and the Acaeno magellanicae-Agrostietum stoloniferae. Communities dominated by *Poa annua* occur largely on sites which are not colonised by indigenous species, notably on strongly compacted soils (cf. Par. 4.2). *Agrostis stolonifera,* however, seems to replace *Acaena magellanica* and other species in the vegetation of comparatively dry drainage lines and embankments along streams, which originally probably were occupied by Acaeno magellanicae-

Brachythecietum rutabuli (cf. Par. 4.3.; Gremmen & Smith 1980).

*Poa pratensis* is found on well-drained soils, usually on slopes. Only a few stands dominated by this grass were found (Table 31) and this species has not spread much beyond its probable sites of introduction. Therefore no separate community was discerned for this type of vegetation. *Festuca rubra*, which was found on still fewer sites, occurs in similar habitats as does *Poa pratensis*. Also *Agropyron repens*, which species is restricted to one locality on Marion, near Ship's Cove, grows on similar well-drained soils. In habitat the stands dominated by these alien grasses are more or less similar to that of the Acaeno magellanicae-Brachythecietum rutabuli (Par. 4.3).

Completely different sites are colonized by *Sagina apetala*. This species, which is restricted to the immediate vicinity of the weather station, but is spreading to other areas, occurs on sites where

*Table 31*. Stands with alien vascular species.

| Relevé number | 559 | 535 | 574 | 558 | 575 | 577 | 580 | 520 | 581 | 585 |
|---|---|---|---|---|---|---|---|---|---|---|
| Elevation (m) | 15 | 130 | 10 | 40 | 15 | 20 | 15 | 200 | 25 | 25 |
| Slope (°) | 25 | 30 | 70 | 55 | 60 | | 30 | 25 | 20 | 5 |
| Aspect | N | ESE | NNE | ENE | E | | N | E | SE | E |
| Depth of groundwater level (cm) | | | | | | | 30 | | 15 | 20 |
| Soil water content (×10%) | 35 | 83 | 79 | 64 | 71 | 96 | 101 | | 80 | 67 |
| Soil pH | 5.6 | 5.5 | 5.2 | 4.4 | 4.6 | 4.6 | 4.0 | | 4.7 | 4.5 |
| Soil loss-on-ignition (%) | 52 | 68 | 77 | 85 | 89 | 83 | 93 | | 76 | 92 |
| Herb layer: cover (%) | 100 | 100 | 100 | 100 | 100 | 100 | 100 | 99 | 100 | 99 |
| height (cm) | 60 | 50 | 35 | 40 | 40 | 30 | 30 | 25 | 7 | 20 |
| Bryophyte layer: cover (%) | | 1 | | 1 | 1 | 1 | 70 | | 1 | 1 |
| height (cm) | | 5 | | 3 | 2 | 2 | 3 | | 1 | 2 |
| Cover of bare soil (%) | | | | | | | | 1 | | 1 |
| Quadrat size ($m^2$) | 25 | 16 | 15 | 50 | 20 | 9 | 15 | 20 | 16 | 25 |
| Number of species | 3 | 5 | 4 | 5 | 6 | 10 | 10 | 5 | 12 | 9 |
| Agropyron repens | 5 | | | | | | | | | |
| Festuca rubra | 1 | 5 | | | | | | | | |
| Poa pratensis | | | 4 | 5 | 5 | 5 | | | | |
| Agrostis stolonifera | | | 4 | | + | | + | | | |
| Rumex acetosella | | | | | | | 5 | | | |
| Stellaria media | | | | | | | | 5 | | |
| Poa annua | | | | | | | | 1 | | |
| Sagina apetala | | | | | | | | | 4 | 5 |
| Cerastium holosteoides | | | | | | | | | | + |
| Acaena magellanica | + | 1 | + | + | + | 1 | | + | | |
| Brachythecium rutabulum | | 1 | | + | | | | | | |
| Blechnum penna-marina | | 1 | | + | | | | | | |
| Poa cookii | | | + | + | + | r | + | + | 1 | |
| Leptodontium proliferum | | | | | | + | 1 | | | |
| Marchatia berteroana | | | | | | | + | | + | |
| Callitriche antarctica | | | | | | | | | 1 | |
| Cotula plumosa | | | | | | | | | 3 | |
| Montia fontana | | | | | | + | | | 2a | 2a |
| Lophocolea randii | | | | | | 1 | 4 | | 1 | 1 |
| Agrostis magellanica | | | | | | 1 | 1 | | + | 1 |
| Peltigera polydactyla | | | | | | r | + | | | |
| Juncus scheuchzerioides | | | | | | | + | | 1 | 1 |
| Ranunculus biternatus | | | | | | | | | 1 | 1 |
| Drepanocladus uncinatus | | 1 | | | | | | | | + |

Additional taxa, recorded once: Azorella selago 577: r Fossombronia australis 581: + Jamesoniella grandiflora 580: 1 Leptoscyphus expansus 577: + cf. Lethocolea radicosa 575: + Plagiochila heterodonta 575: + Pringlea antiscorbutica 520: 1.

trampling has provided bare soil. It occurs most commonly in communities of the *Callitriche antarctica-Poa cookii* complex and in transitional stands between these and communities of the *Juncus scheuchzerioides-Blepharidophyllum densifolium* complex. Usually it does not reach high cover. Only at some sites near the weather station, where the original vegetation has been strongly disturbed by man, does Sagina reach dominance (e.g. relevés 581 and 585; Table 31). This vegetation shows affinity with communities of the *Callitriche antarctica-Poa cookii* complex, although also elements of mire vegetation are present.

*Rumex acetosella,* the only persistent alien vascular species on the islands, occurs in an area near the meteorological station, surrounded by Poo cookii-Cotuletum plumosae and Leptodonito proliferi-Poetum cookii, but the species composition of this stand (Table 31) does not indicate a clear relation to any of the plant communities described in this study.

*Stellaria media* and *Cerastium holosteoides* are frequently found growing in groups on sheltered sites, e.g. at the leeward sides of boulders. Single individuals of both species are frequently found in stands of Acaeno magellanicae-Bracthythecietum rutabuli (cf. Table 25). In this vegetation usually some open soil is available for colonization, while the *Acaena* canopy provides shelter against strong winds.

| | *Cotulo plumosae - Crassuletum moschatae* Crassula-variant | *Cotulo plumosae - Crassuletum moschatae* intermediate variant | *Cotulo plumosae - Crassuletum moschatae* Cotula-variant | *Crassulo moschatae - Clasmatocoleetum vermicularis* | *Crassulo moschatae - Azorelletum selaginis* | *Montio fontanae - Callitrichetum antarcticae agrostietosum magellanicae* | *Montio fontanae - Callitrichetum antarcticae typicum* Montia-variant | *Montio fontanae - Callitrichetum antarcticae typicum* intermediate variant | *Montio fontanae - Callitrichetum antarcticae typicum* Callitriche-variant | *Callitricho antarcticae - Poetum annuae* | *Poo cookii - Cotuletum plumosae* | *Montio fontanae - Clasmatocoleetum vermicularis* | *Leptodontio proliferi - Poetum cookii inops* | *Leptodontio proliferi - Poetum cookii brachythecietosum rutabuli* | *Brachythecietum subplicati philonotidetosum* | *Brachythecietum subplicati inops* | *Brachythecietum subplicati breutelietosum integrifoliae* | *Acaeno magellanicae - Drepanocladetum uncinati breutelietosum integrifoliae* | *Acaeno magellanicae - Drepanocladetum uncinati brachythecietosum rutabuli* | *Acaeno magellanicae - Brachythecietum rutabuli* | *Acaeno magellanicae - Agrostietum stoloniferae* | community of *Juncus scheuchzerioides Campylopus clavatus* variant | community of *Juncus scheuchzerioides Ranunculus biternatus* variant | *Junco scheuchzerioidis - Drepanocladetum uncinati juncetosum scheuchzerioidis* | *Junco scheuchzerioidis - Drepanocladetum uncinati montietosum fontanae* | *Junco scheuchzerioidis - Drepanocladetum uncinati clasmatocoleetosum humilis* |
|---|---|---|---|---|---|---|---|---|---|---|---|---|---|---|---|---|---|---|---|---|---|---|---|---|---|---|
| Number of relevés | 12 | 10 | 11 | 5 | 7 | 6 | 5 | 7 | 7 | 8 | 14 | 7 | 7 | 9 | 2 | 6 | 6 | 5 | 4 | 15 | 5 | 8 | 6 | 6 | 8 | 9 |
| Average number of species per relevé | 4 | 4 | 5 | 4 | 9 | 5 | 4 | 4 | 3 | 5 | 4 | 9 | 7 | 13 | 9 | 5 | 8 | 7 | 7 | 7 | 8 | 4 | 3 | 5 | 6 | 10 |
| Characteristic taxa of communities of the *Crassula moschata* complex | | | | | | | | | | | | | | | | | | | | | | | | | | |
| *Crassula moschata* | V$^5$ | V$^5$ | V$^{2m}$ | V$^3$ | V$^1$ | | | | | | | | | | | | | | | | | | | | | |
| *Eriopus apiculatus* | | | I$^r$ | II$^+$ | | | | | | | | | | | | | | | | | | | | | | |
| *Cotula plumosa* | III$^+$ | V$^3$ | V$^5$ | | V$^+$ | I$^r$ | | | | II$^r$ | V$^5$ | II$^+$ | I$^r$ | I$^r$ | | | | | | | | | | | | |
| *Clasmatocolea vermicularis* | | | | V$^5$ | | | I$^r$ | | | | | V$^5$ | I$^r$ | II$^+$ | | | | | | | | | | | | |
| Characteristic taxa of communities of the *Callitriche antarctica - Poa cookii* complex | | | | | | | | | | | | | | | | | | | | | | | | | | |
| *Montia fontana* | II$^r$ | | IV$^+$ | | I$^r$ | V$^5$ | V$^5$ | V$^3$ | III$^+$ | IV$^{2m}$ | II$^+$ | IV$^{2a}$ | IV$^1$ | V$^1$ | V$^1$ | V$^1$ | V$^1$ | III$^+$ | IV$^+$ | II$^+$ | IV$^+$ | | | II$^r$ | IV$^+$ | III$^+$ |
| *Callitriche antarctica* | I$^r$ | | | V$^+$ | II$^r$ | IV$^1$ | II$^+$ | V$^5$ | V$^5$ | V$^{2a}$ | III$^1$ | IV$^+$ | V$^+$ | IV$^+$ | V$^1$ | | | | | | | | II$^+$ | | | I$^r$ |
| *Poa cookii* | II$^r$ | I$^r$ | II$^r$ | I$^r$ | V$^+$ | I$^r$ | II$^+$ | V$^+$ | III$^+$ | II$^r$ | IV$^1$ | V$^{2a}$ | V$^5$ | V$^5$ | V$^+$ | III$^r$ | I$^r$ | | III$^+$ | IV$^1$ | II$^r$ | | | | | |
| *Poa annua* | | | | | | I$^r$ | I$^r$ | I$^r$ | III$^+$ | V$^5$ | I$^r$ | | | II$^r$ | V$^1$ | | I$^r$ | | | I$^r$ | I$^r$ | | | | | |
| *Marchantia berteroana* | | | | | | | I$^r$ | I$^r$ | | | | III$^1$ | V$^+$ | V$^1$ | | I$^r$ | | | | | I$^r$ | | | | I$^r$ | I$^r$ |
| *Leptodontium proliferum* | | | | | | | | | | | I$^r$ | II$^r$ | III$^r$ | V$^1$ | | | | | | I$^r$ | | | | | | |
| *Mielichhoferia campylocarpa* | | | | | | | | | | | | | I$^r$ | III$^+$ | | | | | | | | | | | | |
| Characteristic taxa of communities of the *Acaena magellanica - Brachythecium* complex | | | | | | | | | | | | | | | | | | | | | | | | | | |
| *Acaena magellanica* | | | | | II$^r$ | I$^r$ | I$^r$ | | | I$^r$ | I$^r$ | I$^r$ | I$^r$ | III$^+$ | | V$^{2b}$ | V$^4$ | V$^3$ | V$^5$ | V$^5$ | V$^4$ | I$^r$ | | II$^r$ | | |
| *Bryum spec.* | | | | | | | | | | | | | | | V$^1$ | | | | | | | | | | | |
| *Philonotis cf. angustifolia* | | | | | | | | | | | | | | | V$^3$ | | I$^r$ | | | | | | | | | |
| *Brachythecium subplicatum* | | | | | | | | | | I$^r$ | | | | | V$^5$ | V$^5$ | V$^5$ | I$^r$ | | | | | | | | |
| *cf. Cratoneuron filicinum* | | | | | | | | | | | | | | | III$^r$ | | IV$^{2m}$ | I$^r$ | | | | | | | | |
| *Brachythecium rutabulum* | | | | | | | | | | I$^r$ | | | | V$^{2m}$ | | | | II$^+$ | IV$^1$ | V$^5$ | V$^1$ | | | | | I$^r$ |
| *Agrostis stolonifera* | | | | | | | | | | I$^r$ | I$^r$ | | | IV$^1$ | | | | | | | V$^5$ | | | | | |
| *Pringlea antiscorbutica* | | | | | I$^r$ | | | | | | | | | I$^r$ | | | | | | I$^r$ | IV$^r$ | | | | | |
| *Stellaria media* | | | | | | | | | | | | | | I$^r$ | | | | | | II$^r$ | II$^r$ | | | | | |
| *Cerastium holosteoides* | | | | | | | | | | | | | | I$^r$ | | | | | | I$^r$ | III$^r$ | | | | | |
| *Drepanocladus uncinatus* | | | I$^r$ | | | II$^r$ | | | I$^r$ | I$^r$ | | I$^r$ | | IV$^+$ | | I$^r$ | V$^+$ | V$^5$ | V$^5$ | III$^+$ | V$^1$ | I$^r$ | I$^r$ | V$^5$ | V$^5$ | V$^5$ |

| Distichophylletum fasciculati | Blepharidophyllo densifolii - Clasmatocoleetum humilis ranunculetosum biternati | Blepharidophyllo densifolii - Clasmatocoleetum humilis clasmatocoleetosum humilis | Blepharidophyllo densifolii - Clasmatocoleetum humilis blechnetosum penna-marinae | Uncinio compactae - Ptychomnietum ringiani | Lycopodio magellanici - Jamesonielletum coloratae ranunculetosum biternati | Lycopodio magellanici - Jamesonielletum coloratae catagonietosum politi | degradation stage of mire vegetation (animal action and erosion) | degradation stage of mire vegetation (inundation) | Jamesonielletum grandiflorae | Bryo laevigati - Breutelietum integrifoliae uncinietosum compactae Breutelia-variant | Bryo laevigati - Breutelietum integrifoliae uncinietosum compactae Riccardia-variant | Bryo laevigati - Breutelietum integrifoliae acaenetosum magellanicae | Isopterygio pulchelli - Blechnetum penna-marinae uncinietosum compactae | Isopterygio pulchelli - Blechnetum penna-marinae brachythecietosum rutabuli | Isopterygio pulchelli - Blechnetum penna-marinae typicum | Jungermannio coniflorae - Racomitrietum crispuli | Andreaeo acutifoliae - Racomitrietum crispuli hymenophylletosum | Andreaeo acutifoliae - Racomitrietum crispuli hypnetosum cupressiformis | sociation os Azorella selago | community of Azorella selago and Acaena magellanica | sociation of Limosella australis | sociation of Potamogeton spec. | | |
|---|---|---|---|---|---|---|---|---|---|---|---|---|---|---|---|---|---|---|---|---|---|---|---|---|
| 10 | 28 | 13 | 8 | 12 | 13 | 9 | 5 | 8 | 6 | 15 | 3 | 5 | 9 | 12 | 12 | 7 | 8 | 4 | 8 | 4 | 1 | 1 | | |
| 7 | 8 | 10 | 16 | 11 | 15 | 16 | 5 | 6 | 13 | 12 | 14 | 10 | 18 | 11 | 10 | 15 | 16 | 19 | 4 | 13 | 1 | 1 | | |
| | | | | | | | | | | | | | | | | | | | | | | | *Crassula moschata* | c 1. |
| | | | | | | | | | | | | | | | | | | | | | | | *Eriopus apiculatus* | c 1.2. |
| | | | | | | | $II^{r}$ | | | | | | | | | | | | | | | | *Cotula plumosa* | d 1.1.;d 2.3. |
| | $I^{r}$ | $I^{r}$ | | $I^{r}$ | $I^{r}$ | | | $I^{r}$ | $I^{r}$ | | | $I^{r}$ | $I^{r}$ | | | | | | | | | | *Clasmatocolea vermicularis* | d 1.2.; d 2.4. |
| | $III^{+}$ | $IV^{+}$ | $II^{+}$ | $III^{r}$ | $III^{+}$ | $II^{r}$ | $V^{2m}$ | | $V^{+}$ | $II^{+}$ | | $III^{+}$ | $III^{+}$ | $I^{r}$ | $II^{r}$ | $II^{r}$ | $II^{r}$ | $IV^{+}$ | $IV^{+}$ | $IV^{+}$ | | | *Montia fontana* | c 2.; cc 2.1. |
| | | | | | | | $II^{1}$ | | | | | | | | | | | | | | | | *Callitriche antarctica* | c 2.; cc 2.1.; cc 2.2.; d 1.2. |
| | $I^{r}$ | $I^{r}$ | | | $I^{r}$ | $II^{r}$ | | | $I^{r}$ | | | | $IV^{+}$ | $II^{r}$ | $III^{r}$ | $II^{r}$ | | $IV^{+}$ | $III^{+}$ | $II^{r}$ | | | *Poa cookii* | c 2.; c 2.5. |
| | | | | | | | | | | | | | | | | | | | | | | | *Poa annua* | cc 2.2. |
| | $I^{r}$ | $II^{r}$ | $II^{r}$ | | | | $I^{r}$ | | $I^{r}$ | | | | | | | | | $II^{r}$ | | | | | *Marchantia berteroana* | c 2.5. |
| | | | | $I^{r}$ | | $I^{r}$ | $I^{r}$ | | | | | | $II^{r}$ | $I^{r}$ | $I^{r}$ | | | | | | | | *Leptodontium proliferum* | c 2.5. |
| | | | | | | | | | | | | | | | | | | | | | | | *Mielichhoferia campylocarpa* | c 2.5.2. |
| | | | | $I^{r}$ | $I^{r}$ | $III^{r}$ | $I^{r}$ | | $II^{r}$ | $I^{r}$ | | $V^{2b}$ | | $V^{2a}$ | $II^{r}$ | $V^{2a}$ | $II^{r}$ | $I^{r}$ | $II^{1}$ | $V^{4}$ | | | *Acaena magellanica* | d 2.5.2.; d4.6.2.; d4.8.2.; d5.1.2.;d6.1.<br>cc 3.2.; cc 3.3.; cc 3.4.; |
| | | | | | | | | | | | | | | | | | | | | | | | *Bryum spec.* | c 3.1.1. |
| | | | | | | | | | | | | | | | | | | | | | | | *Philonotis cf. angustifolia* | c 3.1.1. |
| | | | | | | | | | | | | | | | | | | | | | | | *Brachythecium subplicatum* | c 3.1. |
| | | | | | | | | | | | | | | | | | | | | | | | *cf. Cratoneuron filicinum* | c 3.1.3. |
| | | | | | | | | | | | | $I^{r}$ | | $IV^{1}$ | | | | | | $III^{+}$ | | | *Brachythecium rutabulum* | cc 3.3.; d 2.5.2.; d 5.1.2. |
| | | | | | | | | | | | | | | | | | | | | $II^{r}$ | | | *Agrostis stolonifera* | cc 3.4.; d 2.5.2. |
| | | | | | | | | | | | | | | | $I^{r}$ | | | $III^{+}$ | $II^{r}$ | $III^{+}$ | | | *Pringlea antiscorbutica* | d 6.2.2. |
| | | | | | | | | | | | | | | | | | | | | | | | *Stellaria media* | |
| | | | | | | | | | | | | | | | | | | | | $II^{r}$ | | | *Cerastium holosteoides* | |
| $III^{+}$ | $II^{r}$ | $IV^{+}$ | $IV^{+}$ | $II^{r}$ | $II^{r}$ | $II^{r}$ | $I^{r}$ | $III^{+}$ | $V^{+}$ | $II^{+}$ | $IV^{+}$ | $IV^{+}$ | $III^{+}$ | $V^{2b}$ | $III^{r}$ | | | | | $V^{2a}$ | | | *Drepanocladus uncinatus* | cc 3.2.; cc 4.2.; d 2.5.2.;<br>d 3.1.3.; d 5.1.2. |

| | *Cotulo plumosae - Crassuletum moschatae Crassula-variant*<br>12<br>4 | *Cotulo plumosae - Crassuletum moschatae intermediate variant*<br>10<br>4 | *Cotulo plumosae - Crassuletum moschatae Cotula-variant*<br>11<br>5 | *Crassulo moschatae - Clasmatocoleetum vermicularis*<br>5<br>4 | *Crassulo moschatae - Azorelletum selaginis*<br>7<br>9 | *Montio fontanae - Callitrichetum antarcticae agrostietosum magellanicae*<br>6<br>5 | *Montio fontanae - Callitrichetum antarcticae typicum Montia-variant*<br>5<br>4 | *Montio fontanae - Callitrichetum antarcticae typicum intermediate variant*<br>7<br>4 | *Montio fontanae - Callitrichetum antarcticae typicum Callitriche-variant*<br>7<br>3 | *Callitricho antarcticae - Poetum annuae*<br>8<br>5 | *Poo cookii - Cotuletum plumosae*<br>14<br>4 | *Montio fontanae - Clasmatocoleetum vermicularis*<br>7<br>9 | *Leptodontio proliferi - Poetum cookii inops*<br>7<br>7 | *Leptodontio proliferi - Poetum cookii brachythecietosum rutabuli*<br>9<br>13 | *Brachythecietum subplicati philonotidetosum*<br>2<br>9 | *Brachythecietum subplicati inops*<br>6<br>5 | *Brachythecietum subplicati breutelietosum integrifoliae*<br>6<br>8 | *Acaeno magellanicae - Drepanocladetum uncinati breutelietosum integrifoliae*<br>5<br>7 | *Acaeno magellanicae - Drepanocladetum uncinati brachythecietosum rutabuli*<br>4<br>7 | *Acaeno magellanicae - Brachythecietum rutabuli*<br>15<br>7 | *Acaeno magellanicae - Agrostietum stoloniferae*<br>5<br>8 | *community of Juncus scheuchzerioides Campylopus clavatus variant*<br>8<br>4 | *community of Juncus scheuchzerioides Ranunculus biternatus variant*<br>6<br>3 | *Junco scheuchzerioidis - Drepanocladetum uncinati juncetosum scheuchzerioidis*<br>6<br>5 | *Junco scheuchzerioidis - Drepanocladetum uncinati montietosum fontanae*<br>8<br>6 | *Junco scheuchzerioidis - Drepanocladetum uncinati clasmatocoleetosum humilis*<br>9<br>10 |
|---|---|---|---|---|---|---|---|---|---|---|---|---|---|---|---|---|---|---|---|---|---|---|---|---|---|---|
| Characteristic taxa of communities of the *Juncus scheuchzerioides - Blepharidophyllum densifolium* complex | | | | | | | | | | | | | | | | | | | | | | | | | | |
| *Juncus scheuchzerioides* | I^r | | I^r | II^r | II^r | II^r | I^r | | | II^r | | II^r | | | | I^r | | I^r | | | | V^2b | V^2a | V^2a | III^+ | III^+ |
| *Distichophyllum fasciculatum* | | | | | | | | | | | | | | | | | | | | | | I^r | | | | III^+ |
| *Blepharidophyllum densifolium* | | | | | | | | | | | | II^+ | | | | | | | | | | | | | | III^+ |
| *Clasmatocolea humilis* | | | | | | | | | | | I^r | I^r | | | | | | | | | | | | | | V^+ |
| *Leptoscyphus expansus* | | | | | | | | | | | | I^r | | | | | | | | | | | | I^r | II^r | V^+ |
| *Peltigera polydactyla* | | | | | I^r | I^r | | | | | | II^r | | | | | | | II | | | | | | I^r | II^+ |
| *Lepidozia laevifolia* | | | | | | | | | | | | | | I^r | | | | | | | | | | | | III^+ |
| *Campylopus arboricola* | | | | | | | | | | | | | | | | | | | | | | | | | | |
| *Andrewsianthus lancistipus* | | | | | | | | | | | | I^r | | | | | | | | | | | | | | II^r |
| *Schistochila carnosa* | | | | | I^r | | | | | | | | | | | | | | | | | | | | | |
| *Fossombronia australis* | I^r | I^r | | III^+ | II^r | | I^r | | | | I^r | II^r | II^r | | | | | | | | | | | | I^r | I^r |
| *Jensenia pisicolor* | | | | | | | | | | | | | | | | | | | | | | | | | | |
| *Uncinia compacta* | | | | | | | I^r | | | | | I^r | | | | | | | | | | II^r | | | | II^1 |
| *Ptychomnion ringianum* | | | | | | | | | | | | | | | | | | | | | | I^r | | | | I^r |
| *Jamesoniella colorata* | | | | | | | | | | | | | | | | | | | | | | I^r | | | | |
| *Racomitrium lanuginosum* | | | | | | | | | | | | | | | | | | | | | | I^r | | | | |
| *Lycopodium magellanicum* | | | | | | | | | | | | | | | | | | | | | | | | | | |
| *Jamesoniella grandiflora* | | | | | | | | | | | | II^r | | | | | | | | | | | | | | |
| *Riccardia cf. pinguis* | | | | | | | | | | | | | | | | | | | | | | | | | | |
| *Breutelia integrifolia* | | | | | | | | | | | | | | | | | V^1 | V^1 | | | | I^r | | | II^r | |
| *Bryum laevigatum* | | | | | | | | | | | | | | | | | | | | | | | | | I^r | |
| *Riccardia cf. multifida* | | | | | | | | | | | | | | | | | | | | | | | | | | I^r |
| *Hygroamblystegium filum* | | | | III^+ | | | | | | | | | | I^r | | | | I^r | | | II^+ | | | | | |
| *Acrobolbus ochrophyllus* | | | | | | | | | | | | | | | | | | | | | | | | | | |
| *Symphyogyna marionensis* | | | | | | | | | | | | | | | | | | | | | | | | | | |
| *Distichophyllum imbricatum* | | | | | I^r | | | | | | | | | | | | | | | | | | | | | |
| Characteristic taxa of communities of the *Blechnum penna-marina* complex | | | | | | | | | | | | | | | | | | | | | | | | | | |
| *Blechnum penna-marina* | | | | | | | I^r | | | | | | | I^r | | | | II^r | III^1 | III^1 | | | | I^+ | | II^+ |
| *Plagiothecium platyphyllum* | | | | | | | | | | | | I^r | I^r | II^r | | | | | | | | | | | | I^r |
| *Metzgeria decipiens* | | | | | | | | | | | | | I^r | | | | | | | | | | | | | |
| *cf. Amblystegium* | | | | | | | | | | | | | | | | | | | | II^1 | I^r | | | | | |
| *Isopterygium pulchellum* | | | | | | | | | | | | | | | | | | | | | | | | | | |
| *Peltigera canina* | | | | | | | | | | | | | | | | | | | | | | | | | | |
| *Campylopus polytrichoides* | | | | | I^r | | | | | | | | I^r | II^r | | | | | | | | | | | | |
| *Jungermannia coniflora* | | | | | | | | | | | | | | | | | | | | | | | | | | |
| *Plagiochila heterodonta* | | | | | I^r | | | | | | | | | | | | | | | | | | | | | |
| *Lophozia cylindriformis* | | | | | | | | | | | | I^r | | III^+ | | | | | | | | | | | | |
| *Campylopus clavatus* | | | | | | | | | | | | | | | | | | | | | | II^+ | | | | |
| *Cladonia coll.* | | | | | | | | | | | | | | | | | | | | | | | | | | |
| *Catagonium politum* | | | | | | | | | | | | I^r | | | | | | | | | | | | | | |
| *Colobanthus kerguelensis* | II^+ | I^r | I^r | | V^+ | | | | | | | | | | | | | | | | | I^r | | | | |
| *Azorella selago* | I^r | I^r | I^r | III^+ | V^3 | | | | | | I^r | II^r | II^r | I^r | | | | I^r | II^1 | I^r | I^r | I^r | | | | |

of Marion and Prince Edward Islands (continued).

| | *Distichophylletum fasciculati* | *Blepharidophyllo densifolii - Clasmatocoleetum humilis ranunculetosum biternati* | *Blepharidophyllo densifolii - Clasmatocoleetum humilis clasmatocoleetosum humilis* | *Blepharidophyllo densifolii - Clasmatocoleetum humilis blechnetosum penna-marinae* | *Uncinio compactae - Ptychomnietum ringiani* | *Lycopodio magellanici - Jamesonielletum coloratae ranunculetosum biternati* | *Lycopodio magellanici - Jamesonielletum coloratae catagonietosum politi* | *degradation stage of mire vegetation (animal action and erosion)* | *degradation stage of mire vegetation (inundation)* | *Jamesonielletum grandiflorae* | *Bryo laevigati - Breutelietum integrifoliae uncinietosum compactae Breutelia-variant* | *Bryo laevigati - Breutelietum integrifoliae uncinietosum compactae Riccardia-variant* | *Bryo laevigati - Breutelietum integrifoliae acaenetosum magellanicae* | *Isopterygio pulchelli - Blechnetum penna-marinae uncinietosum compactae* | *Isopterygio pulchelli - Blechnetum penna-marinae brachythecietosum rutabuli* | *Isopterygio pulchelli - Blechnetum penna-marinae typicum* | *Jungermannio coniflorae - Racomitrietum crispuli* | *Andreaeo acutifoliae - Racomitrietum crispuli hymenophylletosum* | *Andreaeo acutifoliae - Racomitrietum crispuli hypnetosum cupressiformis* | *sociation of Azorella selago* | *community of Azorella selago and Acaena magellanica* | *sociation of Limosella australis* | *sociation of Potamogeton spec.* |
|---|---|---|---|---|---|---|---|---|---|---|---|---|---|---|---|---|---|---|---|---|---|---|---|
| | 10 | 28 | 13 | 8 | 12 | 13 | 9 | 5 | 8 | 6 | 15 | 3 | 5 | 9 | 12 | 12 | 7 | 8 | 4 | 8 | 4 | 1 | 1 |
| | 7 | 8 | 10 | 16 | 11 | 15 | 16 | 5 | 6 | 13 | 12 | 14 | 10 | 18 | 11 | 10 | 15 | 16 | 19 | 4 | 13 | 1 | 1 |
| *Juncus scheuchzerioides* cc 4.2.; d 4.2.1. | $V^1$ | $I^r$ | $II^r$ | | $II^r$ | $II^r$ | $II^r$ | $I^r$ | $I^r$ | $III^+$ | $IV^1$ | $V^1$ | $II^r$ | $I^r$ | | | | | | | | | |
| *Distichophyllum fasciculatum* c 4.3.; d 4.2.3. | $V^5$ | $II^+$ | $II^+$ | $IV^+$ | $II^r$ | $I^r$ | $II^r$ | | | | $I^r$ | $II^r$ | | $II^r$ | $I^r$ | $I^r$ | | | | | | | |
| *Blepharidophyllum densifolium* c 4.4.; d 4.2.3.; d 4.4.1. | $IV^1$ | $V^5$ | $V^1$ | $V^5$ | $III^+$ | $V^1$ | $II^r$ | | $II^r$ | $IV^+$ | | $V^1$ | | $II^r$ | $I^r$ | | | | | | | | |
| *Clasmatocolea humilis* c 4.4.; d 4.2.3.; d 4.4.2.; d 4.8.1. + 4.4.3. | $IV^+$ | $V^1$ | $V^5$ | $V^1$ | $IV^+$ | $IV^+$ | $III^+$ | | $III^+$ | $IV^+$ | $III^+$ | $V^1$ | $I^r$ | $IV^+$ | $II^r$ | $I^r$ | $I^r$ | | | | $III^+$ | | |
| *Leptoscyphus expansus* d 4.2.3. | $V^1$ | $II^+$ | $III^+$ | $IV^+$ | $II^r$ | $II^r$ | $II^r$ | | | $V^1$ | $II^r$ | $II^r$ | $I^r$ | $III^+$ | $II^+$ | $IV^+$ | | | | | $II^r$ | | |
| *Peltigera polydactyla* d 4.2.3. | | $II^r$ | $II^+$ | $V^+$ | $III^r$ | $III^r$ | $II^r$ | | | $V^+$ | $I^r$ | | | $II^r$ | $I^r$ | $I^r$ | | | | $I^r$ | $III^r$ | | |
| *Lepidozia laevifolia* d 4.2.3. | | $III^+$ | $IV^+$ | $V^{2m}$ | $V^{2m}$ | $V^{2m}$ | $V^{2m}$ | | | $III^+$ | $I^r$ | $II^r$ | | $V^+$ | $II^+$ | $I^r$ | | $II^r$ | $III^+$ | $I^r$ | $II^r$ | | |
| *Campylopus arboricola* | | $II^+$ | $II^1$ | $II^+$ | $II^r$ | $II^r$ | $II^r$ | | | | | | | $I^r$ | | | | | | | | | |
| *Andrewsianthus lancistipus* | | $I^r$ | $II^r$ | $IV^+$ | $I^r$ | $II^r$ | $II^r$ | | | | $I^r$ | | | | | | | $I^r$ | | | | | |
| *Schistochila carnosa* | | $I^r$ | $II^r$ | | $I^r$ | $I^r$ | | | | | $I^r$ | | | | | | | $II^r$ | | | | | |
| *Fossombronia australis* d 4.4.3.; d 5.1.1. | $I^r$ | | $I^r$ | $III^+$ | $I^r$ | $I^r$ | $I^r$ | $I^r$ | | | | | | $III^+$ | $I^r$ | | | | | | | | |
| *Jensenia pisicolor* d 4.4.3.; d 4.8.1.; d 5.1.1. | | $I^r$ | $I^r$ | $III^+$ | $I^r$ | $IV^1$ | $III^+$ | | $II^+$ | $II^+$ | $III^+$ | $IV^+$ | | $III^+$ | | | $III^+$ | $II^r$ | | | | | |
| *Uncinia compacta* d 4.4.3.; d 4.8.1.; d 5.1.1.; d 6.1. | | $I^r$ | $I^r$ | $IV^1$ | $IV^1$ | $IV^1$ | $IV^+$ | | $IV^+$ | $I^+$ | $V^+$ | $V^+$ | $I^+$ | $V^+$ | $I^r$ | | $III^+$ | | | | | | |
| *Ptychomnion ringianum* c 4.5. | | | $I^r$ | | $V^5$ | $I^r$ | $I^r$ | | $I^r$ | $I^r$ | | $II^+$ | | | $I^r$ | | | | | | | | |
| *Jamesoniella colorata* c 4.6.(tr); c 4.6.1.; d 4.4.3.; d 4.8.1. | | | $I^r$ | $III^+$ | $II^r$ | $V^5$ | $V^1$ | | $III^+$ | | $I^r$ | $V^+$ | | $III^+$ | $I^r$ | $I^r$ | $I^r$ | $II^r$ | | | $III^+$ | | |
| *Racomitrium lanuginosum* c 4.6.(tr); c 4.6.2.; d 4.4.3. | | | $I^r$ | $II^+$ | $I^r$ | $IV^1$ | $V^5$ | | $II^r$ | $I^r$ | $I^r$ | $II^r$ | | $II^r$ | $I^r$ | $I^r$ | $III^1$ | $IV^+$ | $III^r$ | | $II^r$ | | |
| *Lycopodium magellanicum* c 4.6. | | | | | | $I^r$ | $II^r$ | | | | | | | | $I^r$ | | | $I^r$ | | | | | |
| *Jamesoniella grandiflora* c 4.7.; d 4.8.1. | | $I^r$ | $I^r$ | | $I^r$ | $I^r$ | $I^r$ | | $I^r$ | $V^5$ | $I^r$ | $II^r$ | | | $I^r$ | | $II^r$ | $I^r$ | $II^r$ | | $II^r$ | | |
| *Riccardia cf. pinguis* d 4.6. + 4.8.; d 4.8.1. | | | $I^r$ | | $I^r$ | $III^+$ | $I^r$ | | $II^r$ | | $V^+$ | $IV^+$ | | $I^r$ | | | | | | | | | |
| *Breutelia integrifolia* c 4.8.; d 3.1.3.; d 3.2.1. | | | | | $I^r$ | $II^r$ | | | | | $V^5$ | $IV^+$ | $V^5$ | $I^r$ | $I^r$ | | | | $I^r$ | | | | |
| *Bryum laevigatum* c 4.8. | | | | | $I^r$ | $I^r$ | | | | | $V^1$ | $II^r$ | $III^+$ | | $I^r$ | $I^r$ | | | | | | | |
| *Riccardia cf. multifida* c 4.8. | | | $I^r$ | | $I^r$ | $I^r$ | | | $II^r$ | | $III^+$ | $V^5$ | $V^+$ | | | | | | | | | | |
| *Hygroamblystegium filum* d 1.2.; d 4.8.2. | | | | | | | | | | | | | $IV^+$ | | | | | | | | | | |
| *Acrobolbus ochrophyllus* d 6.2.2. | | $I^r$ | $I^r$ | $I^r$ | $I^r$ | $II^r$ | $III^+$ | | | $I^r$ | | | | $II^r$ | $I^r$ | | | $I^r$ | $III^+$ | | | | |
| *Symphyogyna marionensis* d 4.6. + 4.8.; d 5.1.1. | | | | $II^r$ | | $II^r$ | $I^r$ | | | | $II^r$ | $II^r$ | | $III^+$ | | | $I^r$ | $II^r$ | | | | | |
| *Distichophyllum imbricatum* d 5.1.1. | | | | | | $I^r$ | $II^r$ | | | | | | | $III^r$ | | | | | | | | | |
| *Blechnum penna-marina* c 5.1.; d 6.1. | | $II^r$ | $I^r$ | $V^3$ | $IV^+$ | $IV^{2a}$ | $V^3$ | | | $II^+$ | $I^r$ | | $II^r$ | $V^5$ | $V^5$ | $V^5$ | $V^{2b}$ | $II^+$ | $I^r$ | | $III^+$ | | |
| *Plagiothecium platyphyllum* c 5.1.; d 4.6. | | | | | $I^r$ | $I^r$ | $II^r$ | | | | | | | $III^+$ | $III^+$ | $II^+$ | | | | | | | |
| *Metzgeria decipiens* c 5.1.; d 4.6.; d 4.6.2.; d 6.2.2. | | | | $III^+$ | | $I^r$ | $III^+$ | | | | | | | $IV^+$ | $I^r$ | $III^+$ | $III^+$ | | $III^+$ | | $IV^+$ | | |
| *cf. Amblystegium* c 5.1. | | | | $I^r$ | | | $I^r$ | | | $I^r$ | | | | $II^+$ | $III^1$ | $V^1$ | | | | | $II^r$ | | |
| *Isopterygium pulchellum* c 5.1. | | | | | | | $I^r$ | | | | | | | $II^r$ | $II^r$ | $III^+$ | | | | | | | |
| *Peltigera canina* c 5.1. | | | | | | | $I^r$ | | | | | | | $II^r$ | $I^r$ | $II^r$ | | | | | | | |
| *Campylopus polytrichoides* c 5.1. | | | | | | $I^r$ | $I^r$ | $I^r$ | | | | | | $III^+$ | | $II^r$ | $I^r$ | $I^r$ | | | | | |
| *Jungermannia coniflora* d 5.1.1.; d 6.1. | | | | | | | $I^r$ | | | $I^r$ | | | | $II^r$ | | | $III^+$ | $I^r$ | | | | | |
| *Plagiochila heterodonta* d 4.4.3.; d 4.6. | | | $I^r$ | $II^+$ | | $III^+$ | $IV^1$ | | $I^r$ | | $I^r$ | | | $IV^+$ | $I^r$ | $II^+$ | $III^+$ | $V^+$ | $IV^1$ | $I^r$ | $II^r$ | | |
| *Lophozia cylindriformis* d 2.5.2.; d 4.4.3.; d 4.6.2. | | $I^r$ | $I^r$ | $III^+$ | | | $III^+$ | | | $III^+$ | $I^r$ | | | $IV^+$ | $II^r$ | $IV^+$ | | | | | $II^+$ | | |
| *Campylopus clavatus* d 6.2.1. | | | | | | $II^r$ | | | $II^+$ | | | $IV^+$ | | $I^r$ | | | $III^r$ | $IV^+$ | | $I^r$ | | | |
| *Cladonia coll.* d 5.1.1. | | | | | | | $II^r$ | | | | | | | $III^+$ | | $I^r$ | $III^+$ | $IV^+$ | $III^r$ | | | | |
| *Catagonium politum* d 4.6.; d 4.6.2. | | | | | | | $III^+$ | | | | | | | $I^r$ | | $I^r$ | $I^r$ | $II^+$ | $IV^1$ | | | | |
| *Colobanthus kerguelensis* d 1.3.; d 6.2.2. | | | | | | | | | | | | | | | | | | | $IV^+$ | $IV^+$ | | | |
| *Azorella selago* d 1.3. | | $I^r$ | $I^r$ | $II^r$ | $III^+$ | $V^1$ | $IV^1$ | | $I^r$ | $V^1$ | $IV^+$ | $IV^+$ | $II^r$ | $V^{2a}$ | $III^{2a}$ | $III^+$ | $V^{2b}$ | $V^{2a}$ | $V^{2a}$ | $V^5$ | $V^5$ | | |

| | Cotulo plumosae - Crassuletum moschatae Crassula-variant | Cotulo plumosae - Crassuletum moschatae intermediate variant | Cotulo plumosae - Crassuletum moschatae Cotula-variant | Crassulo moschatae - Clasmatocoleetum vermicularis | Crassulo moschatae - Azorelletum selaginis | Montio fontanae - Callitrichetum antarcticae agrostietosum magellanicae | Montio fontanae - Callitrichetum antarcticae typicum Montia-variant | Montio fontanae - Callitrichetum antarcticae typicum intermediate variant | Montio fontanae - Callitrichetum antarcticae typicum Callitriche-variant | Callitricho antarcticae - Poetum annuae | Poo cookii - Cotuletum plumosae | Montio fontanae - Clasmatocoleetum vermicularis | Leptodontio proliferi - Poetum cookii inops | Leptodontio proliferi - Poetum cookii brachythecietosum rutabuli | Brachythecietum subplicati philonotidetosum | Brachythecietum subplicati inops | Brachythecietum subplicati breutelietosum integrifoliae | Acaeno magellanicae - Drepanocladetum uncinati breutelietosum integrifoliae | Acaeno magellanicae - Drepanocladetum uncinati brachythecietosum rutabuli | Acaeno magellanicae - Brachythecietum rutabuli | Acaeno magellanicae - Agrostietum stoloniferae | community of Juncus scheuchzerioides Campylopus clavatus variant | community of Juncus scheuchzerioides Ranunculus biternatus variant | Junco scheuchzerioidis - Drepanocladetum uncinati juncetosum scheuchzerioidis | Junco scheuchzerioidis - Drepanocladetum uncinati montietosum fontanae | Junco scheuchzerioidis - Drepanocladetum uncinati clasmatocoleetosum humilis |
|---|---|---|---|---|---|---|---|---|---|---|---|---|---|---|---|---|---|---|---|---|---|---|---|---|---|---|
| | 12 | 10 | 11 | 5 | 7 | 6 | 5 | 7 | 7 | 8 | 14 | 7 | 7 | 9 | 2 | 6 | 6 | 5 | 4 | 15 | 5 | 8 | 6 | 6 | 8 | 9 |
| | 4 | 4 | 5 | 4 | 9 | 5 | 4 | 4 | 3 | 5 | 4 | 9 | 7 | 13 | 9 | 5 | 8 | 7 | 7 | 7 | 8 | 4 | 3 | 5 | 6 | 10 |
| **Characteristic taxa of communities of the *Andreaea - Racomitrium crispulum* complex** | | | | | | | | | | | | | | | | | | | | | | | | | | |
| *crustose lichens* | | | | | $I^r$ | | | | | | | | | | | | | | | | | | | | | |
| *Racomitrium crispulum* | | | | | $I^r$ | | | | | | | | | | | | | | | | | | | | | |
| *Andreaea acuminata* | | | | | | | | | | | | | | | | | | | | | | | | | | |
| *Herzogobryum vermiculare* | | | | | | | | | | | | | | | | | | | | | | | | | | |
| *Andreaea regularis* | | | | | | | | | | | | | | | | | | | | | | | | | | |
| *Grammitis kerguelensis* | | | | | | | | | | | | | | | | | | | | | | | | | | |
| *Lycopodium saururus* | | | | | | | | | | | | | | | | | | | | | | | | | | |
| *Andreaea acutifolia* | | | | | | | | | | | | | | | | | | | | | | | | | | |
| *Ditrichum strictum* | | | | | | | | | | | | | | | | | | | | | | | | | | |
| *Grimmia kerguelensis* | | | | | | | | | | | | | | | | | | | | | | | | | | |
| *Philonotis scabrifolia* | | | | | | | | | | | | | | | | | | | | | | | | | | |
| *Psilopilum cf. tristaniense* | | | | | | | | | | | | | | | | | | | | | | | | | | |
| *Hymenophyllum peltatum* | | | | | | | | | | | | | | | | | | | | | | | | | | |
| *Bartramia patens* | | | | | | | | | | | | | $I^r$ | $I^r$ | | | | | | | | | | | | |
| *Verrucidens microcarpus* | | | | | | | | | | | | | | | | | | | | | | | | | | |
| *Holodontium pumilum* | | | | | | | | | | | | | | | | | | | | | | | | | | |
| *Hypnum cupressiforme* | | | | | | | | | | | | | | $II^+$ | | | | | | $I^r$ | $I^r$ | | | | | |
| *Ditrichum conicum* | | | | | | | | | | | | | | | | | | | | | | | | | | |
| **other taxa** | | | | | | | | | | | | | | | | | | | | | | | | | | |
| *Limosella australis* | | | | | | | | | | | | | | | | | | | | | | | | | | |
| *Potamogeton spec.* | | | | | | | | | | | | | | | | | | | | | | | | | | |
| *Agrostis magellanica* | $II^r$ | $I^r$ | $II^r$ | $III^+$ | $III^+$ | $V^4$ | $II^r$ | $II^r$ | $III^r$ | $II^r$ | $III^+$ | $V^1$ | $III^+$ | $I^r$ | $V^1$ | $III^+$ | $V^1$ | $V^{2b}$ | $II^r$ | $II^+$ | | $V^1$ | $IV^1$ | $V^{2m}$ | $V^3$ | $V^{2a}$ |
| *Ranunculus biternatus* | $IV^1$ | $V^1$ | $V^1$ | $IV^1$ | $III^+$ | $IV^1$ | $IV^1$ | $II^+$ | $III^+$ | $IV^1$ | | $IV^1$ | | $II^+$ | $V^1$ | $V^+$ | $IV^+$ | $III^+$ | $IV^+$ | $III^+$ | $I^r$ | $II^r$ | $V^1$ | $V^1$ | $V^1$ | $V^1$ |
| *Lophocolea randii* | $I^r$ | $I^r$ | $I^r$ | $I^+$ | $I^r$ | $I^r$ | | | | $I^r$ | $I^r$ | $II^+$ | $III^+$ | $IV^1$ | | $I^r$ | $I^r$ | $I^r$ | $III^1$ | $II^+$ | | | | | $II^+$ | $IV^+$ |
| *Campylopus cavifolius* | | | | | $I^r$ | | | | | | | | | | | | | | | | | | | | | |
| *Temnoma quadripartitum* | | | | $I^r$ | | | | | | | | | | | | | | | | | | | | | | |
| *Brachythecium paradoxum* | | | | | | | | | | $I^r$ | | | | | | | | | | $I^r$ | | | | | | |
| *Cephalozia randii* | | | | | | | | | | | | | $I^r$ | | | | | | | | | | | | | |
| *Plagiochila crozetensis* | | | | | | | | | | | | | | | | | | | | | | | | | | |
| *Campylopus introflexus* | | | | | | | | | | | | | | | | | | | | | | | | | | |
| *Campylopus subnitens* | | | | | | | | | | | | | $I^r$ | | | | | | | | | | | | | |

| *Distichophylletum fasciculati* | *Blepharidophyllo densifolii - Clasmatocoleetum humilis ranunculetosum biternati* | *Blepharidophyllo densifolii - Clasmatocoleetum humilis clasmatocoleetosum humilis* | *Blepharidophyllo densifolii - Clasmatocoleetum humilis blechnetosum penna-marinae* | *Uncinio compactae - Ptychomnietum ringiani* | *Lycopodio magellanici - Jamesonielletum coloratae ranunculetosum biternati* | *Lycopodio magellanici - Jamesonielletum coloratae catagonietosum politi* | *degradation stage of mire vegetation (animal action and erosion)* | *degradation stage of mire vegetation (inundation)* | *Jamesonielletum grandiflorae* | *Bryo laevigati - Breutelietum integrifoliae uncinietosum compactae Breutelia-variant* | *Bryo laevigati - Breutelietum integrifoliae uncinietosum compactae Riccardia-variant* | *Bryo laevigati - Breutelietum integrifoliae acaenetosum magellanicae* | *Isopterygio pulchelli - Blechnetum penna-marinae uncinietosum compactae* | *Isopterygio pulchelli - Blechnetum penna-marinae brachythecietosum rutabuli* | *Isopterygio pulchelli - Blechnetum penna-marinae typicum* | *Jungermannio coniflorae - Racomitrietum crispuli* | *Andreaeo acutifoliae - Racomitrietum crispuli hymenophylletosum* | *Andreaeo acutifoliae - Racomitrietum crispuli hypnetosum cupressiformis* | *sociation of Azorella selago* | *community of Azorella selago and Acaena magellanica* | *sociation of Limosella australis* | *sociation of Potamogeton spec.* | |
|---|---|---|---|---|---|---|---|---|---|---|---|---|---|---|---|---|---|---|---|---|---|---|---|
| 10 | 28 | 13 | 8 | 12 | 13 | 9 | 5 | 8 | 6 | 15 | 3 | 5 | 9 | 12 | 12 | 7 | 8 | 4 | 8 | 4 | 1 | 1 | |
| 7 | 8 | 10 | 16 | 11 | 15 | 16 | 5 | 6 | 13 | 12 | 14 | 10 | 18 | 11 | 10 | 15 | 16 | 19 | 4 | 13 | 1 | 1 | |
| | | | | | | | | | | | | | | | | $V^{+}$ | $IV^{1}$ | $IV^{+}$ | | | | | *crustose lichens* c 6. |
| | | | | | | | | | | | | | | | | $V^{+}$ | $V^{+}$ | $V^{+}$ | $I^{r}$ | | | | *Racomitrium crispulum* c 6. |
| | | | | | | | | | | | | | $I^{r}$ | | | $IV^{1}$ | $II^{+}$ | $III^{+}$ | | | | | *Andreaea acuminata* c 6. |
| | | | | | | | | | | | | | | | | $II^{r}$ | | $III^{+}$ | | | | | *Herzogobryum vermiculare* c 6.; d 6.2.2. |
| | | | | | | | | | | | | | | | | $II^{+}$ | $I^{r}$ | $II^{r}$ | | | | | *Andreaea regularis* c 6. |
| | | | | | | | | | | | | | | | | $II^{r}$ | $I^{r}$ | $II^{r}$ | | | | | *Grammitis kerguelensis* c 6. |
| | | | | | | | | | | | | | | | | $III^{r}$ | | | | | | | *Lycopodium saururus* c 6. |
| | | | | | | | | | | | | | | | | $I^{r}$ | $V^{2m}$ | $IV^{+}$ | | | | | *Andreaea acutifolia* c 6.2. |
| | | | | | | | | | | | | | | | | $I^{r}$ | $V^{1}$ | $IV^{+}$ | | | | | *Ditrichum strictum* c 6.2. |
| | | | | | | | | | | | | | | | | | $II^{+}$ | $II^{r}$ | | | | | *Grimmia kerguelensis* c 6.2. |
| | | | | | | | | | | | | | | $I^{r}$ | | | $IV^{+}$ | $III^{+}$ | | $II^{r}$ | | | *Philonotis scabrifolia* c 6.2. |
| | | | | | | | | | | | | | | | | | $IV^{+}$ | $III^{+}$ | | | | | *Psilopilum cf. tristaniense* c 6.2. |
| | | | | | | $I^{r}$ | | | | | | | | | | $I^{r}$ | $IV^{+}$ | $II^{r}$ | | | | | *Hymenophyllum peltatum* c 6.2. |
| | | | | | | | | | | | | | | | | | $III^{+}$ | $II^{r}$ | | | | | *Bartramia patens* c 6.2. |
| | | | | | | | | | | | | | | | | | $I^{r}$ | $II^{r}$ | | | | | *Verrucidens microcarpus* c 6.2. |
| | | | | | | | | | | | | | | | | | $I^{r}$ | $II^{+}$ | | | | | *Holodontium pumilum* c 6.2. |
| | | | $I^{r}$ | | | | | | | | | | $I^{r}$ | | | | $I^{r}$ | $IV^{+}$ | | | | | *Hypnum cupressiforme* d 6.2.2. |
| | | | | | | | | | | | | | | | | | | $III^{+}$ | | | | | *Ditrichum conicum* d 6.2.2. |
| | | | | | | | | | | | | | | | | | | | | | $I^{2m}$ | | *Limosella australis* c 7.1. |
| | | | | | | | | | | | | | | | | | | | | | | $I^{4}$ | *Potamogeton spec.* c 7.2. |
| $V^{2m}$ | $V^{2b}$ | $V^{2b}$ | $V^{2b}$ | $V^{3}$ | $V^{3}$ | $V^{2m}$ | $V^{5}$ | $V^{5}$ | $V^{4}$ | $V^{3}$ | $V^{4}$ | $V^{2b}$ | $IV^{+}$ | $IV^{+}$ | $IV^{r}$ | $V^{1}$ | $V^{1}$ | $IV^{+}$ | | $III^{1}$ | | | *Agrostis magellanica* d 2.1.1.; d 3.2.1.; d 4.6.1. |
| $III^{+}$ | $IV^{1}$ | $III^{+}$ | $II^{r}$ | $III^{+}$ | $III^{+}$ | $I^{r}$ | $II^{+}$ | | $V^{1}$ | $III^{+}$ | | $III^{+}$ | $II^{r}$ | $I^{r}$ | $II^{r}$ | $II^{r}$ | $I^{r}$ | $II^{r}$ | $I^{r}$ | $V^{+}$ | | | *Ranunculus biternatus* d 4.6.1. |
| $II^{+}$ | $I^{r}$ | $II^{r}$ | | | $I^{r}$ | | | | $II^{+}$ | $I^{r}$ | | $I^{r}$ | $I^{r}$ | $III^{+}$ | $I^{r}$ | | | $I^{r}$ | | $V^{1}$ | | | *Lophocolea randii* d 5.1.2. |
| $I^{r}$ | | | | $I^{r}$ | | | | | | $I^{r}$ | | | | | | | $I^{r}$ | $II^{+}$ | | | | | *Campylopus cavifolius* |
| | | | | $I^{r}$ | | | | | | $I^{r}$ | | | | | | | | | | | | | *Temnoma quadripartitum* |
| | | | | | $I^{r}$ | | | | | | | | $I^{r}$ | $I^{r}$ | $I^{r}$ | | | | | $II^{r}$ | | | *Brachythecium paradoxum* |
| | | | | | $I^{r}$ | | | | | $I^{r}$ | | | | | $I^{r}$ | | | | | | | | *Cephalozia randii* |
| | | | $II^{+}$ | | | | | | | | | | | | $I^{r}$ | | | | | | | | *Plagiochila crozetensis* |
| | | | $I^{r}$ | | | | | | | | | | $I^{r}$ | | | | | | | | | | *Campylopus introflexus* |
| | | | | | $I^{r}$ | | | | | $I^{r}$ | | | | | | | | | | | | | *Campylopus subnitens* |

CHAPTER FIVE

# Ecological outline and concluding remarks

Apart from the very pronounced influence of salt-spray and of trampling and manuring by birds and seals, the most prominent factor determining the distribution of the plant communities on Marion and Prince Edward Islands is the water-regime in the soil. Two aspects of this water-regime are most important:

(1) the amount of water available.
Extreme situations probably affect the vegetation most, viz. inundation in the wetter part of the wet-dry gradient, and at the dry end of the gradient the unavailability of water in the driest periods, which on the islands means after a few days without precipitation.

(2) the occurrence of lateral water movement on the soil surface or through the soil.
Such water flow is alleged to have a marked effect on the nutrient availability for the plants (Chapman 1965; Sparling 1966; Ingram 1967; cf. Par. 4.3 and 4.4).

More or less parallel with the wet-dry gradient often a change from organic to more mineral soils occurs, while frequently also a trend from somewhat sheltered towards strongly exposed conditions coincides with this gradient. Three major types of habitat are discerned along this complex gradient:

a) poorly drained lowland areas, where peat deposits have accumulated. The soils of these mire areas are highly organic and usually very wet, with a groundwater level usually within 40 cm from the soil surface;
b) well-drained lowland areas, not strongly exposed to the wind. These are mostly slopes, with a soil consisting of a layer of organic matter up to c. 30 cm thick overlying deposits of clay;
c) areas strongly exposed to the rigours of the climate. These include nearly all parts of the islands higher than c. 300 m above sea level, as well as those parts of the lowlands which are strongly exposed to the wind. The soils contain little organic matter and usually have a low water content.

Each of these three habitat-types has its own characteristic plant communities. Associated with the islands' mire areas are the communities of the *Juncus scheuchzerioides-Blepharidophyllum densifolium* complex, which are characterized by the prominence of bryophytes in the vegetation. *Agrostis magellanica* and *Uncinia compacta* usually dominate the herb layer (cf. Par. 4.4).

The well-drained slopes are mostly covered by communities of the *Blechnum penna-marina* complex, usually dominated by *Blechnum penna-marina.* The bryophyte stratum in these communities is usually not very prominent, which probably is a consequence of the drier habitat conditions here.

On exposed sites communities of the *Andreaea -Racomitrium crispulum* complex form the vegetation. Cushion plants (both vascular plant species (*Azorella selago*) and many bryophytes) are very prominent. The total cover of the vegetation is not high, commonly less than 10%, but the number of species is usually high (cf. Par. 4.6).

A fourth group of plant communities, the *Acaena magellanica-Brachythecium* complex (cf. Par. 4.3), is not associated with one of the habitat types mentioned above, but forms the plant cover of sites with a more or less strong lateral water flow, such as springs, flushes and drainage lines. These are usually associated with slopes, but are also locally found within the mire complex, e.g. along the edges of streams and in water tracks at the bottom of drainage lines.

Most of the plant communities making up the *Juncus scheuchzerioides-Blepharidophyllum densifolium* complex, the *Blechnum penna-marina* complex and the *Andreaea-Racomitrium crispulum* complex, as well as some communities from the restgroup (cf. Par. 4.7), can be arranged according to their habitat preferences into a single series from wet to dry (Table 33). At the wet end of the gradient the series starts with the *Ranunculus biternatus* variant of the community of *Juncus scheuchzerioides,* which occurs in bog ponds, in 20–80 cm deep water. Along the edges of the ponds and on the wettest sites on peat soils *Drepanocladus uncinatus* enters the vegetation, and the Junco scheuchzerioidis-Drepanocladetum uncinati is the second community in the series. The three subassociations of the Junco-Drepanocladetum occur along the gradient from wet to dry in the following order: juncetosum scheuchzerioidis, montietosum fontanae and clasmatocoleetosum humilis. Next in the series is the Distichophylletum fasciculati, which towards the dry side is followed by the Blepharidophyllo densifolii-Clasmatocoleetum humilis. The subassociation ranunculetosum biternati usually occupies wetter sites than the subassociation clasmatocoleetosum humilis, while the subassociation blechnetosum penna-marinae occurs on drier peat

*Table 33.* Marion and Prince Edward Island plant communities, arranged in series from wet to dry.

| Soil water content (%) | Communities of sites with no pronounced lateral water flow | Communities of sites with a pronounced lateral water flow |
|---|---|---|
| Open water | Comm. of *Juncus scheuchzerioides* (*Ranunculus*-variant) | Brachythecietum subplicati philonotidetosum |
| | Junco–Drepanocladetum juncetosum | |
| 3000 | Junco–Drepanocladetum montietosum | Brachythecietum subplicati inops |
| | Junco–Drepanocladetum clasmatocoleetosum | |
| | | Brachythecietum subplicati breutelietosum |
| | Distichophylletum fasciculati | |
| | | Acaeno–Drepanocladetum breutelietosum |
| 2000 | | |
| | Blepharidophyllo–Clasmatocoleetum ranunculetosum | (Bryo–Breutelietum uncinietosum) |
| | | (Bryo–Breutelietum acaenetosum) |
| | Blepharidophyllo–Clasmatocoleetum clasmatocoleetosum | |
| | Blepharidophyllo–Clasmatocoleetum blechnetosum | |
| 1000 | | Acaeno–Drepanocladetum brachythecietosum |
| | Lycopodio–Jamesonielletum ranunculetosum | |
| | Lycopodio–Jamesonielletum catagonietosum | |
| | | Acaeno–Brachythecietum rutabuli |
| | Isopterygio–Blechnetum uncinietosum | |
| 500 | Isopterygio–Blechnetum brachythecietosum | Acaeno–Agrostietum stoloniferae |
| | Isopterygio–Blechnetum typicum | |
| | Community of *Azorella* and *Acaena* | |
| | Sociation of *Azorella selago* | |
| 100 | Jungermannio–Racomitrietum crispuli | |
| | Andreaeo–Racomitrietum crispuli | |

soils than the two others. The Uncinio compactae-Ptychomnietum ringiani has about the same position as the Blepharidophyllo-Clasmatocoleetum along the wet-dry gradient, but occurs on soils containing more mineral matter than most mire soils. The next association in the series is the Lycopodio magellanici-Jamesonielletum coloratae, which occurs on relatively dry peat soils with a groundwater level usually between 15 and 25 cm below the surface. The subassociation ranunculetosum biternati occupies wetter sites than the subassociation catagonietosum politi. The Jamesonielletum grandiflorae occupies a similar position as the Lycopodio magellanici-Jamesonielletum coloratae in the series from wet to dry, but prefers more mineral soils, and is mostly found along the banks of little streams on deposits of sand or gravel or of scoriae, which are covered by a thin layer of peat. When leaving the Uncinio compactae-Ptychomnietum ringiani and the Jamesonielletum grandiflorae out of consideration, the part of the wet-dry series described sofar can also be interpreted as a successional series of mire communities. The number of species/relevé tends to increase markedly along the gradient from wet to dry (and thus from pioneer communities to communities of a more mature developmental stage), from 3 species in the *Juncus scheuchzerioides* community to on average 16 species/relevé in the Lycopodio magellanici-Jamesonielletum coloratae catagonietosum politi (cf. Table 32).

The series from wet to dry continues with the communities of the *Blechnum penna-marina* complex. This complex contains only one association, the Isopterygio pulchelli-Blechnetum penna-marinae. The subassociation uncinietosum, which contains a number of species which have their optimum in mire vegetation, occurs on comparatively moist slopes, but sometimes also on well-drained peat soils. On drier slopes, usually somewhat influenced by lateral groundwater movement, the subassociation brachythecietosum rutabuli is found, while on still drier slopes the subassociation typicum occurs. In this part of the series the number of species decreases again (cf. Table 32), probably as a consequence of the more unfavourable water regime on the dry slopes.

Towards the dry side of the Isopterygio pulchelli-Blechnetum penna-marinae follows the community of *Azorella selago* and *Acaena magellanica* (Par. 4.7). Floristically this community does not fit well in the series, but according to its habitat preferences it must be placed here. It is followed by the sociation of *Azorella selago*. This community, which is very poor in species, also breaks the clear floristic gradient in this series (cf. Table 32), but again belongs here on the basis of its habitat requirements. It occurs in fjaeldmark areas, on sites where influx of water from surrounding areas and some shelter from the wind cause a somewhat more favourable water-regime.

The end of the series is formed by the fjaeldmark vegetation (*Andreaea-Racomitrium crispulum* complex; Par. 4.6). On exposed sites in the lowlands stands of the Jungermannio coniflorae-Racomitrietum crispuli occur. The vegetation of very strongly exposed lowland sites (e.g. on grey lava ridges) and of more elevated areas consists largely of Andreaeo acutifoliae-Racomitrietum crispuli hymenophylletosum peltati. The soils in these areas contain very little organic matter and commonly consist of a large proportion of rock fragments. Their water retention capacity is small. During the prolonged and frequent periods of rainy weather sufficient water seems to be available even in these areas, but during the occasional dry spells when evaporation is greatly enhanced by the strong winds water shortage becomes an important factor here too. As an adaption to these occasional periods of drought many of the species in the fjaeldmark vegetation possess a cushion-like growth form, while several other species in these communities are largely restricted to strongly sheltered microhabitats such as the leeward sides of boulders and *Azorella*-cushions, along or nearly under the edges of stones, or in cushions of *Azorella* or other species. A special type of such sheltered microhabitats is found in scoriae deposits, where many species of bryophytes find shelter and a more favourable moisture regime in the hollows between the loosely heaped lava fragments. The vegetation of these areas, where an important portion of the plants grow beneath the surface, is formed by the Andreaeo acutifoliae-Racomitrietum crispuli hypnetosum cupressiformis. In the final part of the wet-dry series the number of species/relevé increases again, probably as a result of the large variety in microhabitats available in the fjaeldmark areas (cf. Par. 4.6).

The plant communities of sites with a lateral

water flow at the soil surface or through the soil (*Acaena magellanica-Brachythecium* complex; Par. 4.3) can also be arranged into a series from wet to dry (Table 33). The series starts with the Brachythecietum subplicati, a community occurring in springs and along the edges of streams. The subassociation philonotidetosum is found on mineral soils in shallow, comparatively fast flowing water. The subassociations inops and breutelietosum integrifoliae occur on peat deposits in springs and flushes, the former usually in faster running water than the latter. On somewhat less wet sites the Acaeno magellanicae-Drepanocladetum uncinati is met with, the subassociation breutelietosum integrifoliae on wet peat soils, the subassociation brachythecietosum rutabuli on drier soils. The next community in this series is the Acaeno magellanicae-Bracythecietum rutabuli, which is found on well-drained sites along streams and in drainage lines on slopes. The final community of the series, the Acaeno magellanicae-Agrostietum stoloniferae, is found in comparatively dry habitats, such as drainage lines on steep slopes and gravelly deposits along streams. Considering the species composition of the communities of this series (cf. Tables 25 and 32), a gradual change can be seen when going from wet to dry. In the same direction the initial preponderance of bryophytes in the vegetation changes to a dominance of vascular species in the dry habitats. A similar trend is present in the corresponding part of the wet-dry series of communities from habitats with no marked lateral water flow (cf. Tables 32 and 33). All communities of this series mentioned sofar belong to the *Acaena magellanica-Brachythecium* complex. One association of the *Juncus scheuchzerioides-Blepharidophyllum densifolium* complex, the Bryo laevigati-Breutelietum integrifoliae, is also found in habitats with a strong lateral water flow, notably in water tracks in mire areas, and should find a place in this series. Its position is about the same as that of the Acaeno magellanicae-Drepanocladetum uncinati breutelietosum integrifoliae. This community does not fit in the floristic gradient in the series. Its characteristic habitat, being water tracks in mire areas, is also different from that of the other communities in this series which generally are associated with more minerotrophic conditions.

In saline habitats and in areas affected by trampling and manuring by birds and seals the vegetation differs strongly from that of the other parts of the islands (cf. Table 32).

Only a small number of plant communities occur in areas influenced by seawater or salt-spray. The distribution of these communities, which are grouped together in the *Crassula moschata* complex (Par. 4.1), is largely determined by the intensity of the salt-spray deposition and by the water-regime of the soil. On those sites most strongly affected by seawater and salt-spray the vegetation is Cotulo plumosae-Crassuletum moschatae, notably the variant dominated by *Crassula moschata*. In somewhat less saline habitats soil moisture conditions determine the distribution of the communities to a large extent: on very wet, organic soils the Crassulo moschatae-Clasmatocoleetum vermicularis occurs, on well-drained, organic soils the vegetation is Cotulo-Crassuletum, while on sites with dry, mineral soils, exposed to the winds, the Crassulo moschatae-Azorelletum selaginis is met with. The prominence of cushion plants in the vegetation of these sites (cf. Fig. 27) reflects the severe environmental conditions.

The vegetation of areas strongly influenced by animals shows little relation to the other communities occurring on the islands. The most prominent species in the vegetation of these areas (*Callitriche antarctica-Poa cookii* complex; Par. 4.2) are *Montia fontana, Callitriche antarctica, Poa cookii, Cotula plumosa* and *Poa annua*. The nutrient status of the soils they grow on differs appreciably from that of soils of areas not influenced by animals (cf. Table 9), while frequent trampling in these areas has a considerable, sometimes devastating influence on soil and plant cover. The major factors influencing the distribution of plant communities of these areas strongly influenced by animals are the intensity of the influence of the animals and the water content and content of organic matter of the soil. The relation between habitat conditions and the distribution of the plant communities of this complex has already been described in Par. 4.2, and is represented diagrammatically in Fig. 36.

A final conclusion from the present study concerns the applicability of the Braun-Blanquet approach to vegetation classification in Subantarctic regions. The results of this study show that, despite the relatively small number of species and the wide ecological amplitude of some of them, the differences in habitat-range between the species are pronounced enough to result in a large variation in the plant cover of the islands. This variation allows the distinction of a large number of floristically clearly differentiated plant communities, the distribution of which in nearly all cases could be related to environmental factors.

# Summary

This study gives an account of the plant communities of the Subantarctic Islands Marion and Prince Edward.

Marion and Prince Edward Islands (46° 54′ S, 37° 45′ E) are comparatively young, volcanic islands, associated with the West Indian Ocean Ridge. They form small specs of land in the vast Southern Ocean, far removed from any continent (Africa is 1800 km to the NNW, Antarctica 2300 km to the South). The nearest land is formed by the Iles Crozet, 945 km east of the islands. Marion Island measures 290 km$^2$, Prince Edward 44 km$^2$. The islands lie 22 km apart. They are the summits of shield volcanoes and their age is estimated to be about 0.5 million years. Two periods of volcanic activity have resulted in two distinct series of lavas. Fine-grained, grey basaltic lava originated from eruptions between c. 275 000 and 100 000 years BP. A period of volcanic activity about 15 000 years ago produced strongly vesicular, black lava. Recently (1980) a small eruption took place at Kaalkoppie on Marion Island. Between the two periods of volcanic activity glaciers have covered Marion Island nearly completely. The glaciations are of the same period as the Würm- and Riss-glaciations in Europe. During these ice ages the Antarctic Convergence, the boundary between the cold Antarctic surface water and the c. 2 °C warmer Subantarctic surface water, which is now situated about 250 km to the south of Marion and Prince Edward Islands, had shifted to a position north of the islands.

Marion Island consists of a central mountain area, reaching up to 1230 m above sea level and sloping down in all directions towards the sea. Along the west- and southcoasts this slope abruptly ends in a 200–300 m high escarpment, which steeply descends towards a coastal plain. A radial pattern of elevated grey lava ridges, probably the result of radial faulting and separating low-lying areas covered by black lava, is characteristic of the island. About 130 scoriae cones mark the centres of eruptions of the second volcanic stage. The highland of Prince Edward Island, the summit of which reaches 672 m, gently slopes down towards the East. On the western side it ends abruptly in a c. 400 m high escarpment with descends towards a coastal plain. Towards the north and south the highland ends in high, precipitous cliffs, descending steeply towards the sea.

The climate of the islands is extremely oceanic, with an average annual temperature of 5 °C which shows very little diurnal and seasonal variation, a very high precipitation (2575 mm annually) and humidity, and high wind speeds. Three soil-categories can be discerned: peat soils, soils of lowland slopes consisting of a layer of organic matter overlying a layer of clay, and fjaeldmark soils which consist of clay usually containing many rock fragments. The latter soils are sometimes very shallow. Weathering plays a minor role in supplying nutrients to the soils. The major source of mineral nutrients is the ocean, directly via salt-spray (e.g. Na, Mg and K) or indirectly via excreta and other depositions by sea-going animals (notably N and P).

Marion and Prince Edward Islands are situated within the Subantarctic Zone, i.e. south of the southern limit of tree growth and the occurrence of woody shrubs and north of the southern limit of closed phanerogam communities. The flora of the islands shows a strong affinity to that of the other islands of the Kerguelen Province (Crozet, Kerguelen, Heard and Macdonald Islands). The total number of species is small: 22 indigenous species of vascular plants occur, and 72 mosses, 36 hepatics and 50 species of lichens have been identified. Of a further two species of vascular plants on the islands the status has not been ascertained conclusively. The influence of man on the islands' ecosystems is comparatively small. Although sealing must have had a devastating effect on the population-sizes of Elephant Seal and Fur Seal, these seem to have recovered again. Practically all further influence is indirect; it is a result of, in most cases inadvertent, introductions of plant and animal species to the islands. On Marion 14 species of vascular plants have been introduced, of which four have disappeared again. Three species of vertebrates were introduced: cats, mice and trout. Of these the cats form a pest, heavily influencing the populations of small birds. Some insects appear to have been introduced also. The only alien species recorded from Prince Edward Island is *Poa annua*.

The vegetation of Marion and Prince Edward Islands has been studied using the methods of the Braun-Blanquet School. Forty-one plant communities (on the level of subassociations) were distinguished, and their floristic composition, structure, habitat-characteristics and distribution on the islands has been described. All syntaxa in this study are newly described. As far as the data available in the literature allowed, remarks have been made on the distribution of these communities in other areas. The communities are grouped into six community-complexes which comprise floristically and ecologically related syntaxa.

The first group of communities, united into the *Crassula moschata* complex, forms the vegetation of areas strongly affected by salt-spray. Three associations are discerned: Cotulo plumosae-Crassuletum moschatae, Crassulo moschatae-Clasmatocoleetum vermicularis and Crassulo moschatae-Azorelletum selaginis. The distribution of these associations is largely determined by the intensity of salt-spray deposition and by the soil water regime.

The vegetation of areas strongly influenced by trampling and manuring by animals is formed by the communities of the *Callitriche antarctica-Poa cookii* complex, viz. Montio fontanae-Callitrichetum antarcticae, Callitricho antarcticae-Poetum annuae, Poo cookii-Cotuletum plumosae, Montio fontanae-Clasmatocoleetum vermicularis and Leptodontio proliferi-Poetum cookii. The distribution of these communities is largely dependent on the intensity of the influences by animals and on the water content and the proportion of organic matter in the soil.

Communities dominated by *Acaena magellanica*, united in the *Acaena magellanica-Brachythecium* complex, form the vegetation of springs, flushes, drainage lines and other sites with a pronounced lateral water-flow just above or below the soil surface. The associations which make up this complex form a series from extremely wet habitats (e.g. springs) to comparatively dry sites (e.g. drainage lines on steep slopes) in the following order: Brachythecietum subplicati, Acaeno magellanicae-Drepanocladetum uncinati, Acaeno magellanicae-Brachythecietum rutabuli and Acaeno magellanicae-Agrostietum stoloniferae.

The communities which occur in oligotrophic mires form the *Juncus scheuchzerioides-Blepharidophyllum densifolium* complex. Most of the communities of this complex can be arranged into a single series, ranging in habitat from very wet (bog ponds) to comparatively dry (peat deposits with a groundwater level below 25 cm) in the following order: Community of *Juncus scheuchzerioides* (*Ranunculus biternatus* variant), Junco scheuchzerioidis-Drepanocladetum uncinati, Distichophylletum fasciculati, Blepharidophyllo densifolii-Clasmatocoleetum humilis, Uncinio compactae-Ptychomnietum ringiani and Lycopodio magellanici-Jamesonielletum coloratae. This series can also be interpreted as a successional series. The vegetation on peat soils with a comparatively high proportion of mineral matter often is the Jamesonielletum grandiflorae. In water tracks through the mire expanse the Bryo laevigati-Breutelietum integrifoliae occurs.

On well-drained slopes the vegetation is the Isopterygio pulchelli-Blechnetum penna-marinae, the only association of the *Blechnum penna-marina* complex.

The vegetation of fjaeldmark areas is formed by the communities of the *Andreaea-Racomitrium crispulum* complex: Jungermannio coniflorae-

Racomitrietum crispuli in the lowlands, and Andreaeo acutifoliae-Racomitrietum crispuli at strongly exposed sites in lowland areas and at higher elevations. These communities show a very complex structure and are comparatively rich in species. This is a consequence of the complex nature of the fjaeldmark substrate and the cushion-forming habit of many species in this type of vegetation. In one of the subassociations of the Andreaeo-Racomitrietum plants grow for the most part below the surface in the hollows between the rocks making up the scoriae deposits on which this community occurs. Moss balls, unattached globular colonies of mosses, are locally abundant in fjaeldmark areas.

An ecological outline of the vegetation of Marion and Prince Edward Islands is given in which the most important environmental gradients are described, and the communities are arranged in series along these gradients. The most prominent factor influencing the distribution of the plant communities on the islands is the soil moisture regime. Gradients from organic to mineral soils and from somewhat sheltered to strongly exposed conditions run more or less parallel with the wet-dry gradient. Two series along wet - dry gradients can be formed, one comprising communities from habitats with no pronounced lateral water movement, the other consisting of communities from springs, water tracks and drainage lines, where lateral water-flow is pronounced. The first series contains most communities of the *Juncus scheuchzerioides-Blepharidophyllum densifolium* complex, the *Blechnum penna-marina* complex and the *Andreaea-Racomitrium crispulum* complex, while the second series consists of the communities of the *Acaena magellanica-Brachythecium* complex and one association of the *Juncus-Blepharidophyllum* complex. Only the communities of the *Callitriche antarctica-Poa cookii* complex and the *Crassula moschata* complex, i.e. the vegetation of areas strongly influenced by animals or by salt-spray, do not fit into these series. They differ, floristically, strongly from all other vegetation types on the islands.

# Appendices

**Appendix 1. Infrequent taxa occurring in the relevés of Table 21 (*Crassula moschata* complex)**

| | | | | |
|---|---|---|---|---|
| Acaena magellanica | 261: + | 262: + | Peltigera polydactyla | 262: + |
| Bryum dichotomum | 261: r | | Philonotis scabrifolia | 279: r |
| Campylopus polytrichoides | 600: + | | Pringlea antiscorbutica | 262: + |
| Cladonia coll. | 279: + | | Sagina apetala | 45: r |
| Distichophyllum imbricatum | 505: + | | Schistochila carnosa | 505: + |
| Drepanocladus uncinatus | 67: + | 68: + | Temnoma quadripartitum | 273: + |

**Appendix 2. Infrequent taxa occurring in the relevés of Table 23 (*Callitriche antarctica-Poa cookii* complex)**

| | | | | | |
|---|---|---|---|---|---|
| Andrewsianthus cf. carinatus | 473: + | | Clasmatocolea humilis | 567: + | 560: 1 |
| Andrewsianthus lancistipus | 291: r | | | 419: + | |
| Bartramia patens | 419: 1 | 554: 1 | Hygroamblystegium filum | 428: 1 | |
| Blechnum penna-marina | 554: 2a | | Hypnum cupressiforme | 427: 1 | 428: 1 |
| Blepharidophyllum densifolium | 210: 2b | 291: 1 | | 437: 1 | |
| | 560: 2a | | Jamesoniella grandiflora | 560: + | 175: + |
| Brachythecium paradoxum | 176: + | | | 199: 1 | |
| Brachythecium spec. | 76: + | | Lepidozia laevifolia | 437: + | |
| Bryum amblyolepsis | 584: 2m | | Leptoscyphus expansus | 560: 1 | 419: 1 |
| Bryum argenteum | 427: 1 | | Metzgeria decipiens | 322: + | |
| Bryum dichotomum | 457: + | 421: + | Metzgeria cf. grollei | 427: r | 437: + |
| Bryum spec. | 76: + | | Peltigera polydactyla | 75: + | 210: r |
| Campylopus polytrichoides | 458: 1 | 437: + | Peltigera spec. | 560: + | |
| | 419: 1 | | Plagiothecium platyphyllum | 291: 2a | 129: 2m |
| Campylopus subnitens | 419: + | | | 457: + | 554: + |
| Campylopus spec. | 377: + | 458: 1 | Pohlia nutans | 436: + | |
| Catagonium politum | 291: + | | Pringlea antiscorbutica | 437: r | |
| Cephalozia randii | 377: + | | Sagina apetala | 582: 1 | |
| cf. Ceratodon purpureus | 76: 2b | | Stellaria media | 427: + | 554: + |
| Cerastium holosteoides | 436: r | 437: r | Uncinia compacta | 291: 1 | |

**Appendix 3. Infrequent taxa occurring in the relevés of Table 25 (*Acaena magellanica-Brachythecium* complex)**

| | | | |
|---|---|---|---|
| Brachythecium paradoxum | 387: 1 | 370: 1 | |
| Brachythecium spec. | 508: + | | |
| Breutelia pendula | 459: + | | |
| Bryum laevigatum | 312: + | | |
| Callitriche antarctica | 309: 1 | 310: 1 | |
| Clasmatocolea vermicularis | 252: + | 324: 2a | |
| Cotula plumosa | 252: 3 | | |
| Hypnum cupressiforme | 167: 1 | 426: + | |
| | 370: 1 | | |
| Juncus scheuchzerioides | 380: 2a | 540: + | |
| Leptoscyphus expansus | 508: + | 324: 1 | |
| Marchantia berteroana | 380: + | 219: r | |
| Mielichhoferia campylocarpa | 252: 1 | | |
| Peltigera polydactyla | 557: + | | |
| Philonotis scabrifolia | 508: r | | |
| cf. Riccardia | 324: 1 | | |

**Appendix 4. Infrequent taxa occurring in the relevés of Table 26 (*Juncus scheuchzerioides-Blepharidophyllum densifolium* complex)**

| | | | |
|---|---|---|---|
| Agrostis stolonifera | 214: + | | |
| cf. Amblystegium | 204: + | 545: + | |
| Andrewsianthus marionensis | 162: + | | |
| Andrewsianthus spec. | 217: 1 | 412: + | 346: + |
| | 409: + | | |
| Brachythecium paradoxum | 545: + | | |
| Brachythecium rutabulum | 472: + | 202: r | |
| Brachythecium spec. | 30: + | | |
| Callitriche antarctica | 435: 1 | 564: 2a | 259: 3 |
| | 254: 2a | 606: r | |
| Campylopus cavifolius | 413: + | | |
| Campylopus introflexus | 298: + | | |
| Campylopus polytrichoides | 131: r | 495: + | 687: r |
| | 538: + | | |
| cf. Cephalozia randii | 301: + | 243: + | |
| Colobanthus kerguelensis | 339: + | | |
| Cotula plumosa | 254: + | | |
| Hymenophyllum peltatum | 157: + | | |
| Hypnum cupressiforme | 171: + | 495: + | |
| Isopterygium pulchellum | 495: + | 545: 1 | |
| Jungermannia coniflora | 532: + | 449: + | |
| Leptodontium proliferum | 131: + | 494: + | 98: 1 |
| cf. Lethocolea radicosa | 162: + | | |
| cf. Lophocolea | 485: + | | |
| Metzgeria marionensis | 402: + | 409: + | 404: + |
| Peltigera canina | 415: + | | |
| Plagiochila crozetensis | 298: 1 | 30: + | 461: + |

**Appendix 5. Infrequent taxa occurring in the relevés of Table 28 (*Blechnum penna-marina* complex)**

| | | | |
|---|---|---|---|
| Andreaea acuminata | 441: + | | |
| Brachythecium paradoxum | 269: + | 445: + | 379: + |
| Breutelia integrifolia | 441: + | 464: r | |
| Bryum laevigatum | 268: r | 539: r | |
| Campylopus arboricola | 290: 1 | | |
| Campylopus clavatus | 295: 1 | | |
| Campylopus introflexus | 199: 1 | | |
| Catagonium politum | 552: + | 158: + | 379: + |
| Cephalozia randii | 379: + | | |
| cf. Cephaloziella | 290: + | | |
| Clasmatocolea vermicularis | 295: + | | |
| Cotula plumosa | 132: r | | |
| Distichophyllum fasciculatum | 552: r | 295: + | 290: + |
| | 539: + | | |
| Festuca rubra | 533: (+) | | |
| Jamesoniella grandiflora | 269: 1 | 464: 1 | |
| Juncus scheuchzerioides | 137: 1 | | |
| Lycopodium magellanicum | 285: + | | |
| Marchantia berteroana | 132: + | | |
| Philonotis scabrifolia | 269: r | 544: + | |
| Plagiochila crozetensis | 533: + | | |
| Pohlia nutans | 441: + | | |
| cf. Pohlia | 290: + | | |
| Pringlea antiscorbutica | 539: r | | |
| Ptychomnion ringianum | 269: 1 | 268: 1 | |
| Riccardia cf. multifida | 334: 1 | | |
| Riccardia cf. pinguis | 441: + | | |
| Tylimanthus viridis | 295: 1 | | |

**Appendix 6. Infrequent taxa occurring in the relevés of Table 29 (*Andreaea-Racomitrium crispulum* complex)**

| | | | | |
|---|---|---|---|---|
| Andreaea gainii | 315: r | | Funaria laxa | 443: + |
| Andrewsianthus lancistipus | 451: + | | Lophocolea randii | 313: + |
| Blindia magellanica | 442: + | | Lycopodium magellanicum | 165: + |
| Breutelia integrifolia | 383: + | | Marchantia berteroana | 313: r |
| Campylopus polytrichoides | 163: r | 451: + | cf. Parmelia | 351: r |
| Ceratodon purpureus | 307: + | | Stereocaulon cymosum | 306: + |
| cf. Cheilotheca | 529: + | 165: + | Tylimanthus viridis | 315: + |
| Clasmatocolea humilis | 443: + | | Usnea cf. insularis | 351: + |
| Dicranoweisia brevipes | 372: 1 | | Usnea spec. | 351: r |
| Dicranoweisia breviseta | 281: 1 | 282: + | Willia austro-leucophaea | 281: + |
| Distichium capillaceum | 373: + | | | |

# References


Aarde, R. J. van., 1977. Voeding, habitatsvoorkeur en voortplanting van die wildehuiskat (Felis catus Linnaeus, 1758) op Marioneiland. M. Sc. Thesis, Zoology dept. Pretoria.

Allen, S. E. & Northover, M. J., 1967. Soil types and nutrients on Signy Island. Phil. Trans. Roy. Soc. London, Ser. B, 252: 179–185.

Allison, J. S. & Smith, R. I. L., 1973. The vegetation of Elephant Island, South Shetland Islands. Br. Antarct. Surv. Bull. 33–34: 185–212.

Alston, A. H. G. & Schelpe, E. A. C. L. E., 1957. The Pteridophyta of Marion Island. J. S. Afr. Bot. 23: 105–109.

Anderson, G. D. & Condy, P. R., 1974. A note on the feral house cat and mouse on Marion Island. S. Afr. J. Antarct. Res. 4: 58–61.

Arnell, S., 1953. List of Hepaticae collected in Marion Island by Mr. R. W. Rand. Svensk Bot. Tidskr. 47: 411–425.

Arnell, S., 1963. Hepaticae of South Africa. Stockholm.

Ashton, D. H., 1965. Regeneration pattern of Poa foliosa Hook. f. on Macquarie Island. Proc. Roy. Soc. Victoria 79 (1): 215–233.

Ashton, D. H. & Gill, A. M., 1965. Pattern and process in a Macquarie Island feldmark. Proc. Roy. Soc. Victoria 79 (1): 235–245.

Aubert de la Rüe, E., 1932. La flore et la faune des Iles Kerguelen. La Terre et la Vie 2 (1): 29–51.

Aubert de la Rüe, E., 1962. Quelques particularités des tourbières de l'Archipel de Kerguelen, leur répartition, leur structure et leur destruction. T.A.A.F. 21: 12–30.

Aubert de la Rüe, E., 1964. Observations sur les charactères et la répartition de la végétation des Iles Kerguelen. C. N.F.R.A. -Biologie 1 (1): 1–60.

Aubert de la Rüe, E., 1968. Balles de mousse et mousses vagabondes, curiosités végétales des îles Kerguelen. T.A.A.F. 45: 3–9.

Baker, P. E. & Tomblin, J. F., 1964. A recent volcanic eruption on Bouvetøya, South Atlantic Ocean. Nature 203: 1055–1056.

Ball, D. F., 1964. Loss-on-ignition as an estimate of organic matter and organic carbon in non-calcareous soils. J. Soil Science 15: 84–92.

Barber, H. N., Dadswell, H. E. & Ingle, H. D., 1959. Transport of driftwood from South America to Tasmania and Macquarie Island. Nature 184: 203–204.

Barkman, J. J., Doing, H. & Segal, S., 1964. Kritische Bemerkungen und Vorschläge zur quantitativen Vegetationsanalyse. Acta Bot. Neerl. 13: 394–419.

Barkman, J. J., Moravec, J. & Rauschert, S., 1976. Code of phytosociological nomenclature. Vegetatio 32: 131–185.

Becking, R. W., 1957. The Zürich-Montpellier School of phytosociology. Bot. Rev. 23: 411–488.

Beeftink, W. G., 1965. De zoutvegetatie van ZW-Nederland beschouwd in Europees verband. Thesis Wageningen; Meded. Landbouwhogeschool 65 (1): 1–167.

Beeftink, W. G., 1966. Vegetation and habitat of the salt marshes and beachplains in the South-Western part of the Netherlands. Wentia 15: 83–108.

Beeftink, W. G., 1968. Die Systematik der europäischen Salzpflanzen-gesellschaften. In: R. Tüxen (ed.), Pflanzensoziologische Systematik. Ber. Int. Symp. Stolzenau/Weser 1964: 239–263. Junk, The Hague.

Bellair, P., 1965. Un exemple de glaciation aberrante; les îles Kerguelen. C.N.F.R.A. 11: 3–27.

Benninghoff, W. W., 1955. Jökla mýs. J. Glaciology 2: 514–515.

Berutti, A., Burger, A. E., Williams, A. J. & Siegfried, W. R., 1975. Bird ringing on Marion Island. S. Afr. J. Antarct. Res. 5: 50–52.

Billings, W. D. & Mooney, H. A., 1968. The ecology of arctic and alpine plants. Biol. Rev. 43: 481–529.

Bramwell, D. (ed.)., 1979. Plants and islands. Academic Press, London.

Braun-Blanquet, J., 1928. Pflanzensoziologie. 1. Aufl. Springer, Wien.

Braun-Blanquet, J., 1932. Plant sociology. (transl. G. D. Fuller & H. S. Conard). McGraw Hill, New York.

Braun-Blanquet, J., 1951. Pflanzensoziologie. 2. Aufl. Springer, Wien.

Braun-Blanquet, J., 1964. Pflanzensoziologie. 3. Aufl. Springer, Wien.

Brossard, M. de, 1972. Il y a deux siècles, Marion-Dufresne. Crozet, Kerguelen. T.A.A.F. 58-59: 3-26.

Budd, G. M., 1972. McDonald Island Reconnaissance, 1971. Polar Record 16: 64-67.

Chapman, S. B., 1965. The ecology of Coom Rigg Moss, Northumberland. III. Some water relations of the bog system. J. Ecol. 53: 371-384.

Chastain, A., 1958. La flore et la végétation des Iles Kerguelen. Mem. Mus. Nat. Hist. Nat., B, Botanique 11 (1): 1-136.

Cleef, A. M., 1978. Characteristics of neotropical páramo vegetation and its subantarctic relations. In: C. Troll & W. Lauer (eds.), Geoecological relations between the southern temperate zone and the tropical mountains. Erdwissensch. Forschung XI, Steiner, Wiesbaden: 365-390.

Cleef, A. M., 1981. The vegetation of the páramos of the Colombian Cordillera Oriental. Diss. Bot. No. 61. Cramer, Vaduz.

Clifford, H. T., 1953. The mosses of Macquarie Island and Heard Island. ANARE Sc. Reports (Ser. B) Bot. 2 (11): 1-12.

Cockayne, L., 1909. The ecological botany of the subantarctic islands of New Zealand. In: C. Chilton (ed.), The Subantarctic Islands of New Zealand. Vol. I. Government Printer, Wellington: 182-235.

Cockayne, L., 1928. The vegetation of New Zealand. 2nd. ed. Engelmann, Leipzig.

Condy, P. R., 1978. Distribution, abundance, and annual cycle of fur seals (Arctocephalus spp.) on the Prince Edward Islands. S. Afr. J. Wildl. Res. 8: 159-168.

Condy, P. R., 1979. Annual cycle of the southern elephant seal Mirounga leonina (Linn.) at Marion Island. S. Afr. J. Zool. 14: 95-102.

Costin, A. B., 1957. The high mountain vegetation of Australia. Austr. J. Bot. 5: 173-189.

Costin, A. B., 1965. Long distance seed dispersal to Macquarie Island. Nature 206: 317.

Cour, P., 1958. A propos de la flore de l'Archipel de Kerguelen. T.A.A.F. 4-5: 10-32.

Cour, P., 1959. Flore et végétation de l'Archipel de Kerguelen. II. Végétation. T.A.A.F. 8-9: 3-40.

Croome, R. L., 1972. Preliminary investigations on the nitrogen status of the Marion Island ecosystem. Unpublished expedition report, Dept. of Botany, U.O.F.S. Bloemfontein.

Croome, R. L., 1973. Nitrogen fixation in the algal mats on Marion Island. S. Afr. J. Antarct. Res. 3: 64-67.

Deacon, G. E. R., 1959. The Antarctic Ocean. Science Progress 47: 647-660.

Deacon, G. E. R., 1960. The southern cold temperate zone. Proc. Roy. Soc. London (Ser. B) 152: 441-447.

Deacon, G. E. R., 1964. The Southern Ocean. In: R. Priestly, R. J. Adie & G. de Q. Robin (eds.), Antarctic Research. Butterworth, London.

Dickson, J. H., 1965a. The biological report of the Royal Society Expedition to Tristan da Cunha, 1962. Part I. General introduction. Phil. Trans. Roy. Soc. London (Ser. B) 249: 259-271.

Dickson, J. H., 1965b. The effects of the eruption of 1961 on the vegetation of Tristan da Cunha. Phil. Trans. Roy. Soc. London (Ser. B) 249: 403-424.

Dierschke, H., 1971. Stand und Aufgaben der pflanzensoziologischen Systematik in Europa. Vegetatio 22: 255-264.

Dixon, H. N., 1912. Note on mosses growing unattached. Bryologist 15: 31-32.

Dreux, P., 1964. Observations sur la flore et la végétation de l'Ile aux Cochons (Archipel Crozet). Bull. Soc. Bot. France 111 (7-8): 382-386.

Dreux, P., 1971. Insecta. In: E. M. van Zinderen Bakker Sr., J. M. Winterbottom & R. A. Dyer (eds.), Marion and Prince Edward Islands. Balkema, Cape Town: 335-343.

Ellenberg, H., 1956. Aufgaben und Methoden der Vegetationskunde. In: H. Walter, Einführung in die Phytologie. IV, 1. Ulmer, Stuttgart.

Engler, A., 1964. Syllabus der Pflanzenfamilien. Part II. Borntraeger, Berlin.

Eythórsson, J., 1951. Jökla-mýs. J. Glaciology 1 (9): 503.

Fabricius, A. F., 1957. Climate of the Sub-Antarctic islands. In: M. P. van Rooy (ed.), Meteorology of the Antarctic. Weather Bureau, Dept. of Transport, Pretoria: 111-135.

Falla, R. A., 1960. Oceanic birds as dispersal agents. Proc. Roy. Soc. London Ser. B 152: 655-659.

Fineran, B. A., 1971. A catalogue of bryophytes, lichens and fungi collected in the Auckland Islands. J. Roy. Soc. N. Z. 1 (3-4): 215-229.

Follmann, G., 1965. Una asociación nitrófila de liquenes epipétricos de la Antártica Occidental con Ramalina terebrata Tayl et Hook. como especie caracterizante. Instituto Antártica Chileno. Publicación No 4: 1-8.

Fujiwara, K. & Umejima, S., 1962. On the distribution of windborne salt on the coastal terrace. Res. Bull. Coll. Exp. Forests Hokkaido Univ. 21: 453-464.

Gardner, W. H., 1965. Water content. In: C. A. Black (ed.). Methods of soil analysis. American Society of Agronomy, Madison, Wis.: 82-127.

Gass, I. G., 1963. The Royal Society's Expedition to Tristan da Cunha, 1962. Geogr. J. 129: 281-289.

Gilham, M. E., 1961. Modification of subantarctic flora on Macquarie Island by sea birds and sea elephants. Proc. Roy. Soc. Victoria 74 (1): 1-12.

Gimingham, C. H., 1967. Quantitative community analysis and bryophyte ecology on Signy Island. Phil. Trans. Roy. Soc. London, Ser. B 252: 251-259.

Gimingham, C. H. & Birse, E., 1957. Ecological studies on growth-form in bryophytes. I. Correlations between growth-form and habitat. J. Ecol. 45: 533-545.

Gimingham, C. H. & Smith, R. I. L., 1970. Bryophyte and lichen communities in the Maritime Antarctic. In: M. W. Holdgate (ed.), Antarctic Ecology. Vol. 2. Academic Press, London: 752-785.

Gimingham, C. H. & Smith, R. I. L., 1971. Growth form and water relations of mosses in the Maritime Antarctic. Br. Antarct. Surv. Bull. 25: 1-21.

Girod, M. & Nougier, J., 1972. Volcanism of the Sub-Antarctic islands. In: R. J. Adie (ed.), Antarctic geology and geophysics. Int. Union Geol. Sc. Ser. B 1: 777–788.
Godley, E. J., 1978. Cushion bogs. In: C. Troll & W. Lauer (eds.), Geoecological relations between the southern temperate zone and the tropical mountains. Erdwissensch. Forschung XI, Steiner, Wiesbaden: 141–158.
Good, R., 1974. The geography of the flowering plants. 4th ed. Longman, London.
Greene, S. W., 1964. The vascular flora of South Georgia. Br. Antarct. Surv. Sci. Rep. 45.
Greene, S. W., 1968a. Studies in Antarctic Bryology. I. A basic check list for mosses. Rev. Bryol. Lichenol. 36 (1–2): 132–138.
Greene, S. W., 1968b. Studies in Antarctic Bryology. II. Andreaea, Neuroloma. Rev. Bryol. Lichenol. 36 (1–2): 139–146.
Greene, S. W. & Greene, D. M., 1963. Check list of the subantarctic and antarctic vascular flora. Polar Record 11 (73): 411–418.
Greene, S. W. & Walton, D. W. H., 1975. An annotated check list of the sub-antarctic and antarctic vascular flora. Polar Record 17 (110): 473–484.
Gremmen, N. J. M., 1975. The distribution of alien vascular plants on Marion and Prince Edward Islands. S. Afr. J. Antarct. Res. 5: 25–30.
Gremmen, N. J. M. & Smith, V. R., 1980. Agrostis stolonifera L. on Marion Island (Subantarctic). S. Afr. J. Antarct. Res. 10 (in press).
Gressitt, J. L., 1970. Subantarctic entomology and biogeography. Pacific Insects Monograph 23: 295–374.
Grobbelaar, J. U., 1974. A contribution to the limnology of the sub-antartic island Marion. D. Sc. Thesis, Bloemfontein.
Grolle, R., 1964. Miscellanea hepatologicae (11–20). J. Jap. Bot. 39: 173–178.
Grolle, R., 1969. Grossdisjunktionen in Artarealen Lateinamerikanischer Lebermoose. In: E. J. Fittkau, J. Illies, H. Klinge, G. H. Schwabe & H. Sioli (eds.), Biogeography and ecology in South America. Vol. II. Junk, The Hague: 562–582.
Grolle, R., 1971a. Hepaticopsida. In: E. M. van Zinderen Bakker Sr., J. M. Winterbottom & R. A. Dyer (eds.), Marion and Prince Edward Islands. Balkema, Cape Town: 228–236.
Grolle, R., 1971b. Die Lebermoose der Crozet-Inseln (Subantarktis). Lindbergia 1: 80–82.
Grolle, R., 1972. The hepatics of the South Sandwich Islands and South Georgia. Br. Antarct. Surv. Bull. 28: 83–95.

Hall, K. J., 1977. Some observations on the former sea levels of Marion Island. S. Afr. J. Antarct. Res. 7: 19–22.
Hall, K. J., 1978a. Evidence for Quaternary glaciation of Marion Island (sub-Antarctic) and some implications. In: E. M. van Zinderen Bakker (ed.), Antarctic glacial history and world palaeoenvironments. Balkema, Rotterdam: 137–147.
Hall, K. J., 1978b. Quaternary glacial geology of Marion Island. D. Phil. Thesis, Bloemfontein.
Hartog, C. den & Segal, S., 1964. A new classification of the water-plant communities. Acta Bot. Neerl. 13: 367–393.
Hays, J. D., Lozano, J. A., Shackleton, N. & Irving, G., 1976. Reconstruction of the Atlantic and Western Indian Ocean sectors of the 18 000 BP Antarctic Ocean. Geol. Soc. Am. Mem. 145: 337–372.
Hébrard, J. P., 1970a. Muscinées récoltées aux Iles Kerguelen et Crozet pendant la campagne d'été des Terres Australes et Antarctiques Françaises (1969). Rev. Bryol. Lichenol. 37: 135–162.
Hébrard, J. P., 1970b. Contribution à l'étude de quelques groupements muscinaux des Iles Kerguelen et Crozet. Rev. Bryol. Lichenol. 37: 315–343.
Hedberg, O., 1964. Features of Afroalpine ecology. Acta Phytogeogr. Suecica 49: 1–144.
Hemsley, W. B., 1885. Report on the botany of the Bermudas and various other islands of the Atlantic and Southern Oceans. Second Part. In: C. W. Thompson & J. Murray (eds.), Report on the scientific results of the voyage of H. M. S. Challenger during the years 1873–76. Botany, Vol. 1. H. M. Stationery Office, London: 1–299.
Hill, M. O., 1973. Reciprocal averaging: an eigenvector method of ordination. J. Ecol. 61: 237–249.
Hodgson, E. A., 1962. Hepatics from the subantarctic islands of New Zealand, including the 'Cape Expedition' collections from the Auckland and Campbell Islands. Rec. Dom. Mus. Wellington, N. Z. 4 (11): 101–132.
Holdgate, M. W., 1964. Terrestrial ecology in the Maritime Antarctic. In: R. Carrick, M. Holdgate & J. Prévost (eds.), Biologie Antarctique. Hermann, Paris: 181–194.
Holdgate, M. W., 1967a. Signy Island. Phil. Trans. Roy. Soc. London, Ser. B 252: 173–177.
Holdgate, M. W., 1967b. The Antarctic ecosystem. Phil. Trans. Roy. Soc. London Ser. B 252: 363–383.
Holdgate, M. W., 1967c. The influence of introduced species on the ecosystems of temperate oceanic islands. Proceedings and Papers of the I.U.C.N. 10th technical meeting, I.U.C.N. Publications, New Series 9: 151–176.
Holdgate, M. W., Allen, S. E. & Chambers, M. J. G., 1967. A preliminary investigation of the soils of Signy Island, South Orkney Islands. Br. Antarct. Surv. Bull. 12: 53–71.
Holdgate, M. W. & Wace, N. M., 1961. The influence of man on the floras and faunas of southern islands. Polar Record 10 (68): 475–493.
Holttum, R. E., 1922. The vegetation of West Greenland. J. Ecol. 10: 87–108.
Holub, J. & Jirásek, V., 1967. Zur Vereinheitlichung der Terminologie in der Phytogeographie. Folia Geobot. Phytotaxon. 2: 69–113.
Huntley, B. J., 1970. Altitudinal distribution and phenology of Marion Island vascular plants. Tijdskr. Natuurwet. 10 (4): 255–262.
Huntley, B. J., 1971. Vegetation. In: E. M. van Zinderen Bakker Sr., J. M. Winterbottom & R. A. Dyer (eds.), Marion and Prince Edward Islands. Balkema, Cape Town: 98–160.
Huntley, B. J., 1972. Notes on the ecology of Azorella selago Hook. f. J. S. Afr. Bot. 38: 103–113.
Hydrographic Department. 1948. The Antarctic Pilot. 2nd ed. The Admiralty, London.

Ingram, H. A. P., 1967. Problems of hydrology and plant distribution in mires. J. Ecol. 55: 711–724.

James, T. P., 1876. Musci. Bull. U.S. Nat. Mus., Washington 3: 25–27.

Jeannel, R., 1940. Croisière du Bougainville aux Iles Australes Françaises. I. Partie générale. Mem. Mus. Nat. Hist. Nat., N.S. 14: 1–46.

Johnson, P. N., 1975. Vegetation and flora of the Solander Islands, Southern New Zealand. N. Z. J. Bot. 13: 189–213.

Kable, E. D., Erlank, A. J. & Cherry, R. D., 1971. Geochemical features of lavas. In: E. M. van Zinderen Bakker Sr., J. M. Winterbottom & R. A. Dyer (eds.), Marion and Prince Edward Islands. Balkema, Cape Town: 78–88.

Kessler, A., 1978. Studien zur Klimatologie der Strahlungsbilanz unter besonderer Berücksichtigung der tropischen Hochgebirge und der Kühltemperierten Zone der Südhalbkugel. In: C. Troll & W. Lauer (eds.), Geoecological relationships between the southern temperate zone and the tropical mountains. Erdwissensch. Forschung XI, Steiner, Wiesbaden: 49–61.

Kightley, S. P. J. & Smith, R. I. L., 1976. The influence of reindeer on the vegetation of South Georgia. I. Long-term effects of unrestricted grazing and the establishment of exclosure experiments in various plant communities. Br. Antarct. Surv. Bull. 44: 57–76.

Kok, O., 1975. Verspreiding van veervasgehegte sade deur voëls. Ostrich 46: 261–263.

Kuwahara, Y., 1966. The family Metzgeriaceae in North and South-East Asia. Rev. Bryol. Lichenol. 34: 191–139.

Lacroix, A., 1923. Minéralogie de Madagascar. Vol. 3. Paris.

Lacroix, A., 1940. Les laves des volcans inactifs des Iles Marion et Crozet. Mém. Mus. Hist. Nat. 14: 47–62.

La Grange, J. J., 1954. The South African station on Marion Island. Polar Record 7 (48): 155–158.

La Grange, J. J., 1962. Notes on the birds and mammals on Marion Island and Antarctica (SANAE). J. S. Afr. Biol. Sc. 3: 27–86.

Laing, R. M., 1909. The chief plant formations and associations of Campbell Island. In: C. Chilton (ed.), The Subantarctic islands of New Zealand. Vol 2. Government Printer, Wellington: 482–492.

Langenegger, O. & Verwoerd, W. J., 1971. Topographic survey. In: E. M. van Zinderen Bakker Sr., J. M. Winterbottom & R. A. Dyer (eds.), Marion and Prince Edward Islands. Balkema, Cape Town: 32–39.

Law, P. G. & Burstall, T., 1953. Heard Island. A.N.A.R.E. Interim Reports 7: 1–32.

Lawrence, R. F., 1971. Araneida. In: E. M. van Zinderen Bakker Sr., J. M. Winterbottom & R. A. Dyer (eds.), Marion and Prince Edward Islands. Balkema, Cape Town: 301–313.

Leupe, P. A., 1868. De eilanden Dina en Maerseveen. Verhandelingen en Berigten betrekkelijk het Zeewezen, de Zeevaartkunde, de Hydrographie, de Koloniën en de daarmede in verband staande wetenschappen 3, 2e afd.: 242–253.

Lid, J., 1938. Mosbollar. Nyt Mag. Naturv. 78: 101–104.

Lindeboom, H. J., 1979. Chemical and microbiological aspects of the nitrogen cycle on Marion Island (Sub-Antarctic). Thesis, Groningen.

Lindsay, D. C., 1971. Vegetation of the South Shetland Islands. Br. Antarct. Surv. Bull. 25: 59–83.

Lindsay, D. C., 1976. The lichens of Marion and Prince Edward Islands, Southern Indian Ocean. Nova Hedwigia 28: 667–689.

Maarel, E. van der, Janssen, J. G. M. & Louppen, J. M. W., 1978. TABORD, a program for structuring phytosociological tables. Vegetatio 38: 143–156.

Mägdefrau, K., 1969. Die Lebensformen der Laubmoose. Vegetatio 16: 285–297.

Malloch, A. J. C., 1972. Salt-spray deposition on the maritime cliffs of the Lizard Peninsula. J. Ecol. 60: 103–112.

Marsh, J. H., 1948. No pathway here. Timmins, Cape Town.

Martin, W., 1952. Unattached mosses in the New Zealand flora. Bryologist 55: 65–70.

McDougall, I., 1971. Geochronology. In: E. M. van Zinderen Bakker Sr., J. M. Winterbottom & R. A. Dyer (eds.), Marion and Prince Edward Islands. Balkema, Cape Town: 72–77.

Meurk, C. D., 1977. Alien plants in Campbell Island's changing vegetation. Mauri Ora 5: 93–118.

Moore, D. M., 1963. The subspecies of Montia fontana L. Botaniska Notiser 116 (1): 16–30.

Moore, D. M., 1968. The vascular flora of the Falkland Islands. Br. Antarctic. Surv. Sc. Rep. 60.

Moseley, H. N., 1874. On the botany of Marion Island, Kerguelen's Land and Young Island of the Heard Group. J. Linn. Soc. London, Botany 14: 387–388.

Moseley, H. N., 1877a. Further notes on the plants of Kerguelen, with some remarks on the insects. J. Linn. Soc. London, Botany 15: 53–54.

Moseley, H. N., 1877b. Notes on the flora of Marion Island. J. Linn. Soc. London, Botany 15: 481–482.

Moseley, H. N., 1892. Notes by a naturalist. An account of observations made during the voyage of H.M.S. 'Challenger' round the world in the years 1872–1876. Murray, London.

Mueller-Dombois, D. & Ellenberg, H., 1974. Aims and methods of vegetation science. Wiley & Sons, New York.

Oliver, R. L. & Sorensen, J. H., 1951. Botanical investigations on Campbell Island. Part I. The vegetation. Cape Expedition Series. Bull. No 7. Dept. Sc. Ind. Res., Wellington, N.Z.: 5–24.

Peech, M., 1965. Hydrogen-Ion activity. In: C.A. Black (ed.), Methods of soil analysis. American Society of Agronomy. Madison, Wis.: 914–926.

Rand, R. W., 1954. Notes on the birds of Marion Island. Ibis 96: 173–206.

Rand, R. W., 1955a. The penguins of Marion Island. Ostrich 24:

57–69.
Rand, R. W., 1955b. Marion Island - Home of South Africa's Elephant Seal. Afr. Wildl. 9: 7–9.
Rand, R. W., 1956. Notes on the Marion Island Fur Seal. Proc. Zool. Soc. London 129: 65–82.
Rauh, W., 1978. Die Wuchs- und Lebensformen der tropischen Hochgebirgsregionen und der Subantarktis. Ein Vergleich. In: C. Troll & W. Lauer (eds.), Geoecological relations between the southern temperate zone and the tropical mountains. Erdwissenschaftliche Forschung XI, Steiner, Wiesbaden: 62–92.
Renard, A. F., 1889. Report on petrology of oceanic islands. Rep. Sc. Res. Voyage H.S.S. Challenger 1873–1876. Physics and chemistry 2 (7): 104–107.
Roberts, B. B., 1958a. Chronological list of Antarctic expeditions. Polar Record 9 (59): 97–134.
Roberts, B. B., 1958b. Chronological list of Antarctic expeditions. Polar Record 9 (60): 191–239.
Roskam, E., 1971. Programme ORDINA: multidimensional ordination of observation vectors. Programme Bulletin 16, Psychol. Lab. Nijmegen.
Rothuizen, I. W. J. M., 1977. De geografische verbreiding van de inheemse vaatplanten van de subantarctische floraprovincie Kerguelen. Doctoraal-verslag, Afd. Geobotanie, K.U. Nijmegen (unpublished).

Sainsbury, G. O. K. & Allison, K. W., 1962. Mosses from the Auckland Islands and Campbell Island. Rec. Dom. Mus. Wellington 4 (12): 133–142.
Schalke, H. J. W. G. & van Zinderen Bakker Sr, E. M., 1967. A preliminary report on palynologial research on Marion Island (Sub-Antarctic). S. Afr. J. Sci. 63: 254–259.
Schalke, H. J. W. G. & van Zinderen Bakker Sr, E. M., 1971. History of the vegetation. In: E. M. van Zinderen Bakker Sr., J. M. Winterbottom & R. A. Dyer (eds.), Marion and Prince Edward Islands. Balkema, Cape Town: 89–97.
Schenck, H., 1905. Vergleichende Darstellung der Pflanzengeographie der subantarktischen Inseln, insbesondere über Flora und Vegetation von Kerguelen. Wiss. Ergebn. Deutsche Tiefsee-Exp. Valdivia 1898–1899 2 (1): 1–178.
Schroeder, F.-G., 1969. Zur Klassifizierung der Anthropochoren. Vegetatio 16: 225–238.
Schulze, B. R., 1971. The climate of Marion Island. In: E. M. van Zinderen Bakker Sr., J. M. Winterbottom & R. A. Dyer (eds.), Marion and Prince Edward Islands. Balkema, Cape Town: 16–31.
Segal, S., 1965. Een vegetatie-onderzoek van hogere waterplanten in Nederland. Wetensch. Meded. K.N.B.V. 57.
Séguy, E., 1971. Diptera. In: E. M. van Zinderen Bakker Sr., J. M. Winterbottom & R. A. Dyer (eds.), Marion and Prince Edward Islands. Balkema, Cape Town: 344–348.
Seppelt, R. D., 1977. Studies on the bryoflora of Macquarie Island. I. Introduction and checklist of species. Bryologist 80: 167–170.
Seppelt, R. D. & Ashton, D. H., 1978. Studies on the ecology of the vegetation at Mawson Station, Antarctica. Austral. J. Ecol. 3: 373–388.
Shacklette, T. S., 1966. Unattached moss polsters on Amchitka Island, Alaska. Bryologist 69: 346–352.
Shimwell, D., 1971. Description and classification of vegetation. Sidgewick & Jackson, London.
Skottsberg, C., 1905. Some remarks upon the geographical distibution of vegetation in the colder Southern hemisphere. Ymer 25 (4): 402–427.
Skottsberg, C., 1909. Studiën über das Pflanzenleben der Falklandinseln. Wiss. Ergebn. Schwed. Südpolar-Exp. 1901–1903, 4 (10): 1–58.
Skottsberg, C., 1912. The vegetation of South Georgia. Wiss. Ergebn. Schwed. Südpolar-Exp. 1901–1903, 4 (12): 1–36.
Skottsberg, C., 1913. A botanical survey of the Falkland Islands. (Bot. Ergebn. Schwed. Exp. Patagonien U.D. Feuerlande, 1907–1909, 3) K.Svensk. Vet. Akad. Handl. 50 (3): 1–129.
Skottsberg, C., 1960. Remarks on the plant geography of the southern cold temperate zone. Proc. Roy. Soc. London Ser. B 152: 447–457.
Smith, R. I. L., 1972. Vegetation of the South Orkney Islands, with particular reference to Signy Island. Br. Antarct. Surv. Sc. Rep. 68.
Smith, R. I. L. & Walton, D. W. H., 1975. South Georgia, Subantarctic. In: T. Rosswall & O. W. Heal (eds.), Structure and function of tundra ecosystems. Ecol. Bull. (Stockholm) 20: 399–423.
Smith, V. R., 1976a. Standing crop and nutrient status of Marion Island (Sub-Antarctic) vegetation. J. S. Afr. Bot. 42: 231–263.
Smith, V. R., 1976b. The effect of burrowing species of Procellariidae on the nutrient status of inland tussock grasslands on Marion Island. J. S. Afr. Bot. 42: 265–272.
Smith, V. R., 1976c. The nutrient statuses of Marion Island plants and soils. M. Sc. Thesis, Bloemfontein.
Smith, V. R., 1977a. A qualitative description of energy flow and nutrient cycling in the Marion Island terrestrial ecosystem. Polar Record 18 (115): 361–370.
Smith, V. R., 1977b. Vegetation standing crop of the grey lava flows and of the eastern coastal plain on Marion Island. J. S. Afr. Bot. 43: 105–114.
Smith, V.R., 1977c. Notes on the feeding of Ectemnorrhinus similis Waterhouse (Curculionidae) adults on Marion Island. Oecologia (Berl.) 29: 269–273.
Smith, V. R., 1977d. The chemical composition of Marion Island soils, plants and vegetation. S. Afr. J. Antarct. Res. 7: 28–39.
Smith, V. R., 1978a. Animal-plant-soil relationships on Marion Island (Subantarctic). Oecologia (Berl.) 32: 239–253.
Smith, V. R., 1978b. Soil chemistry on Marion Island (Subantarctic). S. Afr. J. Sci. 74: 174–175.
Smith, V. R., 1979. The influence of seabird manuring on the phosphorus status of Marion Island (Subantarctic) soils. Oecologia (Berl.) 41: 123–126.
Smith, W. A. & Sayers, R. L., 1971. Entomostraca. In: E. M. van Zinderen Bakker Sr., J. M. Winterbottom & R. A. Dyer (eds.), Marion and Prince Edward Islands. Balkema, Cape Town: 361–372.
Snape, C. & Retief, J. A., 1971. Palaeomagnetic study of some recent lavas. In: E. M. van Zinderen Bakker Sr., J. M. Winterbottom & R. A. Dyer (eds.), Marion and Prince Edward Islands. Balkema, Cape Town: 63–71.

Sparling, J. H., 1966. Studies on the relationship between water movement and water chemistry in mires. Can. J. Bot. 44: 747–748.

Taljaard, J. J., 1957. Geographical and hydrological features of the Antarctic. In: M. P. van Rooy (ed.), Meteorology of the Antarctic. Weather Bureau, Pretoria: 1–15.

Taylor, B. W., 1955a. The flora, vegetation and soils of Macquarie Island. A.N.A.R.E. Rep. Ser. B, Vol. 2, Botany.

Taylor, B. W., 1955b. Terrace formation on Macquarie Island. J. Ecol. 43: 133–137.

Tizard, T. H., Moseley, H. N., Buchanan, J. Y. & Murray, J., 1885. Report of the scientific results of the voyage of H.M.S. 'Challenger' 1873–76. Narrative. Vol. 1, first part.

Troll, C., 1944. Strukturboden, Solifluktion und Frostklimate der Erde. Geol. Rundschau 34: 545–694.

Troll, C., 1948. Der asymmetrische Aufbau der Vegetationsstufen auf der Nord- und Südhalbkugel. Jahresber. Geobot. Forschungs-institutes Rübel 1947, Zürich: 46–83.

Troll, C., 1955. Der jahreszeitliche Ablauf des Naturgeschehens in den verschiedenen Klimagürteln der Erde. Studium Generale 8: 713–733.

Troll, C., 1966. Jahreszeitenklimate der Erde. In: H. E. Landsberg, H. Lippmann, K. H. Paffen & C. Troll, World maps of climatology. Springer, Berlin.

Trusswell, J. F., 1965. Marion Island, South Indian Ocean. Nature 205: 64–65.

Vanhöffen, E., 1903. Biologischer Bericht. Heard Insel. Veröff. Inst. Meeresk. Univ. Berlin 5: 126.

Verwoerd, W. J., 1967. Geologiese evolusie van Marion en Prince Edward-eilande. S. Afr. J. Sci. 63: 219–225.

Verwoerd, W. J., 1971. Geology. In: E. M. van Zinderen Bakker Sr., J. M. Winterbottom & R. A. Dyer (eds.), Marion and Prince Edward Islands. Balkema, Cape Town: 40–62.

Vowinckel, E., 1954. Synoptische Klimatologie vom Gebiet Marion Island. NOTOS (Pretoria) 3 (1): 12–21.

Wace, N. M., 1960. The botany of the southern oceanic islands. Proc. Roy. Soc. London Ser. B 152: 475–490.

Wace, N. M., 1961. The vegetation of Gough Island. Ecol. Monogr. 31: 337–367.

Wace, N. M., 1965. Vascular plants. In: J. van Mieghem & P. van Oye (eds.): Biogeography and ecology in Antarctica. Junk, The Hague: 201–266.

Wace, N. M., 1966. Last of the virgin islands. Discovery 27 (2).

Wace, N. M., 1969. The terrestrial biogeography of Antarctica. In: E. M. van Zinderen Bakker (ed.), Palaeoecology of Africa and of the surrounding islands and Antarctica. Vol. 5: 213–238. Balkema, Cape Town.

Wace, N. M. & Dickson, J. H., 1965. The terrestrial botany of the Tristan da Cunha Islands. Phil.Trans. Roy. Soc. London, Ser. B 249: 273–360.

Wace, N. M. & Holdgate, M. W., 1958. The vegetation of Tristan da Cunha. J. Ecol. 46: 593–620.

Walter, H., 1955. Die Klimagramme als Mittel zur Beurteilung der Klimaverhältnisse für ökologische, vegetationskundliche und landwirtschaftliche Zwecke. Ber. Deutsch. Bot. Ges. 68: 331–344.

Walter, H., 1970. Vegetationszonen und Klima. Ulmer, Stuttgart.

Walter, H., Harnickell, E. & Mueller-Dombois, D., 1975. Klimadiagramm-Karten der einzelnen Kontinente und ökologische Klimagliederung der Erde. Fischer, Stuttgart.

Walter, H. & Lieth, H., 1967. Klimadiagramm-Weltatlas. Fischer, Jena.

Walton, D. W. H., 1975. European weeds and other alien species in the Sub-Antarctic. Weed Research 15: 271–282.

Walton, D. W. H., 1977. Studies on Acaena (Rosaceae): Seed germination, growth and establishment in A. magellanica (Lam.) Vahl and A. tenera Alboff. Br. Antarct Surv. Bull. 45: 29–40.

Walton, D. W. H. & Smith, R. I. L., 1973. Status of the alien vascular flora of South Georgia. Br. Antarct. Surv. Bull. 36: 79–97.

Warming, E., 1887. Om Grönlands Vegetation. Meddel. om Grönland 12.

Weather Bureau, 1967a. Report on meteorological data of the year 1965. Government Printer, Pretoria.

Weather Bureau. 1967b. Report on meteorological data of the year 1966. Government Printer, Pretoria.

Weather Bureau. 1968. Report on meteorological data of the year 1967. Government Printer, Pretoria.

Weather Bureau, 1970. Report on meteorological data of the year 1968. Government Printer, Pretoria.

Weather Bureau, 1971a. Report on meteorological data of the year 1963. Government Printer, Pretoria.

Weather Bureau, 1971b. Report on meteorological data of the year 1969. Government Printer, Pretoria.

Weather Bureau, 1972. Report on meteorological data of the year 1970. Government Printer, Pretoria.

Weather Bureau, 1973. Report on meteorological data of the year 1964. Government Printer, Pretoria.

Weather Bureau, 1975a. Report on meteorological data of the year 1972. Government Printer, Pretoria.

Weather Bureau, 1975b. Report on meteorological data of the year 1973. Government Printer, Pretoria.

Werger, M. J. A., 1972. Species-area relationship and plot size: with some examples from South African vegetation. Bothalia 10: 583–594.

Werger, M. J. A., 1973. Phytosociology of the upper Orange River Valley. A syntaxonomical and synecological study. Thesis, Nijmegen, V & R, Pretoria.

Werger, M. J. A., 1974. On concepts and techniques applied in the Zürich-Montpellier method of vegetation survey. Bothalia 11: 309–323.

Werth, E., 1906. Die Vegetation der subantarktischen Inseln Kerguelen, Possession- und Heard-Eiland. 1. Deutsche Südpol.-Exp. 1901–1903, 8, Bot. (1): 125–176.

Westhoff, V., 1979. Phytosociology in the Netherlands: history, present state, future. In: M. J. A. Werger (ed.), The study of vegetation. Junk, The Hague: 81–121.

Westhoff, V. & van der Maarel, E., 1973. The Braun-Blanquet approach. In: R. H. Whittaker (ed.), Ordination and classification of communities. Handbook of vegetation science 5:

617–726. Junk, The Hague.
Westhoff, V. & van der Maarel, E., 1978. The Braun-Blanquet approach. 2nd ed. In: R. H. Whittaker (ed.), Classification of plant communities. Junk, The Hague: 287–399.
White, W. S., 1969. Basic metric surveying. Newnes - Butterworths, London.
Williams, A. J., Burger, A. E. & Berutti, A., 1978. Mineral and energy contributions of carcasses of selected species of sea birds to the Marion Island ecosystem. S. Afr. J. Antarct. Res. 8.
Williams, A. J., Siegfried, W. R., Burger, A. E. & Berutti, A., 1979. The Prince Edward Islands: a sanctuary for seabirds in the Southern Ocean. Biol. Conserv. 15: 59–71.
Winterbottom, J. M., 1971. The position of Marion Island in the Sub-Antarctic avifauna. In: E. M. van Zinderen Bakker Sr., J. M. Winterbottom & R. A. Dyer (eds.), Marion and Prince Edward Islands. Balkema, Cape Town: 241–248.

Young, S. B. & Schofield, E. K., 1973. Palynological evidence for the late-glacial occurrence of Pringlea and Lyallia on Kerguelen Islands. Rhodora 75: 239–247.

Zanten, B. O. van, 1971. Musci. In: E. M. van Zinderen Bakker Sr., J. M. Winterbottom & R. A. Dyer (eds.), Marion and Prince Edward Islands. Balkema, Cape Town: 173–227.
Zanten, B. O. van, 1976. Preliminary report on germination experiments designed to estimate the survival chances of moss spores during aerial trans-oceanic long range dispersal in the southern hemisphere, with particular reference to New Zealand. Journ. Hattori Bot. Lab. 41: 133–140.
Zanten, B. O. van. 1978. Experimental studies on trans-oceanic long-range dispersal of moss spores in the southern hemisphere. Journ. Hattori Bot. Lab. 44: 455–482.
Zinderen Bakker Jr. E. M. van, 1967. Observations on animal life on Marion and Prince Edward Islands. S. Afr. J. Sci. 63: 242–246.
Zinderen Bakker Jr., E. M. van, 1971. Comparative avian ecology. In: E. M. van Zinderen Bakker Sr., J. M. Winterbottom & R. A. Dyer (eds.), Marion and Prince Edward Islands. Balkema, Cape Town: 161–172.
Zinderen Bakker Sr., E. M. van, 1967. The South African biological-geological survey of the Marion and Prince Edward Islands and the meteorological expedition to Bouvet Island. S. Afr. J. Sci. 63: 217–218.
Zinderen Bakker Sr., E. M. van, 1969. Quaternary pollen analytical studies in the Southern Hemisphere with special reference to the Sub-Antarctic. In: E. M. van Zinderen Bakker Sr. (ed.), Palaeoecology of Africa, the surrounding islands and Antarctica 5: 175–212. Balkema, Cape Town.
Zinderen Bakker Sr., E. M. van, 1970. Quaternary climates and Antarctic biogeography. In: M. W. Holdgate (ed.), Antarctic ecology. Academic Press, London: 31–40.
Zinderen Bakker Sr., E. M. van, 1971a. Introduction. In: E. M. van Zinderen Bakker Sr., J. M. Winterbottom & R. A. Dyer (eds.), Marion and Prince Edward Islands. Balkema, Cape Town: 1–15.
Zinderen Bakker Sr., E. M. van. 1971b. The glaciation of Marion Island. S. Afr. J. Antarct. Res. 1: 35.
Zinderen Bakker Sr., E. M. van, 1973. The glaciation(s) of Marion Island (sub-Antarctic). In: E. M. van Zinderen Bakker Sr. (ed.), Palaeoecology of Africa, the surrounding islands and Antarctica 8: 161–178. Balkema, Cape Town.
Zinderen Bakker Sr., E. M. van, 1976. The evolution of late-quaternary palaeoclimates of southern Africa. In: E. M. van Zinderen Bakker Sr. (ed.), Palaeoecology of Africa, the surrounding islands and Antarctica 9: Balkema, Cape Town.
Zinderen Bakker Sr., E. M. van, 1978. Geoecology of the Marion and Prince Edward Islands. In: C. Troll & W. Lauer (eds.). Geoecological relations between the southern temperate zone and the tropical mountains. Erdwissensch. Forschung XI. Steiner, Wiesbaden: 495–515.
Zinderen Bakker Sr., E. M. van, Verwoerd, W. J. & Langenegger, O., 1967. Marion and Prince Edward Islands. Nature 213: 230–232.
Zinderen Bakker Sr. E. M. van, Winterbottom, J. M. & Dyer, R. A. (eds.), 1971. Marion and Prince Edward Islands. Balkema, Cape Town.

# Subject index

# Syntaxonomic index

# Systematic index

# Samenvatting

Marion en Prince Edward zijn twee relatief jonge, vulkanische eilanden, gelegen in de Zuidelijke Oceaan, ver verwijderd van enig ander land. De Iles Crozet vormen het dichtstbijzijnde land, op 945 km ten oosten van Marion en Prince Edward. Het meest nabije continent is Afrika, 1800 km in noordnoordwestelijke richting gelegen, terwijl Antarctica 2300 km naar het zuiden ligt. Marion en Prince Edward meten respektievelijk 290 en 44 $km^2$ en zijn 22 km van elkaar verwijderd.

Twee fasen van vulkanische aktiviteit zijn te onderscheiden in de ontwikkelingsgeschiedenis van de eilanden. Gedurende de oudste fase, c. 275 000–100 000 jaar geleden, werd grijze lava gevormd, terwijl de lava van de jongste vulkanische fase zwart van kleur is. Nog tijdens een eruptie in 1980 op Marion werd zwarte lava gevormd. Tussen de twee perioden van vulkanische aktiviteit heeft er een ijstijd plaatsgevonden, gedurende welke het grootste deel van Marion door gletsjers was bedekt.

Het klimaat van de eilanden is extreem oceanisch en wordt gekenmerkt door een lage, zeer gelijkmatige temperatuur (jaargemiddelde 5 °C), door een hoge neerslag (2575 mm jaarlijks) en door sterke winden.

Marion en Prince Edward liggen in de subantarctische zone. Hun flora vertoont grote gelijkenis met die van de overige eilanden van de Kerguelen Provincie (Crozet, Kerguelen, Heard en Macdonald). Het aantal soorten is laag: 23 inheemse soorten vaatplanten, 72 soorten mossen, 36 levermossen en 50 lichenen. Daarnaast zijn 14 soorten vaatplanten door de mens ingevoerd, waarvan er inmiddels weer 4 zijn verdwenen.

De vegetatie van Marion en Prince Edward is onderzocht met behulp van de methodieken van de Braun-Blanquet school. Op het niveau van de subassociatie werden 41 plantengemeenschappen onderscheiden. Al deze gemeenschappen worden hier voor het eerst beschreven. Ze zijn verenigd in complexen, d.w.z. groepen van gemeenschappen met een duidelijke floristische en oecologische verwantschap.

De belangrijkste faktor die de verspreiding van de plantengemeenschappen op de eilanden bepaalt is het waterregime van de bodem. Twee series van plantengemeenschappen kunnen langs nat-droog gradienten worden onderscheiden, een op standplaatsen met geen of nauwelijks zijwaartse waterbeweging door de bodem, een andere op plaatsen met een duidelijke zijwaartse beweging van water door de bodem of aan het oppervlak. De eerste serie bestaat uit de vegetatie van plasjes, venen, hellingen en tenslotte hooggelegen, sterk aan de wind blootgestelde plaatsen, terwijl de tweede serie wordt gevormd door plantengemeenschappen van stroompjes, bronnen, watergeulen in het veen, drainage geulen op hellingen en beekoevers. De vegetatie van gebieden met een sterke invloed van zout, of van dieren sluit niet bij deze series aan. Deze verschilt floristisch sterk van alle andere vegetatietypen op de eilanden.

# Curriculum vitae

Nicolaas Johannes Marie Gremmen werd geboren te Schijndel op 13 juli 1948. Hij doorliep de lagere school te Schijndel en de middelbare school te 's Hertogenbosch, alwaar hij in 1966 het diploma gymnasium $\beta$ behaalde. In ditzelfde jaar begon hij aan de studie biologie aan de Katholieke Universiteit te Nijmegen. Hij behaalde het kandidaatsdiploma in 1969. In 1971 behaalde hij de graad van doctorandus (cum laude), met als hoofdvak vegetatiekunde (onder begeleiding van Prof. Dr. V. Westhoff en Dr. H. Strijbosch) en als bijvakken dieroecologie (onder begeleiding van Dr. H. C. J. Oomen en Dr. J. J. van Gelder) en genetica (onder begeleiding van Prof. Dr. S. J. Geerts en Dr. C. Stumm). In 1971 was hij als vegetatiekundig onderzoeker in dienst van de Provinciale Planologische Dienst van Drente. Vervolgens was hij wetenschappelijk medewerker aan de afdeling Geobotanie van de Katholieke Universiteit te Nijmegen. Van eind 1973 tot 1977 was hij als vegetatiekundig onderzoeker verbonden aan de Universiteit van de Oranje Vrijstaat te Bloemfontein, Zuid Afrika. Gedurende deze periode nam hij deel aan twee expedities naar de subantarctische eilanden Marion en Prince Edward. Gedurende een deel van 1978 was hij werkzaam als vegetatiekundig onderzoeker bij de Provinciale Planologische Dienst van Overijssel. In 1979 werd hij aangesteld als wetenschappelijk medewerker aan de afdeling Geobotanie van de Katholieke Universiteit te Nijmegen, alwaar hij op dit moment werkzaam is.

GPSR Compliance

*The European Union's (EU) General Product Safety Regulation (GPSR) is a set of rules that requires consumer products to be safe and our obligations to ensure this.*

*If you have any concerns about our products, you can contact us on ProductSafety@springernature.com*

In case Publisher is established outside the EU, the EU authorized representative is:

Springer Nature Customer Service Center GmbH
Europaplatz 3
69115 Heidelberg, Germany

**Batch number: 10809836**

Printed by Printforce, the Netherlands